Applied Pascal
for Technology

MERRILL PUBLISHING COMPANY
A Bell & Howell Information Company
Columbus Toronto London Melbourne

Published by Merrill Publishing Company
A Bell & Howell Information Company
Columbus, Ohio 43216

This book was set in Garamond Light.

Administrative Editor: Steve Helba
Production Coordinator: Linda Kauffman Peterson
Art Coordinator: Lorraine Woost
Cover Designer: Brian Deep

Library of Congress Catalog Card Number: 88–062362
International Standard Book Number: 0–675–20771–1
Printed in the United States of America
1 2 3 4 5 6 7 8 9—92 91 90 89

MERRILL'S INTERNATIONAL SERIES IN
ELECTRICAL AND ELECTRONICS TECHNOLOGY

GAONKAR   *Microprocessor Architecture, Programming, and Applications with the 8085/
                       8080A, Second Edition,* 20675–8
                       The Z80 Microprocessor: Architecture, Interfacing, Programming, and Design,
                       20540–9

GILLIES   *Instrumentation and Measurements for Electronic Technicians,* 20432–1

HUMPHRIES   *Motors and Controls,* 20235–3

KULATHINAL   *Transform Analysis and Electronic Networks with Applications,* 20765–7

LAMIT/LLOYD   *Drafting for Electronics,* 20200–0

LAMIT/WAHLER/   *Workbook in Drafting for Electronics,* 20417–8
HIGGINS

MARUGGI   *Technical Graphics: Electronics Worktext,* 20311–2

McINTYRE   *Study Guide to accompany Electronics Fundamentals,* 20676–6

MILLER   *The 68000 Microprocessor: Architecture, Programming, and Applications,*
                       20522–0

MONACO   *Introduction to Microwave Technology,* 21030–5
                       Laboratory Exercises in Microwave Technology, 21031–3

NASHELSKY/   *BASIC Applied to Circuit Analysis,* 20161–6
BOYLESTAD

QUINN   *The 6800 Microprocessor,* 20515–8

REIS   *Electronic Project Design and Fabrication,* 20791–6

ROSENBLATT/   *Direct and Alternating Current Machinery, Second Edition,* 20160–8
FRIEDMAN

SCHOENBECK   *Electronic Communications: Modulation and Transmission,* 20473–9

SCHWARTZ   *Survey of Electronics, Third Edition,* 20162–4

SORAK   *Linear Integrated Circuits: Laboratory Experiments,* 20661–8

STANLEY, B. H.   *Experiments in Electric Circuits, Third Edition,* 21088–7

STANLEY, W. D.   *Operational Amplifiers with Linear Integrated Circuits, Second Edition,* 20660–X

TOCCI   *Fundamentals of Electronic Devices, Third Edition,* 9887–4
                       Electronic Devices: Conventional Flow Version, Third Edition, 20063–6
                       Fundamentals of Pulse and Digital Circuits, Third Edition, 20033–4
                       Introduction to Electric Circuit Analysis, Second Edition, 20002–4

WEBB   *Programmable Controllers: Principles and Applications,* 20452–6

YOUNG   *Electronic Communication Techniques,* 20202–7

Preface

This text is designed to be used in a first or second programming course for technology students. The first two chapters cover introductory topics, and they neither require nor assume any prior programming experience or knowledge of computers. Chapter Three begins the coverage of material based on information learned in the earlier chapters. The mathematical requirements for the text's programs do not go beyond elementary algebra. The text adhered to a careful, disciplined approach to programming in Pascal, emphasizing immediate, practical applications. The goal of this text is to teach a stepwise approach to technical problem solving using the applied programming method (APM), featuring top-down design and the systematic development of algorithms.

To facilitate this approach, the Boehm and Jacopini completeness theorem is introduced early, along with the development of Pascal procedures. Action blocks, branch blocks, and loop blocks are introduced as needed, allowing the student sufficient time and practice to understand these universal programming concepts.

Most of the chapters include a case study of the programming of a technical problem. Here, the technology student has the opportunity to experience the development of a technical program from conception to final design and testing. Each case study demonstrates top-down design, algorithm development, and various types of program documentation resulting in a substantial program with multiple procedures. Each case study is followed by a self-test to guide the student through the key parts of the program. The use of Pascal in interactive programming is emphasized throughout this text. Turbo Pascal 4.0 is used as the operating system; it is presented in a logical step-by-step fashion to give the student full benefit of this new industry standard.

Some important features of this text include the following:

1. **Early introduction of the completeness theorem**. This theorem is introduced in Chapter Two so that the student understands the three fundamental programming structures that set the foundation for the rest of the text.

2. **Spiral approach.** This method can be used because of the early introduction of the completeness theorem; the concepts of action, loop, and branch blocks are presented throughout the text with increasingly more detail, as needed.

3. **Use of Turbo 4.0 Pascal environment.** The student is immediately exposed to the most widely used Pascal system. Each chapter of the text contains an important element of this powerful system, starting with simple editing commands and progressing up to the construction of a Pascal library using the Turbo UNIT.

4. **Early introduction of the applied programming method.** The concepts of top-down design and a systematic approach to algorithm development are presented in Chapter Two, allowing this process to be used in all the chapters following.

5. **Emphasis on modern technology.** All major programming examples and problems are presented in the environment of a technology discipline. Thus, the student has the advantage of seeing immediate practical applications for new Pascal commands. Students are encouraged to think in concrete technical terms; the programming jargon common to many programming texts is avoided.

6. **Early introduction of procedures.** Procedures are presented in Chapter Two, including modular programming with top-down design. The early emphasis on good programming practices helps the student in the programming activities that follow.

7. **Case studies.** Case studies from different areas of technology are presented in key chapters. These large programs are carefully developed, using only material that the student has already learned. Applied programming method tables are used to develop necessary algorithms, data types, procedures, functions, and program documentation.

8. **Variety of programming formats.** This variety of styles affords the student valuable experience that will prove useful in real-world settings. The style used to present and demonstrate new information is consistent throughout the text to facilitate learning, but the programming style may vary for the case studies.

Pedagogical Features

1. **Chapter objectives.** Clearly stated at the beginning of each chapter, the objectives give the student a quick overview of what to expect.

2. **Key terms.** Key terms at the beginning of each chapter include all new terms that will be presented and defined in the chapter.

3. **Glossary.** A glossary of all key terms appears at the end of the book and provides a handy reference for the beginning student.

4. **Program implementation and debugging.** The sections in each chapter on program implementation using Turbo Pascal 4.0 and debugging help the beginning student, who needs guidance to prevent potential problems and correct existing ones, progress through the course.

5. **Self-check review questions.** These questions at the end of each chapter section are open-ended to stimulate class discussion on key topics.
6. **Interactive exercises.** The unique interactive exercises section in each chapter encourages student participation at the computer console. Programming activities are selected to give immediate feedback and demonstrate the unique properties of the particular computer being used by the student, at the same time emphasizing new concepts presented in the chapter.
7. **Self-test.** These tests, which question the student about the Pascal program case studies, may require programming debugging as well as the careful analysis of programs.
8. **New command summaries.** At the end of each chapter, the summary of all new Pascal commands aids in reviewing as well as test preparation.
9. **Varied problem sets.** End-of-chapter problems are divided into major areas. General concepts test the student's general knowledge about the chapter. Program analysis consists of programming excerpts to analyze and correct for potential problems. Program design includes problems carefully selected to fit into representative areas of technology, so that the technology student may work in an area of interest and see the programming commonalities of other technology disciplines.
10. **Pascal dialect.** Turbo Pascal 4.0 is presented as the standard Pascal used in this text.
11. **Introduction to DOS.** The DOS operations described in an appendix are important for beginning students using the IBM PC, or compatible, with Turbo Pascal.
12. **Reference Appendices.** There are also appendices that present the forms and tables used with the applied programming method and suggested Pascal formats.
13. **Instructor's Diskette.** The instructor's diskette, available with an Instructor's Guide, contains "bugged" Pascal programs for each chapter in the text. These programs may be distributed to students as part of class assignments. This diskette also contains the source code of all major programs used in the text.

The author would like to thank the following reviewers for their many suggestions and guidance during the development of this text:

Stephen Aninye	ITT Technical Institute
	Arlington, Texas
Bill Champion	DeVry Institute of Technology
	Irving, Texas
Lall Comar	Florida Community College
	Jacksonville, Florida
William Diman	DeVry Institute of Technology
	Kansas City, Missouri
Don Distler	Belleville Area College
	Belleville, Illinois
Tom Eichler	DeVry Institute of Technology
	Lombard, Illinois

Charles Goodspeed	Cabrillo Community College
	Aptos, California
Walter Hedges	Fox Valley Technical College
	Appleton, Wisconsin
Michael Karamolengos	DeVry Institute of Technology
	Chicago, Illinois
Robert Lambiase	Suffolk Community College
	Selden, New York
Paul Ross	Millersville University
	Millersville, Pennsylvania

Finally, I would like to express my appreciation to the people at Merrill Publishing. In particular, I would like to thank Steve Helba, administrative editor, for his help and encouragement in this project. Also, a special thanks goes to Anne Daly and Linda Peterson, production editors; Lorraine Woost, art coordinator; Lois Porter and Elizabeth Wong, freelance copyeditors; and Catherine Parts, editorial assistant.

T. Adamson

Contents

Glossary   625

Answers   633

1 Getting Started

Objectives

This chapter gives you the opportunity to learn:

1 How to benefit from the features in this text that help you learn, reinforce, and retain new material.
2 The major sections of a computer and their purposes.
3 What a computer language is and why there are different computer languages.
4 The definition of *structured programming* and why such programming can be very important to you and your present or future employer.
5 Why Pascal is used as a programming language and what is necessary to use Pascal in a computer.
6 Some idea of Pascal.
7 An introduction to the Turbo environment.

Key Terms

Computer
Hardware
Processor
Memory
Peripheral Devices
Monitor
Keyboard
Disk Drive
Printer
Interface

Software
Program
Read-Only Memory (ROM)
Firmware
System Software
Applications Software
Data
Integrated Circuit
Chip
Microprocessor

Read-Write Memory (RAM)
Resident Monitor Program
Disk Operating System (DOS)
Floppy Disk (Diskette)
Initialize (Format)
Sectors
Instruction
Computer Language
Machine Language
Language Level
Character Set
High-Level Language
Pascal
Low-Level Language
Hexadecimal Number
Machine-Language Program

Interpreter
Compiler
Computer Science
Editor
Object Code
Pascal Operating System
Structure
Program Structure
Unstructured Program
Structured Program
Block
Block Structure
Programmer's Block
Remarks
Top-Down Design

Outline

This chapter presents an introduction to

1. This text.
2. The computer.
3. Computer languages.
4. Structured programming.
5. The programming language called Pascal.
6. What Pascal looks like.

If you have already had some exposure to computers, this chapter will be light reading. Even if you are an experienced programmer, you should still read the section on how to use this text.

If you are new to programming, this chapter is for you. You'll find it an easy chapter, and beginning programming information is fun to learn, making your first experiences with a computer memorable. Remember that all programmers start out knowing as much about programming as you do now. All expert programmers have done what you are doing now—taking that important first step.

Let's get started.

1–1   How to Use This Text

Welcome

This text is written for the beginning programmer, but even if you have had some experience in programming, you will find the material in this chapter an interesting review. This material is presented using the best available learning aids. Take advantage of them; you'll find they will help you learn and are fun as well.

Chapter Organization

The organization of each chapter in this book is given in Table 1–1, listing each feature and the reason for that feature.

Table 1–1   Chapter Organization

Feature	Purpose
Chapter Objectives	Presents what you should expect to know once you have completed the chapter.
Key Terms	Lets you see all of the terms that will be formally defined within the chapter.
Chapter Introduction	Shows why the chapter was written.
Program Debugging and Implementation	Introduces you to the Turbo Pascal 4.0 operating system and gives valuable hints on debugging your programs.
Case Studies	Some of the advanced chapters present case studies of actual Pascal programs with technical applications, giving you the opportunity to follow the creation of large, complex programs in a technological arena.
Section Reviews	Short questions at the end of each section that test your understanding of the material just presented.
Interactive Exercises	Gives you the opportunity to practice what you have learned on your own computer and learn the specifics of your system.
Pascal Commands	All new Pascal commands introduced in the chapter are summarized for you.
Self-Test	Presents needed practice in the analysis of large programs with technical applications.
Problems	End-of-chapter problems presented in three areas: General Concepts Program Analysis Program Design
Answers	Answers to section reviews, self-tests, and selected end-of-chapter problems are found at the end of the text.
Introduction to DOS	Appendix A presents an introduction to the disk operating system.

It's recommended that you read the material in each section, and then try the section review and check your answers. Doing this helps identify problem areas you may have and prevents you from reading new material with a poor understanding of old material.

The interactive exercises are intended for interaction between you and your computer, and will help introduce you to your Pascal system. Don't do these exercises unless you are no more than three feet from your computer.

It's a good idea to do the *self-test* as soon as you have read the chapter. These tests are carefully designed to point out key features presented in the chapter. Taking the test while the material is still fresh makes it fun and helps you retain the information longer.

The list of Pascal commands is handy to review before a chapter quiz, or to jog your memory of material in a previous chapter.

Ideally, you should select end-of-chapter problems in an area of your interest, but this may not always be possible in a classroom environment. However the problems are assigned, you should find many applicable to other courses or your own personal needs.

Conclusion

This section does not contain a section review, but the next section does; you should find it enjoyable as well as informative.

1–2   How the Computer Works

This section will introduce you to the main elements of a computer. Because microcomputers are so popular, they will be presented here. The components are likely to be similar to the ones you will be using with this text.

Hardware

Computer **hardware** consists of the computer and all the physical things attached to it. Figure 1–1 shows the major elements of computer hardware. First there is the computer itself, which includes the **processor** and **memory** (both explained later in this section). Then there are the **peripheral devices**, which include the **monitor** (the television-like screen), the **keyboard** (lets you put information into the computer), the **disk drive** (where you can save your information and get information from others), and the **printer** (where you get a "hard copy"). These peripheral devices are connected to the computer through special circuits called **interfaces**.

Software

Computer **software** is a list of instructions, called a **program**, that the computer processor will perform. One kind of software is stored in the computer's **Read-Only**

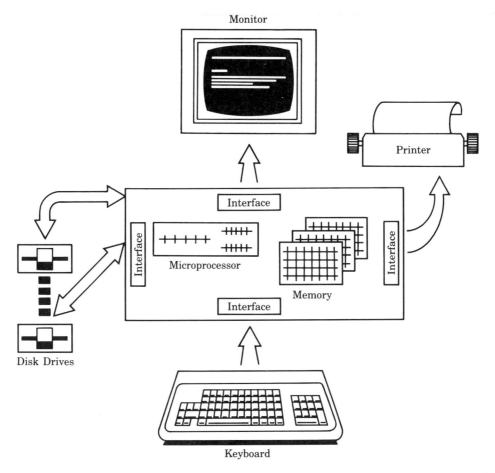

Figure 1-1 Computer Hardware

Memory (ROM). These are programs installed inside the computer at the factory, and you, the computer operator, cannot change them. For this reason, they are called **firmware** or **system software** programs. These programs are necessary to instruct the processor how to read the keyboard, display things on the screen, and make the printer operate.

Applications Software

Applications Software consists of instructions (programs) that are brought into the main memory from a peripheral device such as the disk drive to allow you to perform specialized tasks with the computer. These programs allow the computer to act as a word processor, spelling checker, video game, or one of many other applications.

Information

For the purpose of this text, **information** will be defined as what you type on the computer keyboard. As you will see, the computer will then use this information in many different ways, depending on what you have entered. Information of this kind is also referred to as **data**.

The Processor

The processor for a microcomputer consists of an **integrated circuit**, sometimes referred to as a **chip**. The chip shown in figure 1–2 is called a **microprocessor** because

Figure 1–2   Typical Microprocessor Chip (Courtesy Motorola Inc.)

it is a physically small processor. But don't let the size fool you! These microprocessors are more powerful than many of their giant grandfathers.

The microprocessor interacts with memory and does a series of **procedures**. All processors have a limited number of procedures. What makes them appear so powerful is their ability to do these procedures quickly (over a million each second).

Main Memory

The **main memory** is the region inside the computer where the processor stores information. This memory is called **RAM**. RAM stands for **Randomly Accessible read-write Memory**. This means that, unlike ROM (where information can only be read), new information can be entered into RAM as well as read from it. Another difference is that when the computer is turned off, everything inside RAM is forgotten. So if you want to save information for later use, you must store what is in RAM some other place, usually the disk drive or printer, before leaving the computer.

Resident Monitor Program

The **resident monitor program** is a set of instructions inside the computer's ROM that starts up the computer when you turn it on.

Disk Operating Systems

A **disk operating system (DOS)** is a program that can be loaded into the computer (usually through the disk drive) which allows you to work easily with the disk drive system. This program contains instructions to the processor on how to copy information from one disk to another, transfer information to other devices, and many other useful functions. It enables you to give a simple instruction, such as COPY, and have the information contained on one disk automatically copied on another. Figure 1–3 shows some of the functions performed by a disk operating system.

The Disk

You will store many of your programs on your own **disk**. This is sometimes referred to as a **floppy disk**, because it is made from a flexible material enclosed in a protective jacket. The disk is actually another form of memory.

Before you can use a new disk to save your information, you must first **initialize**, or **format**, the new disk by placing the new disk inside your computer disk drive and following the instructions for your own system.

When you format a disk, any information that was on the disk will be erased, so you usually only perform this operation when you first get the disk. It's very important to remember this: **When you initialize a disk, you erase everything on it—your homework, your best program, your favorite game—and you can never get them back from that disk again—ever!**

Formatting the disk causes the disk operating system (DOS) to place concentric **tracks** on the disk and divide the disk into **sectors**. Information that you later store on

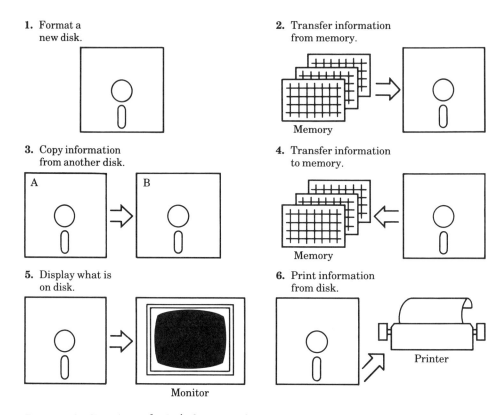

1. Format a
new disk.

2. Transfer information
from memory.

Memory

3. Copy information
from another disk.

A   B

4. Transfer information
to memory.

Memory

5. Display what is
on disk.

Monitor

6. Print information
from disk.

Printer

Figure 1–3   Functions of a Disk Operating System

the disk will, with the assistance of the DOS, be placed in one or more of these sectors. Such an arrangement is illustrated in figure 1–4.

Conclusion

This section introduced you to the basic parts of computer hardware and software. No matter what computer system you will be using, the basic concepts presented in this section will apply. Test your new knowledge in the following section review.

1–2 Section Review

1 Describe the major parts of a computer system.
2 How would you explain the difference between hardware and software?
3 List some examples of a peripheral device.
4 Explain the purpose of applications software.
5 Describe the properties of the microprocessor inside a microcomputer.
6 State the difference between RAM and ROM.
7 Explain the purpose of a resident monitor program.

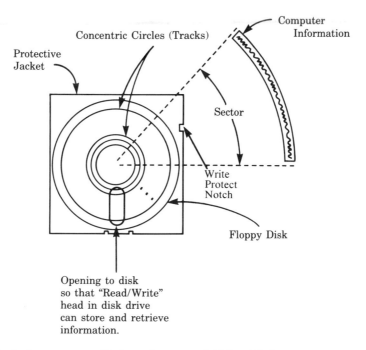

Computer Information

Concentric Circles (Tracks)

Protective Jacket

Sector

Write Protect Notch

Floppy Disk

Opening to disk so that "Read/Write" head in disk drive can store and retrieve information.

Figure 1–4 Arrangement of Information on an Initialized Disk

8 Describe the purpose of a disk operating system.
9 State the functions that can be performed by DOS.
10 Describe how information is stored on a floppy disk.

1–3 Computer Languages

What the Computer Understands

All computers understand only two things: **ON** or **OFF**. These ONs and OFFs are fed into the computer using a 1 for an ON and a 0 for an OFF. As an example, the sequence

10110010

is an **instruction** to the microprocessor within the IBM PC that causes the microprocessor to move information into itself. This process is easy for a computer to understand, but it requires a great deal of training and practice for people to understand. In the "old days" this is how a computer was programmed. To program a computer this way today means either that you are a masochist or you have a very old computer! Even though the computer you use today still operates with ONs and OFFs, we people no longer have to.

Computer Language

Remember that the microprocessor inside your computer understands only one **computer language**, the language of 1s and 0s, called **machine language**. Since this is the most fundamental language understood by a computer, it is called a **low-level language**.

What People Understand

Rather than using a machine language of 1s and 0s, it would be easier to give instructions and get answers back using symbols that we already understand. The easiest **set of characters** that we can use is that which we are now using to communicate. It would be very easy to program if you could simply tell the computer

> Figure out my income tax and let me know when you're finished so I can tell you where to mail it.

Language Levels

Although computers can't quite yet follow those instructions, a program can be written in symbols and letters that people easily understand; such a program is written in what we call a **high-level language**. The language presented in this text, **Pascal**, is a high-level language. You should be happy to know that! Figure 1–5 illustrates the different levels of some of the major computer languages.

Any language that is higher in level than machine language requires a program to convert that language into machine language. For example, the next level understood by the microprocessor above the 1s and 0s is the **hexadecimal number system**, a number system to the **base 16**. Using this system, the instruction

`10110010`

can be entered into the computer as

`B2`

—two keystrokes instead of eight!

Programming a computer this way is still called machine-language programming by many programmers. Now, however, there must be a program inside the computer to **interpret** the instruction `B2` and convert it to `10110010`. The program that does this is called an **interpreter**. Thus, for any computer to be programmed in a language higher in level than the 1 and 0 machine language, the computer must have an interpreter stored in its memory. Most microcomputers have an interpreter program installed inside their ROM at the factory that will interpret the programming language called BASIC. The action of an interpreter is shown in figure 1–6 (p. 12).

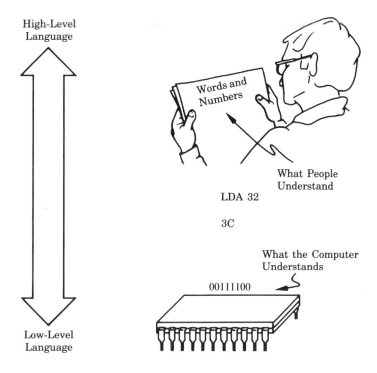

Figure 1–5   Different Levels of Computer Languages

Why Other Languages?

The main reason there are so many different programming languages is an attempt to make things easier for the programmer. Table 1–2 lists some of the more popular languages and why they were invented.

Table 1–2   Some Programming Languages and Their Advantages

Programming Language	Reason for Development
APL	(A Programming Language) A very powerful language for investigative work.
Assembly language	A low-level programming language that allows direct control of the microprocessor.
BASIC	Easy to learn.
COBOL	For business-oriented programming.
LOGO	A simple programming language originally designed to teach children how to use computers.
Pascal	To teach good programming habits.
PILOT	To develop computer-aided instruction.
C	To easily control the computer.
FORTRAN	To solve mathematical formulas.

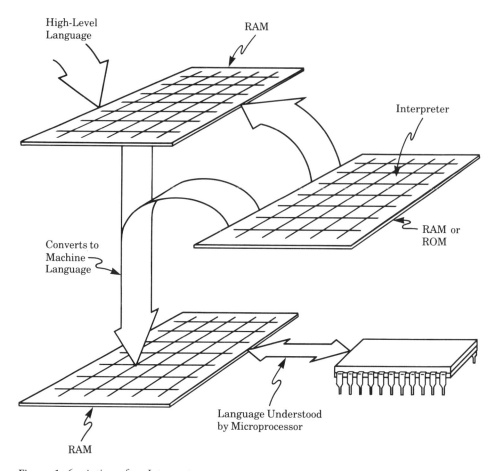

Figure 1–6   Action of an Interpreter

There are other programming languages; new ones are being developed almost every year. Those listed in table 1–2 were selected to give you an idea of the reasons for different languages.

Compilers

Pascal is a **compiled** language. The action of a **compiler** is similar to that of an interpreter. It is the same in that it is used to convert instructions into those the computer understands, but there are also some major differences. The compiler is a program that comes on a disk that must be loaded into your computer. The compiler then takes what you enter into the computer and converts it into another program that can be understood by the computer. This new program can be used over and over again, never needing to be recompiled. With an interpreter, however, the program is interpreted every time it is executed. Figure 1–7 shows the action of a compiler.

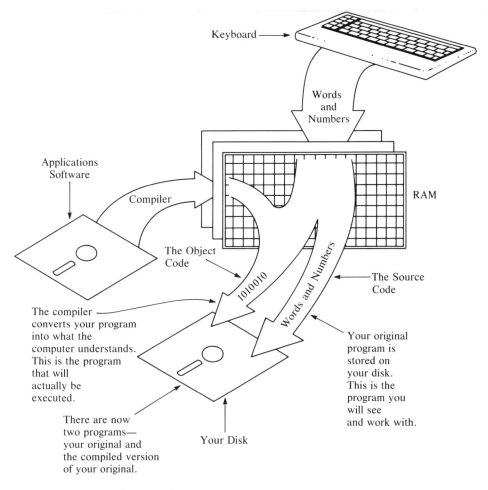

Keyboard

Words
and
Numbers

Applications
Software

Compiler

RAM

The Object
Code

1010010

Words and Numbers

The Source
Code

The compiler
converts your program
into what the
computer understands.
This is the program
that will
actually be
executed.

Your original
program is
stored on
your disk.
This is the
program you
will see
and work with.

There are now
two programs—
your original and
the compiled version
of your original.

Your Disk

Figure 1–7   Action of a Compiler

Conclusion

Some fundamental concepts about computer languages were presented in this section. The study of these languages is in itself the major discipline of **computer science**. Your task will be to use Pascal in a manner that assists you in developing and using your technical knowledge.

1–3 Section Review

1   Explain what is meant by a computer language.
2   Describe the two things that computers "understand."
3   State the name of the computer language that is used by the microprocessor in your computer.

4   Explain what is meant by language levels concerning computer languages.
5   Describe the function of an interpreter as applied to computers.
6   State what is meant by a BASIC interpreter.
7   What must your computer have before it can use any high-level language?
8   Give the names of some of the most commonly used higher-level languages and the reasons for their use.
9   What is a compiler?

1–4   The Pascal Operating System

Overview

An operating system ideally provides the programming tools needed to work with a particular programming language. When you program in the language called BASIC, the environment is usually built into the microcomputer (in its ROM). In this case, all that is usually necessary is to turn the computer on, type in BASIC commands, and instruct the computer to interpret the program as it runs. But this is not the case with other programming languages. This section presents what is called the **Pascal operating system**. (Section 1–7 will later introduce the Turbo Pascal* operating system.)

Using the Pascal Operating System

There are many manufacturers who provide software on a floppy disk that will allow you to program your computer using the programming language called Pascal. The operating system used with this text is **Turbo Pascal**. This system is used because it is inexpensive, readily available, easy to use, and widely accepted in industry as well as the classroom. The Program Debugging and Implementation sections of the beginning chapters show you how to get started using the Turbo Pascal operating system. This section will give you a brief introduction to what the Turbo Pascal operating system will do for you.

There are three fundamental steps in using a Pascal operating system.

1. Use a program called an **editor**. This program allows you to type in Pascal commands from the computer keyboard.
2. Submit the typed program to another program called the compiler. The compiler will check your typed Pascal program for errors and let you know if there are any and what they are. If there are no errors, the compiler will translate your Pascal **source code** (you are the source) to the computer **object code** (the computer is the object).

---

*The Turbo Pascal operating system and the name Turbo Pascal are copyrighted by Borland International Inc.

3. You now have two programs. One program is the original you wrote (containing symbols you understand and entered from the keyboard), and the other is the compiled program (containing the code the computer understands). You will execute the compiled program.

This information is presented in detail in the Program Debugging and Implementation sections of the beginning chapters. You will also need some kind of **disk operating system** to assist you in saving your programs. You can refer to appendix A for an introduction to the disk operating system. The main parts of the Pascal operating system are shown in figure 1–8.

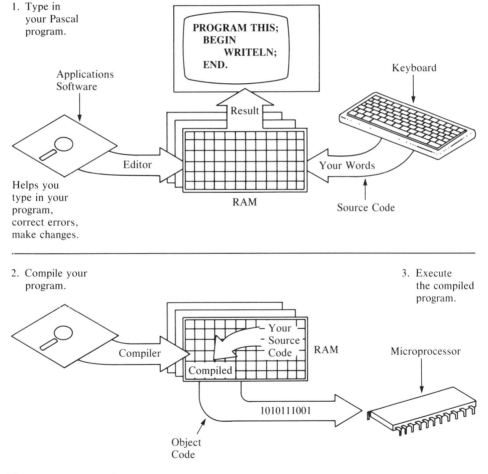

Figure 1–8 A Pascal Operating System

Conclusion

This section introduced a Pascal operating system to give you an idea of how to enter the program into your computer and get the program to do what you want (executing the program). In the next section, you will learn some of the reasons Pascal is such a useful programming language and how you can use it in technology.

1–4 Section Review

1   State the purpose of an editor. Why is it needed?
2   Give the reason for using a compiler.
3   Explain the action of a compiler.
4   What is the difference between source code and object code?

1–5   Why Pascal?

Background

In 1965, Professor Niklaus Wirth of the Swiss Technical Institute in Zurich, Switzerland presented a new language as an enhancement and replacement of another programming language called ALGOL 60. At that time, ALGOL 60 was the most popular language in Europe for teaching programming. Dr. Wirth developed the Pascal language for two primary reasons:

> "The development of the language Pascal is based on two principal aims. The first is to make available a language suitable to teach programming as a systematic discipline based on certain fundamental concepts clearly and naturally reflected by the language. The second is to develop implementations of this language which are both reliable and efficient on presently available computers."*

Reasons for Pascal

The reasons for the popularity of Pascal and their meaning to you as a beginning programmer are summarized in table 1–3.

What Pascal Looks Like

Program 1–1 is a Pascal program. Note that the material between the brackets is disregarded by the compiler; this is a form of internal documentation.

After being compiled and executed, program 1–1 will display

```
Send me a 10 ohm resistor.
```

---

*PASCAL, User Manual and Report, Kathleen Jensen, Niklaus Wirth, 1978, Springer-Verlag, New York

Table 1–3   The Power of Pascal

Advantage	What this means to you
Designed for top-down programming.	Your programs will be easier to design.
Designed for structure.	Your programs will be easier to read and understand.
Allows modular design.	Enhances the programs' appearance so others can easily follow and modify. Makes it easier for you to debug.
An efficient language.	More compact, quicker-running programs.
Portability.	A program you write on one computer will operate on another computer system with little or no change.

Program 1–1

```
{ This is a Pascal program.  It will print a message on
   the computer screen.                                 }
PROGRAM First;
 BEGIN
    WRITELN('Send me a 10 ohm resistor.');
 END.
```

Observe that what was entered between the { and } was not displayed on the screen when the program was executed.

Note that no one doubts what program 1–1 will do, because its **structure** makes it easy to read, even though there are some "strange" symbols (like the { and }) that will be explained shortly. What you see in this program is the source code that you, the source, would type in.

In program 1–1, the items required by Pascal are in **boldface** and CAPITALIZED. The items put in by choice are not in boldface and are in lowercase letters. You can thus conclude that such words as **PROGRAM** and **END.** (followed by a period) are required in every Pascal program.

The Different Parts

As outlined in table 1–4, all Pascal programs must have certain parts.

Don't worry about the other items (such as the **WRITELN** and the ') used in program 1–1. They help get the sentence displayed on the monitor; you will learn their meaning and uses later.

Table 1–4   Major Parts of a Pascal Program

Item	Purpose
PROGRAM	This marks the point where the Pascal program really starts. Required for all Pascal programs.
First	Every Pascal program must be given a name, but you can make it up yourself. There are some rules to follow (for example, you cannot start the name with a number), which you will soon be learning.
BEGIN **END.**	The word **BEGIN** indicates where a Pascal program begins and the work **END** followed by a period (.) tells where it ends. All of the program instructions must go between the **BEGIN** and the **END.**
{ }	These symbols are optional and are used to enclose comments, which are remarks used to help clarify the program for people but ignored by the computer.
;	The semicolon (**;**) plays an important role in all Pascal programs. It must be used at the end of most programming lines.

Conclusion

Congratulations! You have just seen your first Pascal program, and it displayed a message on the screen. The program was made as simple as possible to enable you to concentrate on the main idea of a Pascal program. In the next section, you will learn about structure and why you should use it.

1–5 Section Review

1   State, in your own words, the two principle aims of the Pascal programming language.
2   What is meant by portability in a programming language?
3   What must all Pascal programs start with?
4   What two words must all Pascal program instructions go between?
5   Give an example of a comment in a Pascal program.
6   What is the purpose of the brackets { } in a Pascal program?

1–6   Program Structure

All Pascal programs in this text will conform to a specific structure. You can think of a **program structure** as the format you use when entering the program. The first Pascal program had a structure.

When compiled and executed, program 1–1 would cause the monitor to display (on the first line starting at the left of the screen)

Send me a 10 ohm resistor.

but program 1–1 could just as well have been written with a different structure.

Program 1—1

```
{ This is a Pascal program.  It will print a message on
    the computer screen.                                    }

PROGRAM First;

 BEGIN

    WRITELN('Send me a 10 ohm resistor.');

 END.
```

Program 1–2

```
    { This is a Pascal program.  It will print a message on
         the computer screen. } PROGRAM First; BEGIN WRITELN('Send
me a 10 ohm resistor.'); END.
```

When compiled and executed, program 1–2 would display exactly what program
1–1 displayed. The only difference is that the structure has been changed, and the
program has become a little harder to read. You could also have written

Program 1–3

```
PROGRAM First; BEGIN WRITELN('Send me a 10 ohm resistor.'); END.
```

and again, when compiled and executed, the program would have shown the same
display. The point to notice is that good program structure follows rules of appearance
that make the reading and understanding of programs easier for people (not
computers). Program structure may seem to make very little difference for the simple
program introduced here, but program structure becomes very important for larger
programs.

Structured vs Unstructured Programming

One of the measures of a "good" computer program is that it can be read and
understood by anyone—even if they do not know how to program. For example, the

first Pascal program presented in this book was easy to understand. Admittedly it didn't do much, but if good structure is employed, a program will be easier to understand, modify, and debug when necessary. An **unstructured program** gives little or no effort to help people read and understand it. Remember that the structure of a program makes no difference to the computer—only to the people who have to work with it.

Program Blocks

You can think of a **structured program** as having certain parts of the program located at particular positions in the program document. These positions can be thought of as **blocks** of information. A letter also contains blocks of information.

> John Student
> 123 Page Mill Road
> Programville, USA
> 01234
>
> The Resistor Company
> 321 Mill Page Road
> Sourcecode, USA
> 43210
>
> Dear Sir,
>
> Send me a 10 ohm resistor.
>
> Sincerely,
>
> John Student
>
> PS Thanks for the prompt delivery of my last order.

John Student's letter could just as well have been written without spacing.

John Student 123 Page Mill Road Programville, USA 01234 The Resistor Company 321 Mill Page Road Sourcecode, USA 43210 Dear Sir, Send me a 10 ohm resistor. Sincerely, John Student PS Thanks for the prompt delivery of my last order.

The difference between the two letters is that one is structured; the other isn't. The unstructured letter would take less time to write, less time to print, use less paper, and might even save you some postage. But such a letter is difficult to read, not because of its writing style, but because of its structure.

Let's look at John Student's letter again, emphasizing its **block structure** this time.

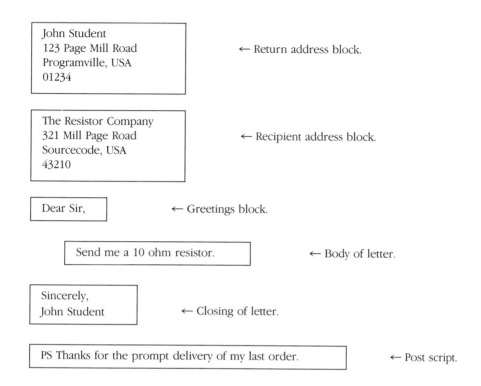

Just as the block structure of the letter makes it easy to read, when block structure is used in a program, it, too, is easier to read. There is an accepted block structure for letters, and in a professional programming environment there will be an accepted block structure for Pascal programs. Blocks required in most professional and technical programming situations will be used in this text, preparing you to understand, use, modify, and update Pascal programs in your work environment.

The Programmer's Block

All Pascal programs start with a block called the **programmer's block**. It consists of **remarks** containing the following information:

1. Program name
2. Developer and date
3. Description of the program
4. Explanation of all variables
5. Explanation of all constants

You already know enough about Pascal to develop a Pascal program that would solve for the voltage drop across a 10 ohm resistor with a specified amount of current. The formula for this relationship is

$$V = I \times R$$

where   V = the voltage across the resistor measured in volts
         I = the current in the resistor measured in amps
         R = the value of the resistor measured in ohms
        The Pascal program would start with the programmer's block, as illustrated in
program 1–4.

Program 1–4

```
{   ******************************************************
    *                  Voltage Solver                   *
    ******************************************************
    *          Developed by:  A. Good Structure         *
    *          September 1993                           *
    ******************************************************
    *   This program will solve for the voltage drop*
    * across a 10 ohm resistor.  The program user   *
    * must enter the value of current.              *
    ******************************************************
    *           Constants used:                         *
    *---------------------------------------------------*
    *         R = 10   (A 10 ohm resistor)              *
    ******************************************************
    *           Variables used:                         *
    *---------------------------------------------------*
    *         V = Voltage across resistor.              *
    *         I = Current in resistor.                  *
    ******************************************************     }

PROGRAM Voltagesolver;

   BEGIN

     { Body of the program to do the above.   }

   END.
```

   Program 1–4 will compile and execute, but nothing will happen, because no
commands were put into the program. The program consists only of remarks and the
essential elements of a Pascal program, but the programmer's block is complete. It
tells you exactly what the program will do, defines the variables (V and I), and the
value of the constant R that will be used in the program. **Top-down design**, as
presented in chapter 2, will show you that stating the problem in words is the essential
first step in programming design. Selecting variables and defining them is a second
important step. Now you know what must be included in a programmer's block and
how to write such a block using Pascal.

Note that the program name is `Voltagesolver`. Recall that you can give a Pascal program any name you want. You will learn what you can and cannot use in a program name, but for now, any words made of letters of the alphabet but no spaces can be used as names except **PROGRAM, BEGIN,** and **END.** These few rules will keep you out of trouble most of the time.

Advantages and Disadvantages of Structured Programming

The main disadvantage of structured programming is for those who have been used to unstructured programming. Old habits die hard. If you've never programmed before, structured programming will not have this disadvantage for you. The only other disadvantage of structured programming is that it makes short, unstructured programs longer.

In the past, computer memory was relatively expensive and limited. Thus it did, at one time, pay to conserve computer memory space by making programs as brief as possible. Today, even pocket computers have more memory than many of the older, larger machines, so there is no longer the necessity to be brief in programming. There is, however, the necessity to be clear in programming and develop good programming habits that result in a completed program that is easy to understand, easy to modify and easy to correct. That's the purpose of this book.

What you learn in this text can be applied to any structured language. When you complete this text you will be proficient in creating programs using structured programming techniques.

Conclusion

In this section you were shown the general difference between structured and unstructured programming. You will learn much more, including the power of Pascal as a programming language. Test your new skills in the following section review.

1–6 Section Review

1 Define the term structure as used in programming.
2 Is it necessary to give a structure to Pascal for it to compile without errors?
3 Describe a block structure, and give an example.
4 Explain the reason for structured programming.
5 Describe a programmer's block, and state what it must contain.

1–7 Program Debugging and Implementation— Starting with Turbo

To get full benefit from this book, you will need to have Turbo Pascal 4.0, produced and copyrighted by Borland International. The owner's handbook that accompanies the software has detailed information for using this integrated environment. Here, and in the other program debugging and implementation sections you will find the basic information that will help you get started with this extensive and powerful operating system. This text is not intended to replace the over-600-page owner's handbook supplied with the Turbo Pascal 4.0 system.

First Things

Be sure to make backup copies of your Turbo Pascal disks. Refer to Appendix A (Introduction to DOS) if you are not sure how to do this.

After making the backup copies, put your original Turbo Pascal disks away in a safe place. Now make a **working disk** that contains the following programs:

```
TURBO.EXE
TURBO.TPL
TURBO.HLP
```

Label your working disk

Turbo Pascal 4.0
Working Disk
Your Name   Date

Be sure to use a soft felt-tip marker when writing on the disk label. The pressure needed to use a ball point pen or lead pencil may require you to press so hard that the disk becomes unusable.

Getting Started

If you used the procedure recommended in Appendix A for making your working copy, once your system is booted and with your working disk in drive **A:**, simply type the word **TURBO** from the DOS prompt **A>**. After a few seconds of disk use, your computer screen will appear as shown in figure 1–9. Included will be a copyright notice. Simply press the space bar and the copyright notice will disappear.

Tour of the Main Turbo Screen

At the top line of the screen you will see five **menu items**:

File Edit Run Compile Options

Each of these menu items represents a distinct part of the Turbo Pascal environment. The functions of these menu items will be covered later as needed. For now, just know that the top line of this screen is called the **main menu** and contains five menu items.

The very last line of the screen is referred to as the **bottom line**. This bottom line tells you which of the function keys (such as F1 or F2) on your keyboard will cause something to happen. Again, don't worry about what they do for now; just know they are there and that on this screen they are referred to as the bottom line.

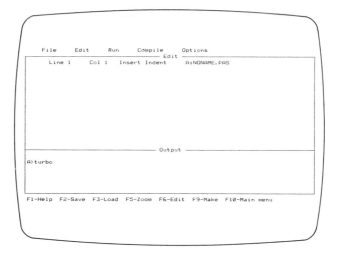

Figure 1–9   Main Turbo Screen

The next thing to note is that the screen has two major sections, divided by a horizontal line containing the word **OUTPUT**. The blank screen above this line is referred to as the **edit screen**. The edit screen will contain the Pascal source code (usually entered by you). The blank screen area below this line, called the **output screen**, will contain the results a Pascal program in the edit screen may have on the screen when it is executed.

Entering the Edit Screen

Note that in the main menu, one of the menu items will be highlighted. When the Turbo system is first used, `File` will be highlighted. You can change the highlighted menu item by pressing the left or right arrow keys on the keyboard.

To begin, make sure that the highlighted menu item is `File` and then press the RETURN/ENTER key; the **file menu** will then appear as shown in figure 1–10.

Such a display is called a **pull-down window** because you essentially "pull it down" when you activate it. This particular pull-down, the file menu, consists of nine items each starting with a boldfaced letter.

Note that now one of these nine menu items will be highlighted. The up and down arrow keys on your keyboard will cause different menu items to be highlighted. Using these arrow keys, highlight the menu item called `New` and press RETURN/ENTER. The file menu will now disappear and you will see a blinking cursor at the top left of the edit screen. You are in the Turbo Pascal editor and anything you type on the keyboard will now appear on the edit screen. Try it by typing

```
hello
```

What you have just typed is not, of course, Pascal, but it does confirm that you got into the editor. Note that at the top of the edit screen you see

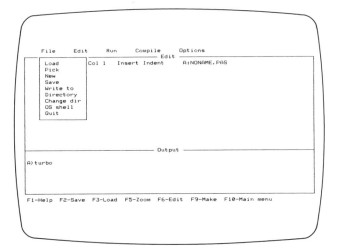

Figure 1–10   The File Menu

`Line 1 Col 6  Insert Indent   A:NONAME.PAS`

The first two items (Line and Col) show you where your cursor is located on the screen (it should now be at line 1, column 6). The other items will be explained later.

Now you know how to enter a word into the Turbo editor. To remove the word, use the **backspace key** (the one at the top of the keyboard with the big left arrow on it). Keep depressing this key and note that the cursor will move to the left, erasing everything in its path.

Putting in Your First Program

While you are in the Turbo editor, type in the following Pascal program.

```
program first;
begin
write('my first program.');
end.
```

Press the RETURN/ENTER key at the end of each line. Don't worry about capital letters for now, but do make sure that all of your punctuation is exactly as shown in the above program or it will not compile successfully. Make sure you use the ' and the ; as well as the period at the end of `end`. If you make any mistakes entering the program, use the backspace key to erase the mistakes, then type what should have been there.

Running Your Program

Once you're sure the program is entered correctly, you are ready to leave the editor. Look in the bottom line of the screen for `F10:Main Menu`. Press F10 and you will leave the editor and return to the main menu. Note that even though you are no

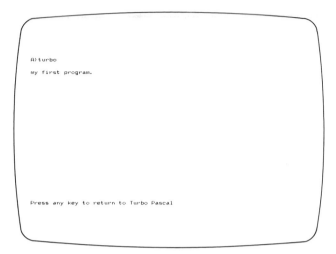

```
A) turbo

my first program.




Press any key to return to Turbo Pascal
```

Figure 1–11   Effects of First Program on Turbo Screen

longer in the editor, what you typed in is still there. The **File** selection will now be highlighted. Use the right arrow key to highlight the **Run** option of the main menu and then press RETURN/ENTER.

Now your program will automatically begin to compile. If you didn't make any errors in entering it, you screen will appear as shown in figure 1–11.

If you did make an error, the compiler will stop and automatically return you to the editor. An error message will be displayed and the cursor will be placed at or near the location of the error. (Note: No compiler is perfect in finding all possible errors. You will get its best guess—it's up to you to find and correct the actual error!) Correct your mistake (the first key you press will automatically erase the error message). Then repeat the process of leaving the editor (**F10**) and selecting the **Run** option to try your program again.

Leaving Turbo

You could, at this point, simply turn off the computer, remove the disk, and go home. However, the preferred method of exiting the Turbo environment is to make sure you are at the main menu (**F10** always gets you there). Use the arrow keys to highlight the **File** option, and press RETURN/ENTER once you are there. The file menu will appear again with its nine options. Use the arrow keys to select the **Quit** option, and press RETURN/ENTER.

A message will appear asking if you want to save **NONAME.PAS.** Enter **N** for no. Your system will be returned to the DOS prompt. Now you can remove your working disk, turn off the system and the monitor (some monitors will turn off automatically with the computer), and you're finished. Your program was not saved, but you will learn how to do that in the next Program Debugging and Implementation section.

Conclusion

In this section you were introduced to the Turbo Pascal 4.0 environment, and given the minimal commands necessary to enter a Pascal program, compile, and run it. In the next chapter, you will learn how to save your programs to the disk and how to retrieve programs from the disk. For now, test your understanding of this section by trying the following section review.

1–7 Section Review

1 State one of the first things you should do before using your new Turbo Pascal operating system.
2 What is the main menu?
3 Explain how to enter the editor.
4 What key gets you back to the main menu?
5 State the preferred method of leaving the Turbo environment.

Summary

1 This text contains many different learning features, including the chapter introduction, objectives, section reviews, summary, interactive exercises, self-tests, new commands, chapter problems, and answers to section reviews and self-tests.
2 Hardware can be thought of as the tangible parts of the computer.
3 Software can be thought of as the instructions used to operate the hardware.
4 The computer consists of many different parts which may include the processor and peripheral devices such as the monitor, disk drives, and the printer.
5 Applications software is a program that allows you to perform a specific task with your computer.
6 The main memory of the computer consists of RAM and ROM.
7 RAM means Randomly Accessible (read-write) Memory. It is used for the temporary storage of software.
8 ROM means Read Only Memory. It is used for the permanent storage of software usually installed at the factory.
9 The Disk Operating System (DOS) is a program that is loaded into the computer to give it instructions on how to operate the disk drives. It allows you to store, retrieve, and copy programs from the computer to the disk and between disks.
10 A new disk must first be initialized before it can be used on your computer system. This process erases any information that may have been on the disk. It should only be done once, when the disk is first used.
11 A computer language can be thought of as a set of characters that form symbols and the rules for using these symbols so that the programmer and the computer can communicate.
12 The computer understands a very fundamental process that consists of a series of electrical ons and offs. Mathematically this is represented as 1 for on and 0 for off.
13 A low-level language is a programming language that is close to what the computer understands.
14 A high-level language is a programming language that is close to what people understand.

15 An interpreter is a program that converts a high-level language into the low-level on/off code that the computer understands, every time the program is executed.

16 There are many different programming languages. Each was designed to fill a need seen by the developer of the language.

17 A compiler is a program that converts a high-level language into the low-level on/off code that the computer understands. But, unlike when an interpreter is used, the high-level language is compiled only once and need not be compiled every time the program is executed.

18 The Pascal programming language requires its own operating system that is usually different from the operating system that is built into your microcomputer.

19 The main parts of a Pascal operating system are a program called an editor, and a program called a compiler used in conjunction with some kind of disk operating system.

20 There are many reasons for the popularity of Pascal over that of other programming languages. Among these are ease of program design, efficiency, and portability.

21 Pascal requires all programs to start with the word `PROGRAM.`

22 Comments are an optional but important part of programming. In Pascal comments go between the `{` and the `}`.

23 The words `BEGIN` and `END.` are used in Pascal in order to indicate the beginning and the end of the program instructions.

24 The Turbo Pascal operating system contains an editor, filer and other systems designed to help you in creating, saving, modifying, and debugging your Pascal programs.

Interactive Exercises

Directions

These exercises require that you have access to a computer and software that supports Pascal, specifically the **Turbo Pascal Development Environment**, Version 4.0 (or higher), Borland International. The exercises will give you valuable experience and immediate feedback on what the concepts and commands introduced in this chapter will do. They are also fun. Note: If you are a first-time user, it is suggested that you start with Appendix A for an introduction to DOS.

Exercises

1 Find the brightness control on your monitor. What happens when you turn the brightness control fully counterclockwise? What happens when you turn it fully clockwise? About where do you place it for the most comfortable viewing? (NOTE: You really need to adjust this control to achieve the least amount of eye strain; this requires sharp characters with a maximum dark background.)

2 Find the contrast control on your monitor. What happens when you turn the contrast control fully counterclockwise? What happens when you turn it fully clockwise? About where do you place it for the most comfortable viewing (least eye strain)?

3 How many disk drives does your computer have? In which drive do you place the DOS when starting up the system?

4 How do you evoke the editor for your Pascal system?

5 What happens when you compile the Pascal program in program 1–5?

6 Do you have two programs when program 1–5 is compiled? What are these two programs called?

7 What happens when you execute program 1–5?

8 Enter program 1–5 again, but this time omit the programmer's block and enter

Program 1–5

```
{ *********************************************
  *            Program demonstration           *
  *********************************************
  *       Developed by:  Your Name              *
  *       Date:                                 *
  *********************************************
  *     This program prints a message on the    *
  * the computer screen.  It is part of the     *
  * interactive exercises for Chapter 1.  See   *
  * if you can predict what the message will be. *
  *********************************************
  *            Variables used:                  *
  *-------------------------------------------*
  *          No variables for this program.     *
  *********************************************
  *            Constants used:                  *
  *-------------------------------------------*
  *          No constants for this program.     *
  *********************************************  }

PROGRAM Thisone;

  BEGIN

    WRITELN('This is a Pascal program for technology.');

  END.
```

```
PROGRAM Thisone; BEGIN WRITELN('This is a Pascal program for
technology.');END.
```

9 What happens when you compile the unstructured version of program 1–5? Is this any different from what happens when the program is structured?

10 What happens when you execute the unstructured program? Is this any different from when the program was structured?

11 Enter and try to compile program 1–6.

Program 1–6

```
{ *********************************************
  *     To enclose comments start with a { and *
  * end with a } .                             *
  *********************************************  }
  PROGRAM Anotherone;
    BEGIN
    END.
```

11   Enter and try to compile program 1–6.

12   What happens when you try to compile program 1–6? Why does this happen?

Pascal Commands

PROGRAM Name;   In Turbo Pascal this is optional, but its use will be required in this text because of its universal acceptance in Pascal programs. Every Pascal program must start with the **reserved word PROGRAM** followed by an **identifier** consisting of a letter or underscore followed by any combination of letters, digits, or underscores to a maximum length of 127 characters. Turbo Pascal does not make a distinction between upper and lower case letters. The identifier must end with a semicolon (**;**).

BEGIN   END.   All Pascal commands must go between these two reserved words. The **END** must be immediately followed by a period (**.**).

WRITELN(' Any characters you want.');   A **standard identifier** used in Pascal. The characters between the (**'** and **'**) will appear on the screen. For now, it must end with a semicolon (**;**).

{Any characters you want.}   Brackets **{ }** are used to enclose comments to make the program easier to understand. Anything that appears between them is ignored by the interpreter; they have no effect on the operation of the Pascal program. Asterisks (***** and *****) can also be used in the same way, but in this text, brackets **{ }** will be used.

Self-Test

Directions

Program 1–7 was developed for construction technology; it could be used in the design of heating/cooling systems or in the design of the total structure. It is used to calculate the volume of a room. The programmer is using top-down design; the program is now in an intermediate stage before program code has been entered. There are, however, some errors in the program. See if you can spot them. Answer the following questions on page 33 by referring to program 1–7.

Program 1–7

```
}  *******************************************
   *   This program computes the volume of a  *
   * room.  The program user must enter the   *
   * value of the room dimensions.            *
   *******************************************
   *           Constants used:                *
   *------------------------------------------*
   *         There are no constants used.     *
   *******************************************
   *           Variables used:                *
   *------------------------------------------*
```

Program 1–7 *continued*

```
     *        H = Height of room.              *
     *        L = Length of room.              *
     *        W = Width of room.               *
     *********************************************   }

     BEGIN

     {| Explain program to user; |
      ----------------------------------------------------------}

       WRITELN('This program will compute the volume of a room.');
       WRITELN('You need to enter the value of the room height,');
       WRITELN('width, and length, and the computer will do the rest.');

      {----------------------------------------------------------
       | End explain program to user; |}

      {| Get values from user; |
       ---------------------------------------------------------}

         {  Ask for value of room height;  }
         {  Get value of room height;      }

         {  Ask for value of room length;  }
         {  Get value of room length;      }

         {  Ask for value of room width;   }
         {  Get value of room width;       }

    {-------------------------------------------------------------
      | End get values from user; |}

      {| Calculate volume of the room; |
       --------------------------------------------------------}

         {  V = H × L × W;  }

    {-------------------------------------------------------------
      | End calculations; |}

      {| Display the answer; |
       ---------------------------------------------------------}

        {  The room volume is V;  }

    {---------------------------------------------------------
      | End display answer; |}

   {|| Main Programming Sequence; ||
      ===========================================================}
```

Program 1–7 *continued*

```
{ Explain program to user;  }
{ Get values from user;     }
{ Calculate volume of room; }
{ Display the answer;       }

END.
{=======================================================
|| End of main programming sequence. || }
```

Questions

1   Describe what the above program will do if it compiles correctly and is executed.
2   If the program compiled correctly, what would be displayed on the monitor?
3   List the variables used in the program. State the purpose of each.
4   Is the programmer's block complete? If not, what is missing?
5   Does the program contain any errors that would prevent it from compiling? If it does, what are they?
6   What will the program do when it is completed?

Problems

General Concepts

1   Sketch the major parts of a computer system. Be sure to label all of the parts. [Section 1–2]
2   Name three different types of application software. If possible use the trade name of the software. [Section 1–2]
3   State what the resident monitor program causes your system to do. Be sure to include all of the sequences of operations. [Section 1–2]
4   Write the DOS commands used by your system that do the following. [Section 1–2]
    A. Get a program from the disk.
    B. Copy a program from one disk to the other.
5   Which programming language is the easiest to learn? Which one was designed to teach good programming habits and which was developed for business applications? [Section 1–3]
6   State the difference between an interpreter and a compiler. [Section 1–4]
7   Explain the main parts of the Pascal operating system. [Section 1–4]
8   Describe all of the required parts of a Pascal program as presented in this chapter. [Section 1–5]
9   Write a Pascal program that will display the following message on the monitor. Be sure to include the programmer's block. [Section 1–6]

Applied Pascal for technology.

Program Analysis

10   Will the following program compile? If not, why not?

```
{ ******************************
  * This is a Pascal program. *
  ****************************** }
```

11   When the following program is compiled and executed, what will the program cause to be displayed on the screen?

```
PROGRAM Pascal;
BEGIN
WRITELN('{ This is a Pascal program.}');
END.
```

12   What will be the screen display for the following program, once compiled and executed?

```
PROGRAM Pascal;
BEGIN
{This is also a Pascal program.}
END.
```

13   Correct any errors in the following program so it will compile correctly.

```
PROGRAM Pascal;
BEGIN
This is a comment;
END.
```

14   The following program has some errors. Correct them so it will compile correctly.

```
PROGRAM Name { This is the name of the program. }

   BEGIN  { The program begins here }
     WRITELN('END.');
{END.} {This is where the program ends.}
```

Program Design

Each problem is presented in a particular area of technology. The program you are to develop is to consist of a programmer's block and the minimum requirements for a Pascal program. No other programming code is required or encouraged.

```
PROGRAM Name;
BEGIN
END.
```

Manufacturing Technology

15   Develop a Pascal program that will compute the wage of a mechanic if the hourly rate is $30.00 per hour. The input variables are the number of hours worked and the hourly pay.

Drafting Technology

16   Develop a Pascal program that computes the area of a rectangle. The input variables are the height and width of the rectangle.

Computer Science

17   Develop a Pascal program that returns the cube of a number. The input variable is the number to be cubed.

Construction Technology

18   Develop a Pascal program that computes the sales tax on lumber. Assume the sales tax is 6%. The input variable is the cost of the lumber; the program constant is the sales tax.

Electronics Technology

19   Develop a Pascal program that computes the power dissipation of a resistor when the value of the resistor current and resistor voltage are known. The formula for power dissipation in a resistor is the product of the voltage and the current. Input variables are the voltage and the current.

Agriculture Technology

20   Develop a Pascal program that computes the volume of grain in a cylindrical silo. The formula for the volume of a cylinder is

$$V = \pi r^2 h$$

where   V = volume of cylinder
π = the constant 3.14159 . . .
r = radius of the cylinder
h = height of the cylinder

The input variables are the radius and the height. Program constant is π.

Health Technology

21   Develop a Pascal program that computes the number of hospital patient hours served by a doctor in one week. The input variables are the number of patients seen by the doctor and the amount of time, in hours, spent with each patient.

Business

22   Develop a Pascal program that computes the amount of withholding for an employee retirement plan that requires a 12% deduction for the gross weekly income. The input variable is the employee's weekly income. The program constant is 12%.

2 What Computers Can Do (Using Pascal to Do It)

Objectives

This chapter will give you the opportunity to learn:

1 The fundamental processes that computers are good at.
2 The fundamental processes that people are good at.
3 A theorem that formally states all of the processes a digital computer is capable of performing.
4 The difference between and the meanings of sequence, decision, and loop processes.
5 How to read basic flowcharts and use them in problem solving.
6 The concept of top-down design and how to apply it when developing a computer program.
7 How to convert a technology problem into a Pascal program using the applied programming method.
8 What program blocks are and how to develop them using Pascal.
9 A specific format to be used in the development of any Pascal program.

Key Terms

Completeness Theorem
Action Block
Branch Block
Open Branch
Closed Branch
Loop
Counting Loop

Sentinel Loop
Flowcharts
Top-Down Design
Main Programming Block
Procedure
Applied-Programming Method
Guide

Problem Statement Guide   Real Number
Problem Solution Guide   Integer Number
Algorithm   Character
Algorithm Development Guide   Boolean
Control Blocks Guide   String
Constants and Variables Guide   Pascal Design Guide
Type

Outline

Many beginners to programming are surprised to find that there are actually only three distinct processes any computer can do, no matter what computer language is used.

This chapter is an important chapter because it lays the groundwork for the chapters to follow, as well as for other programming languages you may learn that work with sequential digital computers. Here you will discover what computers and people are good at, and how to maximize the abilities of both.

Here you will learn how to approach a problem in technology and convert it into a computer program using top-down design. You will also be introduced to some very important problem-solving tools that will give you a step-by-step procedure for converting technology problems into Pascal programs. You will also learn the concept of block structure and how to apply it to Pascal. When you finish this chapter, you will be better at solving technical problems because you will have some new and powerful problem-solving techniques.

2–1   What Computers and People Do Best

Introduction

Many beginners approach programming without understanding just when a computer program should (or should not) be used. In this section you will learn what computers do well and what they don't do well, to help you decide how to best use a computer.

Table 2–1   What a Computer is Good at Doing

Activity	Meaning	Example
Repetition	Any task that has to be repeated many times.	Solving the same formula for many different values.
Arithmetic	Any task that requires a computation.	Addition, subtraction, multiplication, and division.
Logical Decision	Any task that requires a decision based on a yes or no answer.	Allowing the program user to repeat the program by typing in a "yes" or else end the program.
Sorting Information	Any task that requires the arranging of information from one form to another.	Alphabetizing or sorting numbers.
Graphical Representation	Any task that requires the display of pictures related to technical information.	Developing a graph from numerical information supplied to the computer.

What Computers are Good at Doing

The computer is a tool. More specifically, for the technician, it is a tool to be used for analyzing, monitoring, or problem solving. Just like any other tool you have, such as your pocket calculator, pencil, or paper, it is good at some tasks and very poor at other tasks. Table 2–1 summarizes the tasks a computer performs well.

One reason the computer performs the above tasks well is that it can perform them very quickly—much more quickly than people can. This text will examine the items listed in table 2–1, and you will learn how to get the computer to do what it does well using the programming language called Pascal.

What's Hard for the Computer

Since the computer is a machine, it cannot "think" in the sense that you or I think, but by programming very carefully, you can make the computer respond in such a way that it appears to be thinking. This is one promise of artificial intelligence, but at this time the "thinking" is illusion, not reality. Things that a computer does poorly are summarized in table 2–2.

These activities that a computer performs poorly are activities that people perform well. You can see a pattern here: tasks that computers perform well, people generally perform poorly, and vice versa. You can think of the combination of the human mind and the machine called a computer as a powerful mental system. Just as the airplane allows us to travel in the air, and the submarine gives us life underwater, the computer gives the human mind an extraordinary mental vehicle, allowing the combination of the two to transcend the ability of either one alone.

Table 2–2 What a Computer is Not Good at Doing

Activity	Meaning	Example
Patterns	Being able to determine a continuing pattern based upon limited or new information.	Determining what will follow in a series such as 3, 6, 9, 12, 15 . . . Recognizing the face of a friend.
Judgement	Deciding for yourself what to do in a situation which may or may not be familiar.	Deciding if you want to repeat a computer program over again.
Adaptation	Using information from past experience when facing a new situation.	Substituting a pair of pliers for a wrench when the directions for assembling a bike tell you to use a wrench.
Incomplete Information	Performing a complex task based on incomplete or partial instructions.	Successfully completing a homework assignment.

Conclusion

This section compared the abilities of computers and people. You saw many important activities that computers are good at performing, as well as what people are good at doing. In this text, you will see how to put the human mind and computers together using Pascal to create a powerful mental vehicle that will assist in the analysis and solution of technology problems. Test your understanding of this section by trying the following section review.

2–1 Section Review

1 State three activities computers are good at doing.
2 Name an activity that a computer is not good at doing.
3 What is the difference between logical decision-making and judgement?

2–2   No Matter What, This Is All a Computer Can Do

Introduction

There are only three ways a computer can approach a problem; this is true no matter what programming language is used. Everything else done in programming is a refinement of these three. Pascal makes it easier for the programmer to understand these three approaches.

An Important Theorem

In 1966, two computer scientists, Boehm and Jacopini, proved what is called the **completeness theorem**, that *any program logic, no matter how complex, can be* *resolved into action blocks, branch blocks and loop blocks.* This section will demonstrate what each type of block means.

Going From One Place to Another

You can think of an **action block** as a straight sequence of computer instructions. Think of a road map; an action block is similar to traveling along a road, going from one town to another, each town representing a computer instruction. This is illustrated in figure 2–1.

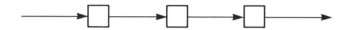

Figure 2–1   Illustration of an Action Block

You can think of an action block as instructions saying

```
DO THIS, THEN DO THAT, THEN THIS . . . ALWAYS
```

Deciding Where to Go Once You're There

The second kind of programming block is a **branch block**. Using the road map again, this is equivalent to driving to a town, then deciding which other town to go to next. This is illustrated in figure 2–2.

As shown in figure 2–2, there are two types of branch blocks. An **open branch** gives the chance to do something else before continuing, or just continue; a **closed branch** will always do one of two unique possibilities and then continue on. You can think of an open branch as instructions saying

```
IF THIS, DO THAT, BUT ALWAYS CONTINUE ONWARD
```

Using an open branch is similar to traveling to Beachtown and deciding on the way what other town to go to depending on the weather. If it's raining, you'll go to the movies in Movietown and to Beachtown the next day. If it's not raining, you'll go straight to the beach in Beachtown.

Keep in mind that these analogies are being made for a machine called a computer. When a branch block is programmed into a computer, it will do one thing or the other depending upon a very specific condition given in the program. For the computer there is no such thing as a cloudy day—it will either be raining or it will be sunny; a computer cannot exercise judgement. It will not turn around and go back home because of a cloudy day or a bad movie.

Using a closed branch is similar to going to Movietown if it's raining, or Beachtown if it's not raining; in either case you will go to Nexttown afterward. Another way of looking at the closed branch block is

```
IF THIS, THEN DO THAT, OR ELSE DO SOMETHING DIFFERENT,
     THEN CONTINUE ONWARD
```

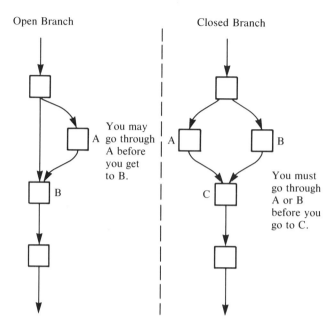

Open Branch

Closed Branch

You may A go through A before you get to B.

You must go through A or B before you go to C.

Figure 2–2   Illustration of a Branch Block

Going Back Where You Came From

The third kind of programming block is called a **loop**. You can think of this as going back and repeating a process. The road map analogy here is similar to a race track. Traveling along the race track causes you to pass by the same places over and over again. In this case, each place represents a computer instruction. This is illustrated in figure 2–3.

Loop blocks are used in programming when the same processes are to be used over and over again. There are two kinds of loop blocks; one kind is used when you know ahead of time how many times you want to perform the loop; the other is used when you don't know how many times the loop will be performed. This concept is illustrated in figure 2–4.

When you know ahead of time how many times the loop is to be repeated, you can think of it as

REPEAT THE LOOP UNTIL A CERTAIN CONDITION IS COMPLETED

This is called a **counting loop**.

When you don't know ahead of time how many times the loop is to be repeated, you can think of it as

WHILE SOMETHING IS HAPPENING DO REPEAT THE LOOP

This is called a **sentinel loop**.

There are many times in programming that you will want the same process repeated an exact number of times; you may want to solve a formula for ten values—

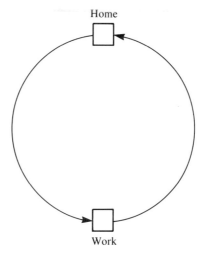

Figure 2–3   Illustration of a Loop Block

no more, no less. There will be other times when you don't know how many times a process will need to be repeated. You may be writing a program that will allow the program user to repeat the program as many times as the program user wants.

Sample Problem

Before starting the example, it will be helpful to summarize the three types of programming blocks available. This is done in table 2–3.

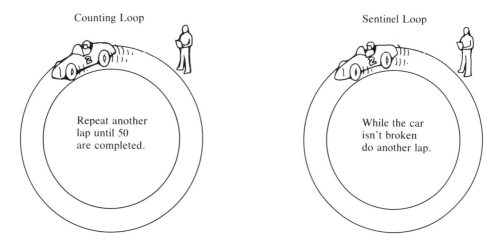

Figure 2–4   Concept of Two Different Kinds of Programming Loops

Table 2–3   Three Types of Programming Blocks

Type	Meaning	Example
Action	Do this, then do that.	Get a number, then double it.
Branch		
(Open)	If this, do that, but always continue onward.	If a number is odd, double it, and always continue onward.
(Closed)	If this, do that, or else do something different, and continue onward.	If a number is odd, double it, or else triple it, and continue onward.
Loop		
(Counting)	Repeat the loop until a certain condition is completed.	Solve the same formula for 10 different values.
(Sentinel)	As long as a certain thing is happening, repeat the loop.	As long as the program user types "YES", repeat the program.

Program 2–1 is the outline of a sample program.

Analysis of Sample Program

Program 2–1 is written in Pascal, and even though you know very little Pascal programming code, you can read through program 2–1 and get a very good idea of what the program is supposed to do and how it will do it. The only thing left to do for the above program is to put in the programming code. That's the easy part—the hard part, designing the program, (the thinking part) has already been done.

Program 2–1 will compile but little will be accomplished other than having some text appear on the screen when the program is executed. It is important to see how program 2–1 is divided into program blocks, and what each block will do. Note from figure 2–5 (p. 47) the block structure of program 2–1. The programming blocks of a Pascal program make the program easy to follow and understand.

Conclusion

This section presented a very important programming theorem. The idea that all your computer programs can be written using just three processes is a major step in applying Pascal to the problems of technology.

In the next section, you will see how to develop flow charts that represent these major concepts. For now, test your understanding of this section by trying the following section review.

Program 2–1

```
{   ****************************************************************
    *                 Program block demonstration                 *
    ****************************************************************
    *             Developed by:  An Important Theorem             *
    *             Date:  September 18, 1999                        *
    ****************************************************************
    *    This program will solve for the voltage drop            *
    * across a given value of resistor for ten different         *
    * values of current.  The program user must supply           *
    * value of the resistor.  The values of current used         *
    * are from 1 amp to 10 amps in steps of 1 amp each.          *
    *    The program will tell the program user if the           *
    * voltage drop across the resistor ever becomes              *
    * larger than 100 volts.                                     *
    ****************************************************************
    *              Constants used:                                *
    *------------------------------------------------------------*
    *        Start = 1   Beginning value of the current.          *
    *          End = 10  Ending value of the current.             *
    *         Step = 1   Value to increase current each           *
    *                    time.                                    *
    * Maxvoltage = 100 If voltage across resistor                 *
    *                    exceeds this value, tell the             *
    *                    program user.                            *
    ****************************************************************
    *              Variables used:                                *
    *------------------------------------------------------------*
    *            V = Voltage across resistor.                     *
    *            R = Value of resistor.                           *
    *            I = Value of current.                            *
    *     Response = If program is to be repeated.                *
    ****************************************************************   }

PROGRAM Blockdemo;

    BEGIN

{| Explain program to user; | <== THIS IS AN ACTION BLOCK
    ------------------------------------------------------------}
    WRITELN('This program will give the voltage drop across');
    WRITELN('a resistor for ten values of current ranging');
    WRITELN('from 1 amp to 10 amps in steps of 1 amp each.');
    WRITELN('All you need to do is supply the value of the');
    WRITELN('resistor in ohms.');
    WRITELN('The program will tell you if the value of the');
    WRITELN('voltage drop ever exceeds 100 volts.');
    {---------------------------------------------------
    | End explain program to user; | }

{| Get values from user; | <== THIS IS AN ACTION BLOCK
    ------------------------------------------------------------}
      { Ask for value of the resistor in ohms;   }
    {---------------------------------------------------
    | End get values from user; | }
```

Program 2–1 *continued*

```
{| Do computations; |   <== THIS IS A COUNTING LOOP BLOCK
---------------------------------------------------------------}
   { I = Start; }  { Loop starts at 1 amp. }
   {REPEAT}  { Repeat the following: }
    {.....................................................}
       { V = I × R; }     { Calculate voltage drop. }
                     { A closed branch block follows. }
          {.....................................................}
          {IF V > 100 THEN WRITELN(' Over 100 volts!');
             ELSE WRITELN('100 volts or less.');}
          {.....................................................}
          WRITELN('Value of voltage = ',V);
       { I = I + Step; }  { Increase current by 1 amp. }
    {UNTIL I = 10; }  { Stop the loop. }
  {-----------------------------------------------------------
   | End do computations; | }

{| Ask for program repeat; |  <== THIS IS AN ACTION BLOCK
---------------------------------------------------------------}
   WRITELN('Do you want to repeat the program?');
   WRITELN('Type Y or N and press RETURN/ENTER.');
      { Get response from user; }
  {-----------------------------------------------------------
   | End ask for program repeat; | }

 {|| Main Programming Sequence; ||  <== THIS IS A SENTINEL LOOP
=============================================================}
{ Explain program to user; }
   { WHILE Response is not N DO }
     { BEGIN loop here }
      { Get values from user; }
      { Do computations; }
      { Ask for program repeat; }
     { END loop here; }
  END.
   {=============================================================
    || End main programming sequence. || }
```

2–2 Section Review

1 Explain the importance of the completeness theorem.
2 Describe the difference between an action block, branch block, and loop block.
3 Explain the two different types of branch blocks.
4 Describe the two different kinds of loop blocks.

2–3   Using Road Maps—The Flowchart

Introduction

Flowcharting is one of the classical ways to begin to design a new program. Many text books on programming describe how programmers first write complete **flowcharts** and then start working on the actual program, but you will find that very few practicing

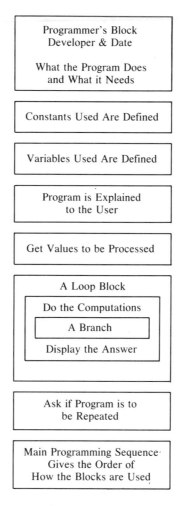

Figure 2–5   Block Structure of Sample Program

programmers actually do this. Flowcharting is now more of a text-book exercise for students than a requisite for designing computer programs. Flowcharting will thus be used in this text primarily to help illustrate and clarify explanations given in the text.

Flowchart Symbols

There is no agreed-upon standard in programming text books for the symbols used in flowcharting. Some of the flowchart symbols that will be used in this text are shown in Figure 2–6, and explained in table 2–4.

The Completeness Theorem

Recall that section 2–2 stated that any programming logic, no matter how complex, could be resolved into three types of blocks: action, branch, and loop. The flowchart symbolism for each of these blocks will now be explained.

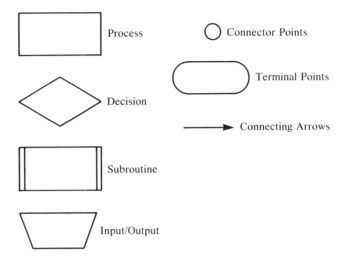

Figure 2–6   Symbols Used in Flowcharting

Table 2–4   Flowchart Symbols

Type	Discussion
Process	This represents an activity such as adding two numbers. The process operation usually indicates an action.
Decision	This is used in a branch block or a loop block.
Subroutine	This is a part of the program that is subordinate to the main program and may be routinely used, hence the name subroutine. In Pascal, you will see these are called procedures and functions.
Input/Output	This symbol represents getting information into the computer (input) from an external device such as the keyboard, or sending information out (output) to a device external to the computer such as the printer.
Connector Points	Allows the flowchart to be connected with a minimum number of connecting lines.
Terminal Points	Shows where the program starts and where it ends.
Connecting Arrows	The symbols are connected together with connecting arrows. By following the direction of the connecting arrows you follow the structure of the program.

Action Block

The flowchart in figure 2–7 shows us that an action block consists of simply going from one program step to another. This is very similar to the road map illustration of the action block used in section 2–2.

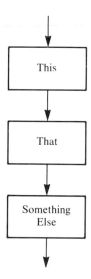

Figure 2–7   Flowchart of an Action Block

Branch Block

Figure 2–8 shows a flowchart of the two types of branch blocks. Recall that the difference between these two types of branch blocks is that the open branch may do something, then continue with the program, while the closed branch will do one thing or another thing, and then continue with the program. Again, both of these are very similar to the road map explanation of the branch used in the last section.

Loop Block

Figure 2–9 (p. 51) shows a flowchart of the two different types of loop blocks. Observe from figure 2–9A that the sentinel loop block tests the condition before the process is done; if this kind of loop is used, the process may not be performed at all. Figure 2–9B shows counting a loop, where the process is done first, then the condition is tested; if this type of loop is used, the process will always be performed at least once.

Difference Between a Loop and a Branch

Beginning students often question the difference between a loop and a branch, feeling that they are the same. The difference is clear: a loop block can go back and repeat a process; a branch block always goes forward.

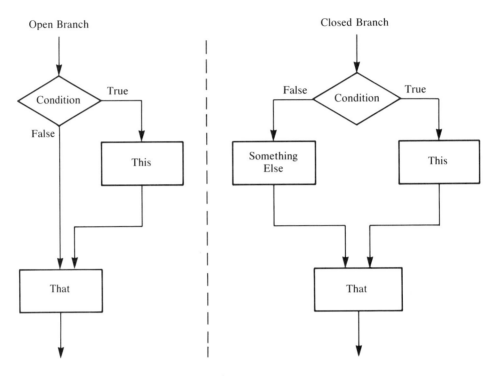

Figure 2–8   The Two Types of Branch Blocks

Flowchart Example

A flowchart of the sample program, program 2–1, is shown in figure 2–10 (p. 52). As you can see, the flowchart helps you get a different perspective on the program, allowing you to follow the action blocks, different loop blocks, and the branch block.

Advantages of Flowcharting

There are some advantages to flowcharting; one is that flowcharting gives a pictorial representation of the program, which may be useful in program analysis. As seen in figure 2–10, the flowchart representation of a program may give more of an insight into the program's structure.

Another advantage of flowcharting is that it is computer-language independent. A person need not know any programming language to follow the purpose and structure of the program; this is similar to one of the advantages Pascal offers.

Flowcharting is widely used in areas other than computer programming. For example, it is common practice to flowchart technical processes, professional operations, and decision-making in addition to computer programs when they are completed to help people understand them. Understanding how to use flowcharting could be beneficial in other areas of interest.

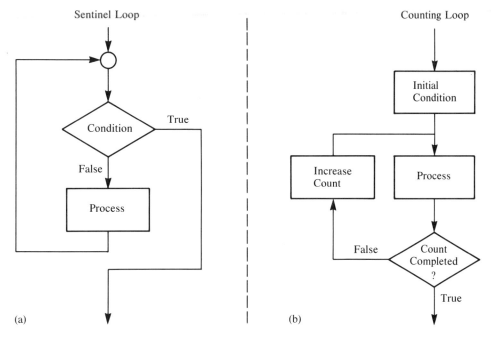

Figure 2–9   Two Types of Loop Blocks

Conclusion

Flowcharting is used in this text primarily to illustrate the action of different programming blocks. Because flowcharts can help explain how a section of a program works, they are widely used in programming text books and the documentation of many programs used in industry.

2–3 Section Review

1   State the main use of flowcharting in computer programming.
2   Identify each of the flowchart symbols shown in figure 2–11 on page 53.
3   Explain what is meant by a process block flowchart symbol. Give an example of such a process.
4   State the use of a decision logic block flowchart symbol.
5   State what an input/output block flowchart symbol represents. Give an example.
6   Explain the purpose of a terminal point in flowcharting.

2–4   Top-Down Design

Introduction

There have been many approaches to developing computer programs to solve technology problems; over the years the best have emerged to the surface. The

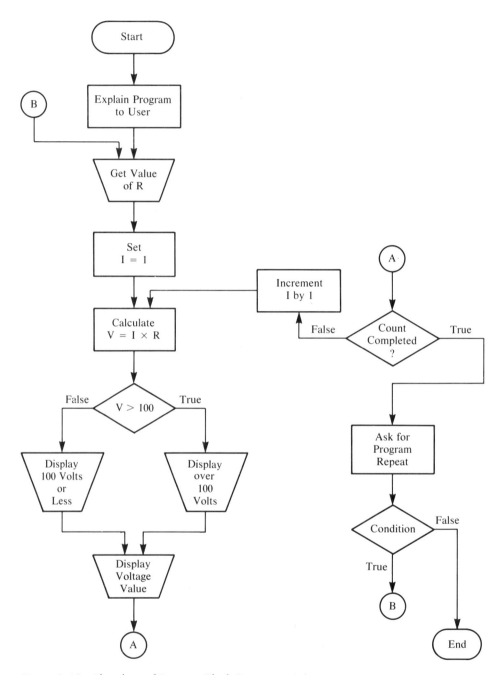

Figure 2–10   Flowchart of Program Block Demonstration

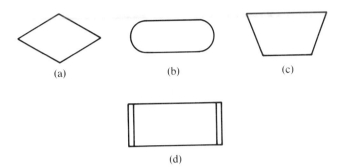

Figure 2–11 Flowchart Symbols for 2–3 Section Review

method that seems to work best for most programmers is called **top-down design**. This section will give you the tools necessary to approach a technology problem using top-down design.

Approach to Developing Programs

As a technician attempting to apply Pascal to a particular technology problem, your initial exposure to the problem will usually be in the form of words or ideas. The basic concept of top-down design is to start with the most abstract description of the problem and then work from the top down. In other words, the last thing you will do is enter program code.

Think of how a newspaper article is presented. First, there is the headline. This is the most generalized idea of what is in the article—in fact, a well-written headline will tell the whole story even though it gives no details. The lead paragraph of the newspaper story contains a little more detail; each sentence imparts an important facet of the story to follow. The remaining paragraphs will fill in the details of the story.

A good writer will work hardest on the headline and the lead paragraph. Once this is done, the rest of the story will flow easily. In the same way, a good programmer will spend the majority of programming time developing the major ideas of the program, without using program code other than remarks and just enough code to get the program to compile. Only when this has been done, is the program code entered. Entering program code, then, is the easy part. The hard part—the thinking part—is what top-down design will help you do.

Steps in Top-Down Design

The first step in top-down design is to set down in writing just what the program will do. You have seen this in the opening comments of the sample program, program 2–1.

```
{ ********************************************************
  * This program will solve for the voltage drop        *
  *across a given value of resistor for ten different    *
  *values of current. The program user must supply       *
  *value of the resistor. The values of current used    *
  *are from 1 amp to 10 amps in steps of 1 amp each.    *
  * The program will tell the program user if the        *
  *voltage drop across the resistor ever becomes         *
  *larger than 100 volts.                               *
  ******************************************************** }
```

Beginning programmers tend to omit this first step and begin by entering program code. They often run into one of two major problems: either the program doesn't do what it was supposed to do, or it does more. Either way, much time is wasted; since the programmer never had a clear idea of what the program should do, neither will another programmer who is trying to modify the program to make it work. There is another old saying in computer programming: "If you don't understand it, don't try to solve it."

The next step in top-down design is to see if the problem you have just stated in writing will fit the following programming sequence.

1. Explain program to user.
2. Get values from user.
3. Do computations.
4. Display answer(s).
5. Ask for program repeat.

You will find that most technology problems fit into this format, or variations of this format. Once the format is decided, you have created the **main programming block** in Pascal. All of this can be entered in Pascal using comments.

```
{|| Main Programming Sequence; ||
=============================================================}
  { Explain program to user; }
  { Get values from user; }
  { Do computations; }
  { Display answers; }
  { Ask for program repeat; }
{=============================================================
|| End main programming sequence. || }
```

The next step is to create program blocks that will do each of the operations outlined in the main programming block. All of these are to be put ahead of the main programming block.

```
{| Explain program to user; |
------------------------------------------------------------}
{------------------------------------------------------------
  | End explain program to user; |}
```

```
{| Get values from user; |
-----------------------------------------------------------}
{-----------------------------------------------------------
 | End get values from user; | }

{| Do computations; |
-----------------------------------------------------------}
{-----------------------------------------------------------
 | End do computations; |}
{| Display answer; |
-----------------------------------------------------------}
{-----------------------------------------------------------
 | End display answer; |}

{|| Main Programming Sequence; ||
  =========================================================}
    { Explain program to user; }
    { Get values from user; }
    { Do computations; }
    { Display answers; }
    { Ask for program repeat; }
{=========================================================
 || End main programming sequence. || }
```

You now have the major ideas down, and the only program code you have used is the **{ . . . }**. Everything you have entered is comments. The program won't compile because you need the reserved words **BEGIN** and **END.** But you have now used top-down design to give your program direction (and also limitations).

Developing Procedures

A Pascal procedure is nothing more than a program block that begins with the reserved word **PROCEDURE**, followed by a name given by you. The name of the procedure should be descriptive of what it does. Play it safe for now; if you use a name that contains letters of the alphabet and at least two words separated by an underscore, you won't wind up using any of Pascal's other reserved words. A procedure must also contain the reserved words **BEGIN** and **END.** A sample procedure is shown below.

```
PROCEDURE This_is_a_Sample;
   BEGIN
     WRITELN('This is from the sample procedure');
   END;
```

Note that the **END** of a procedure is followed by a **;** and not a period. Each Pascal programming block is a separate procedure and its name can be used by the main programming sequence. Thus, once you have written a program block, you will not need to rewrite it in that program, but need only use the name you gave it in the main programming block. You can use a procedure in any order you wish in the main programming block. The compiler distinguishes the main programming block from the others by noting that the **END** statement of the main programming block ends with a period.

Structure of a Typical Pascal Program

Program 2–2 shows the structure of a typical Pascal program.

The general Pascal structure shown in program 2–2 is also found in Appendix G for your convenience.

You will see more applications of procedures in this chapter. For now, know that the program blocks you have been reading about actually do exist in Pascal coding; they are called procedures and you can give them names with meaning.

Conclusion

This section presented the first important steps in top-down design; next, you will see how to increase the detail of this important problem-solving process. For now, test your understanding of this section by trying the following section review.

2–4 Section Review

1 State the main idea of top-down design.
2 What is the first step in top-down design? Why is this important?
3 Describe the programming sequence most commonly found in computer programs used to solve technology problems.
4 Define a Pascal procedure. What programming code must it contain? How does it end?
5 Define the main programming sequence block. What does it use to execute the Pascal program?

2–5   The Applied Programming Method

Discussion

The **Applied Programming Method (APM)** is an exacting, step-by-step approach to converting a problem into a Pascal program. Knowing programming code alone is not enough to develop successful computer programs. This section will present a process you can use to approach a technical problem and convert it into a Pascal program.

Using APM

APM uses seven phases in the development of a Pascal program, as presented in table 2–5 (p. 59).

Program 2–2

```
PROGRAM Example_One;

    {**********************************************************
            Developed by:  Name of program developer.
                    Date:  Date of development.
        ***********************************************************
        Program description:
        What the program will do, what is needed for input
        and what will be given for output.
        ***********************************************************
                        Constants used:
        ----------------------------------------------------------
                Constant1 = Value          [Type]
                Constant2 = Value          [Type]
                    .          .              .
                    .          .              .
                    .          .              .
                ConstantN = Value          [Type]
        ***********************************************************
                        Variables used:
        ----------------------------------------------------------
            Variable1 = Explain what variable is.   [Type]
            Variable2 = Explain what variable is.   [Type]
                .                 .                 .
                .                 .                 .
                .                 .                 .
            VariableN = Explain what variable is.   [Type]
        **********************************************************}

    PROCEDURE First_One;
    {-------------------------------------------------------------
        Explain what this procedure does.
        [Action, looping, and/or branching]
        -------------------------------------------------------------
        Constants used:
        Variables used:                                          }

    BEGIN

        { Step one.....;}
        { Step two.....;}
        {     .        ;}
        {     .        ;}
        { Step N.......;}
    {  ------------------------------------------------------------}
     END;   {of Procedure First_One;}

    PROCEDURE Second_One;
    {  -----------------------------------------------------------
        Explain what this procedure does.
        [Action, looping and/or branching]
        -------------------------------------------------------------
        Constants used:
        Variables used:                                          }
```

Program 2–2 *continued*

```
   BEGIN

      { Step one.....;}
      { Step two.....;}
      {     .        ;}
      {     .        ;}
      {     .        ;}
      { Step N.......;}
  { --------------------------------------------------------------}
     END;    {of Procedure Second_One;}

     PROCEDURE Last_One;
  { ---------------------------------------------------------------
      Explain what this procedure does.
      [Action, looping and/or branching]
      ---------------------------------------------------------------
      Constants used:
      Variables used:                                          }

     BEGIN

      { Step one.....;}
      { Step two.....;}
      {     .        ;}
      {     .        ;}
      {     .        ;}
      { Step N.......;}
  { --------------------------------------------------------------}
     END;    {of Procedure Last_One;}

  { || Main Programming Sequence ||
    ========================================================}
  BEGIN
      { These may be in any order you need. }
          First_One;
          Second_One;
              .
              .
              .
          Last_One;
  { ========================================================}
     END.   {Of main programming sequence;}
```

The first six steps each have a **guide** which is a document that outlines exactly what you should do for that phase. An explanation of each of these guides and how to use them will be presented in this section. The first six phases of APM will thus be covered in detail in this section; the seventh step, entering programming code, will be covered in the remaining chapters of this book. There is another old saying in programming: "The sooner you start to enter program code, the longer it will take to complete the program." There are usually good reasons for old sayings, so please be patient.

Table 2–5 Seven Phases of APM

Phase	Action	Discussion	Guide
1	Understand the problem.	This is the most important step in the solution of any problem.	Problem Statement
2	Solve the problem by hand.	This confirms that you understand the problem, and helps you identify all of the steps necessary for the solution of the problem. You may of course use a pocket calculator or even the computer in the solution.	Problem Solution
3	Record all of the necessary steps needed for solving the problem by hand.	Here you are creating a recipe, called an algorithm, of what to do. It lists, step-by-step, each part of the process you went through to solve the problem. This is really the first phase of developing the computer program.	Algorithm Development
4	Decide which of the three control methods will be needed.	Recall the Boehm and Jacopini Theorem of action blocks, branch blocks, and loop blocks. In this step you are documenting which of these blocks you will need in order to replicate the algorithm from step 3.	Control Blocks
5	List and define all program constants and variables.	You must have a clear and well documented record of all formulas, as well as what each part of each formula means. This is very important for all computer languages, but especially so for Pascal.	Constants and Variables
6	Write the Pascal program just using procedures and enough code to compile.	This is the all-important documentation that all good programs require. In top-down design, the program documentation is done first and the coding is done last.	Program Design Steps
7	Enter the Pascal program code and test the program.	If the first six steps have been done well, this step will require very little time.	

Using the APM Phases

The first phase of APM is to understand the problem. The **Problem Statement Guide** is used to help you. This guide and all of the other guides presented in this section are included together in Appendix E. Use them to assist you in applying Pascal to your technical problems.

Suppose the problem you had to program was to solve for the power dissipated in a resistor using the formula: $P = I \times E$, where P = power in watts, I = current in Amps, and E = voltage in volts. The program user would input the value of the current and the voltage, and the program would find the solution and display the answer. Here is how you would use the Problem Statement Guide.

PROBLEM STATEMENT GUIDE

1. What will the program do?
   Solve for the power dissipation in a resistor using
   P = I × E.
2. What is needed for input?
   The value of I and E.
3. What will be given as output?
   The power P.

Once you have defined what your program will do, you can go on to the second phase, solving the problem by hand.

Solving the Problem by Hand

Being able to solve the problem yourself helps you identify all of the steps involved. It also helps you make sure you have all of the information needed to solve the problem. The **Problem Solution Guide** helps you identify the steps needed to solve the problem.

PROBLEM SOLUTION GUIDE

Formula:
Values Used:

Step	Computation	What did you actually do?
1		

Formula:
Values Used:

Step	Computation	What did you actually do?

Observe the use of this guide for the solution of the resistor power problem.

PROBLEM SOLUTION GUIDE

Formula: $P = I \times E$
Values Used: $I = 2$ **Amps** $E = 3$ **Volts**

Step	Solution process	What did you actually do?
1	$P = 2 \times 3$	Entered the value of I and E.
2	$P = 6$	Multiplied and set P equal to the result.

Now that you have worked out the problem by hand, you are ready for the third phase, developing an algorithm.

Developing an Algorithm

Do you remember the programming sequence most commonly used to solve technology problems? This is where the **Algorithm Development Guide** is helpful.

Look at your Problem Solution Guide for the resistor power problem. Under the column "What did you actually do?" are the steps you took to solve the problem. You have already developed a major part of your recipe, or algorithm. Since most technology problems have many procedures in common, the following Algorithm Development Guide is helpful.

ALGORITHM DEVELOPMENT GUIDE

  I. Explain program to user.
    a. What does the program do?
      [The process]
    b. What does the program user need to do?
      [The input]
    c. What will be displayed?
      [The output]
  II. Get values from user.
 III. Do computations.
 IV. Display answer(s).
  V. Ask if program is to be repeated.

Now, apply this Algorithm Development Guide to the resistor power dissipation problem.

ALGORITHM DEVELOPMENT GUIDE

  I. Explain program to user.
    a. What does the program do?
      [The process]
      Computes the power dissipated in a resistor.

b. What does the program user need to do?
   [The input]
   Input the value of I and E.
c. What will be displayed?
   [The output]
   The power dissipation in watts.
 II. Get values from user.
   a. **Get the value of I.**
   b. **Get the value of E.**
III. Do computations.
   a. **P = I × E**
IV. Display answer(s).
   a. **P =**
 V. Ask if program is to be repeated.
    (Use only when necessary.)

Control Methods

This is the fourth phase of APM. Here, you are actually constructing your program blocks, Pascal procedures. Because the sample resistor power program has no loop blocks (nothing is to be repeated) or branch blocks (there is no decision to be made), this program will be composed of action blocks alone.

If, however, the power dissipated by the resistor was to be computed for a whole range of current and voltage values, a loop block would be necessary to solve the formula over and over with the different values. Only if the program could do something different depending on a previous solution or user selection would a branch block be required.

To illustrate, suppose the resistor power dissipation problem required the display of the amount of power dissipated by the resistor for different values of current at a voltage value selected by the program user. This would require a loop block. Also suppose that if the power dissipation of the resistor exceeds 100 watts the program is required to display a warning message—a branch block would now be required.

The following **Control Blocks Guide** is helpful in organizing the programming blocks needed for your program.

CONTROL BLOCKS GUIDE (PROCEDURES)

Action Blocks Number	Action Taken
1	[PROCEDURE Name;]

Counting
Loop Blocks

No:	What is repeated: Starting Value: Increment: Ending Value:	`[PROCEDURE Name;]`
No:	What is repeated: Starting Value: Increment: Ending Value:	`[PROCEDURE Name;]`

Sentinel
Loop Blocks

No:	What is repeated: Under what condition:	`[PROCEDURE Name;]`
No:	What is repeated: Under what condition:	`[PROCEDURE Name;]`

Branch
Blocks

No:	Condition for branch: If condition is met: If condition is not met:	`[PROCEDURE Name;]`
No:	Condition for branch: If condition is met: If condition is not met:	`[PROCEDURE Name;]`

You can also add an option to your Control Blocks Guide that will allow the program user to repeat the program.

CONTROL BLOCKS GUIDE (PROCEDURES)

Action Blocks Number	Action Taken	
1	Explain program to user.	`[Explain_It;]`
2	Get value of E from user.	`[Get_Value;]`
3	Ask user if program is to be repeated.	`[Program_Repeat;]`

Counting
Loop Blocks

No: **1** What is repeated: **P = I × E, Display P, Branch block #1.**
Starting value: **I = 1, E = Selected by user.**
Increment: **I by + 1**
Ending Value: **I = 5**

`[Compute_and_Display;]`

Sentinel
Loop Blocks

No: **1** What is repeated: **The main programming sequence.**
Under what condition: **Desire of program user.**
`[MAIN PROGRAM]`

Branch
Blocks

No: **1** Condition for branch: **If P is greater than 100.**
If condition is met: **Display warning—continue program.**
If condition is not met: **Continue program.**
[Part of `Compute_and_Display Block`]

Now you have developed an outline of all the Pascal procedures you will use in your program. Notice the names given to the different procedures. Each name is descriptive of what the block does, and uses at least two words, each separated by an underscore __. This insures that you are not using any reserved words for the name of your procedure.

You now have the programming almost completed; you are spending so much time putting items into words because Pascal is a language of words, meant to read like paragraphs in a book. As you translate your ideas into words, you will find that you are not far from putting them into the actual code used by Pascal, which, you will find, will be very similar to the words you have already used.

Program Constants and Variables

In phase five, as you will see, Pascal is very fussy about defining all program variables and constants before they are used in the program; this makes sure that you, yourself, know the purposes of the constants and variables. This will help clarify your thinking process when you begin entering the program code, and prevent you from using the same variable for two different things. Here, then, is the **Constants and Variables Guide.**

CONSTANTS AND VARIABLES GUIDE

Formula	Variables and Constants	Meaning	Type					Block(s) Used
			R	I	C	B	S	

This guide has five main columns and five subcolumns under **Type**. For now, think of these five subcolumns as

- R = **Real number**—any number that uses a decimal point (such as 1.24, 0.5, or 3.0).
- I = **Integer number**—any number that does not use a decimal point (such as 124, 58, or 2).
- C = **Character**—any single keyboard character (such as a, b, or c).
- B = **Boolean**—any variable that can only be true or false.
- S = **String**—any variable that will be words (such as HELLO!). The words don't have to make sense (they could be OpW$!!xPz).

The Constants and Variables Guide would be used in this way for the sample problem.

CONSTANTS AND VARIABLES GUIDE

Formula or Other	Variables and Constants	Meaning	Type					Block(s) Used
			R	I	C	B	S	
P = I × E	P [**VAR**]	Power dissipated by resistor	X					`Compute_and_ Display;`
	I [**VAR**]	Current in resistor	X					`Compute_and_ Display;`
	E [**VAR**]	User input	X					`Get_Value; Compute_and Display;`
Program loop	I1 = 1 [**CONST**]	Beginning value of loop		X				`Compute_and Display;`
	Inc = 1 [**CONST**]	Increment value of loop		X				
	I2 = 10 [**CONST**]	Ending value of loop		X				
Branch	Pmax = 100 [**CONST**]	Branch condition	X					`Compute_and Display;`
[Other]	[Answer] [**CHAR**]	Response from program user to repeat pgm.			X			`Program_ Repeat;`

The constants and variables used in the power formula were all selected to be real numbers, because when dealing with technical information, you may need to use a decimal point. It's a good rule for now to make all formula variables real numbers when you select the type of constant or variable for Pascal. The type selected for the counting loop variables was integer, because none of the counting numbers use fractions, which is usually the case with counting loops.

Note that a character type was chosen for the response from the program user. The program will ask the user to input the character "Y" (for Yes) if the program is to be repeated, or the character "N" (for No) if the program is not to be repeated.

First Program Writing

Now you have reached the last phase, and are at last ready to develop the Pascal program—your first chance to sit at the computer keyboard and enter Pascal code. The **Pascal Design Guide** is used to help you do this.

PASCAL DESIGN GUIDE

Step 1. Give the program a descriptive name, ending with a **;**
  Example: **PROGRAM** `Resistor_Power;`
      (Note: You are safe if you use at least two words separated by an underscore.)

Step 2. Give your name and the development date using comments.
    Example: **{ ***
            Developed by:
                Date:
        * }

Step 3. State what the program is to do and what is needed.
    Example: Refer to your Problem Statement Guide.

Step 4. Define all constants and variables.
    Example: Refer to your Constants and Variables Guide.

Step 5. Write a program block for each of your procedures. The generalized format of a program block (procedure) is:
      PROCEDURE `Name_of_Procedure;`
    `{------------------------------------------------`
      `Purpose of procedure:`
      `Constants used:`
      `Variables used:`   `}`

      BEGIN

```
{List procedure steps you will want the procedure
    to do.
    .
    .
    .                                                                }
{---------------------------------------------------------}
END; { End of Name_of_Procedure }
```
Observe that the **END** for a procedure finishes with a **;** not a period!

Step 6. Write the main programming sequence using the names of each of your procedures. Put this between the Pascal **BEGIN** and **END.** statements. Note there must be a period following the reserved word **END.**

Example:
```
{ || Main Programming Sequence ||
  ===========================================}
        BEGIN
            Name_of_a_Procedure;
            Name_of_Anotherone;
            Name_of_the_Last_One;
  {===========================================}
        END. {of main programming sequence }.
```

Step 7. Compile the program, correct any errors, save it on your disk, and make a printed copy. You are now ready to enter program code. The design phase of your program is completed.

This process is illustrated in figure 2–12.

Conclusion

The last section of this chapter will take you step-by-step through the design of a Pascal program; information presented in this section will be applied there.

Congratulations! You have come a long way in your understanding of what it takes to apply the Pascal language (or any programming language) to the solution of technology problems. Test your understanding of this section by trying the following section review.

2–5 Section Review

1 State, in your own words, the general idea of APM.
2 Describe the first six steps used in APM.
3 State the purpose of each of the following APM guides:
   A) Problem Statement Guide
   B) Problem Solution Guide
   C) Algorithm Development Guide
   D) Control Blocks Guide
   E) Constants and Variables Guide
   F) Pascal Design Guide
4 State the steps used in Pascal program design.

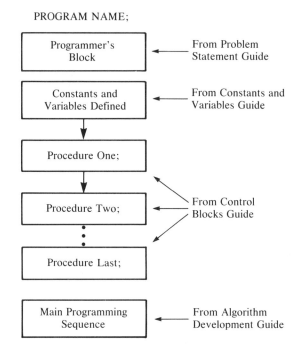

Figure 2–12   First Steps in Program Writing

2–6   Developing a Pascal Program

Introduction

Recall that you can write a Pascal program with one coded instruction immediately following another. You can also write a letter in one continuous line, but a letter written this way is difficult to read. A Pascal program written with instructions all jumbled together is also very difficult to read.

This section takes you step-by-step through the development of a Pascal program using APM. You will refer to this section many times as you develop your programming skills for the solution of technical problems.

Sample Program

You will now follow the complete development of a sample technology problem into a Pascal program. The only step to be omitted will be that of entering Pascal code. The only Pascal code you will use is:

{ }   For comments.

;   Required by Pascal, used after the **END** at the end of each Pascal procedure.

PROGRAM   A reserved word.

PROCEDURE Used to define a Pascal procedure.
BEGIN
        END Defines program blocks.
            • Used at the end of the last **END.**

Problem Statement

Show all of the steps in the development of a Pascal program that will solve for the area of a circle. The program user must enter the radius and the computer will display the area. No program repeat is required. (Formula: $A = \pi r^2$)

Solution
Use the seven phases of APM. First, use the Problem Statement Guide to define the problem.

PROBLEM STATEMENT GUIDE

---

1. What will the program do?
   Compute the area of a circle.
2. What is needed for input?
   Value of the radius.
3. What will be given as output?
   Area of the circle.

Next, use the Problem Solution Guide to solve the problem by hand and write down each step you used in the process.

PROBLEM SOLUTION GUIDE

Formula: $A = \pi r^2$
Values Used: $r = 3$

Step	Computation	What did you actually do?
1	$A = 3.14159 \times 3^2$	Entered the value of 3, set π equal to a constant.
2	$A = 28.27$	Squared 3 and multiplied by π, set A equal to the result.

Now you are ready for the Algorithm Development Guide, where you will actually develop the sequence of your program.

ALGORITHM DEVELOPMENT GUIDE

I. Explain program to user.
   a. What does the program do?
      Solve for the area of a circle.

    b. What does the program user need to do?
      Enter the value of the radius.
    c. What will be displayed?
      The value of the area.
 II. Get values from user.
    a. Value of radius (r)
III. Do computations.
    a. A $= \pi r^2$
IV. Display answer(s).
    a. Area equals A
 V. Ask for program repeat.
    a. Not used in this example.

Using this algorithm, you can start the development of the Pascal blocks. In the fourth phase of APM you use the Control Blocks Guide to actually lay out your Pascal program.

CONTROL BLOCKS GUIDE

Action Blocks Number	Action Taken	
1	Explain program to user.	[Explain_It;]
2	Get value from user.	[Get_Value;]
3	Do Computations.	[Compute_It;]
4	Display answers.	[Show_Answer;]

There are no loops or branches, so your next step is to use the Constants and Variables Guide, to help you carefully define all of the constants and variables you will use in your program. As you will see when you begin Pascal coding, this is very important.

CONSTANTS AND VARIABLES GUIDE

Formula	Constant or Variable	Meaning	R	I	C	B	S	Block(s) Used
A $= \pi r^2$	A	Area of						Compute_It;
	[VAR]	circle.	X					Show_Answer;
	π	3.14159						Compute_It;
	[CONST]	Value of π	X					
	r	Radius of						Get_Value;
		circle	X					Compute_It;

You have now reached the sixth phase of APM, and can do the first steps in program writing. The Pascal program will be written out using only the Pascal code you have learned up to this point, using the first six steps in the Pascal Design Guide.

PASCAL DESIGN GUIDE

Step 1. Give the program a descriptive name.

```
PROGRAM Circle_Area;
```

Step 2. Name and development date, using comments.

```
{***********************************************************
                Developed by: Your Name Here
                Date: Month, Day, Year
***********************************************************
```

Step 3. State what the program is to do and what is needed.
                (This is from your Problem Statement Guide.)

```
***********************************************************
   This program computes the area of a circle. Program
user must input value of the radius. The computer will
display the area of the circle.
***********************************************************
```

Step 4. Define all constants and variables. Use the information from your Constants and Variables Guide.

```
***********************************************************
                     Constants used:
----------------------------------------------------------
          Pi = 3.14159              Real
***********************************************************
                     Variables used:
----------------------------------------------------------
          A = Area of circle        Real
          r = Radius of circle      Real
*********************************************************}
```

Step 5. Write a program block for each of your procedures. The details of this are contained in the Algorithm Development Guide. These procedures will be placed ahead of the Main Programming Sequence.

```
PROCEDURE Explain_It;
{----------------------------------------------------------
   This procedure explains the program to the user.
   Action block.
----------------------------------------------------------}
   No constants or variables in this procedure.
```

```
BEGIN
   { This program will solve for the area of a circle;}
   { You only need enter the value of the radius;}
   { The program will then display the area of the circle;}
{---------------------------------------------------------------}
   END; {of procedure Explain_It}

 PROCEDURE Get_Value;
{---------------------------------------------------------------
     This procedure gets the value of the radius from user.
     Action block.
-----------------------------------------------------------------
     Variables: r [REAL]                                         }

   BEGIN

     { Enter the value of the radius;}
     { Get value of r;}
{------------------------------------------------------------------}
   END; {of procedure Get_Value;}

 PROCEDURE Compute_It;
{------------------------------------------------------------
     This procedure computes the area of the circle.
     Action block.
------------------------------------------------------------
     Constants: PI     ; REAL
     Variables; r      ; REAL
                A      ; REAL                                   }
   BEGIN

     {A = PI × r × r;}
{------------------------------------------------------------------}
   END; {of procedure Compute_It;}

 PROCEDURE Show_Answer;
{------------------------------------------------------------------
     This procedure displays the area of the circle.
     Action block.
-----------------------------------------------------------------
     Variables: A      ; REAL                                   }

   BEGIN

     { The area of the circle is A; }
{------------------------------------------------------------------}
   END; {of procedure Compute_It;}
```

Step 6. Write the main programming sequence using the names of each of your program blocks (procedures). You get this information from the Control Blocks Guide.

```
{|| Main Programming Sequence. ||
=====================================================}
  BEGIN
        Explain_It;
        Get_Value;
        Compute_It;
        Show_Answer;
{=====================================================}
END. {of main programming sequence}
```

Putting it all Together

Program 2–3 shows the completed design of the Pascal program.

Program 2–3

```
   PROGRAM Circle_Area;
   {***********************************************************
           Developed by:  Your Name Here
                Date:  Month, Day, Year
   ***********************************************************
       This program computes the area of a circle.  Program
   user must input value of the radius.  The computer will
   display the area of the circle.
   ***********************************************************
                      Constants used:
   ----------------------------------------------------------
           Pi = 3.14159              Real
   ***********************************************************
                      Variables used:
   ----------------------------------------------------------
           A = Area of circle        Real
           r = Radius of circle      Real
   ***********************************************************}

   PROCEDURE Explain_It;
   {----------------------------------------------------------
       This procedure explains the program to the user.
       Action block.
   ----------------------------------------------------------
       No constants or variables in this procedure.          }

     BEGIN

        { This program will solve for the area of a circle;}
        { You only need enter the value of the radius;}
        { The program will then display the area of the circle;}
   {----------------------------------------------------------}
      END;  {of procedure Explain_It}
```

Program 2–3 *continued*

```
  PROCEDURE Get_Value;
{-----------------------------------------------------------
      This procedure gets the value of the radius from user.
      Action block.
      -----------------------------------------------------
      Variables: r     [REAL]                                }

    BEGIN

        { Enter the value of the radius;}
        { Get value of r;}
{----------------------------------------------------------}
    END;  {of procedure Get_Value;}

  PROCEDURE Compute_It;
{-----------------------------------------------------------
      This procedure computes the area of the circle.
      Action block.
      -----------------------------------------------------
      Constants: PI   : REAL
      Variables: r    : REAL
                 A    : REAL                                 }

    BEGIN

        { A = PI × r × r;}
{----------------------------------------------------------}
    END;  {of procedure Compute_It;}

  PROCEDURE Show_Answer;
{-----------------------------------------------------------
      This procedure displays the area of the circle.
      Action block.
      -----------------------------------------------------
      Variables: A    : REAL                                 }

    BEGIN

        { The area of the circle is A; }
{----------------------------------------------------------}
    END;  {of procedure Compute_It;}

{ ‖ Main Programming Sequence. ‖
    ========================================================}
  BEGIN

          Explain_It;
          Get_Value;
          Compute_It;
          Show_Answer;
{=========================================================}
  END.  {of main programming sequence;}
```

Conclusion

This section followed the step-by-step development of a problem in technology—finding the area of a circle. You could have found it more easily using a pocket calculator, but of course the intent was to show you practical development tools that you can apply to creating Pascal programs to solve very complex problems in technology.

The sample program contained only action blocks. This is as it should be; even though you realize that there can be loop blocks and branch blocks, you should spend some practice time getting used to using Pascal development tools. Working with technical problems that only require action blocks and omitting the program repeat step allows you to concentrate on the fundamental techniques of APM. Now, test your understanding of this section by trying the following section review.

2–6 Section Review

1   State the first three phases of APM.
2   What guides are used for the first three phases of APM?
3   Name the fourth through sixth phases of APM.
4   What guides are used for phases four through six?

2–7 Program Debugging and Implementation— Saving and Retrieving Your Program

Introduction

In chapter one you learned how to enter a Pascal program and execute (run) it using the Turbo Pascal 4.0 environment. However, you found that what you entered was lost after you left the Pascal system. In this section, you will see how to save your program to the disk so that you can use it again. You will also see how to retrieve a program that is already on the disk so you can execute it as well as modify it.

Creating a Data Disk

In this section you will see how to save your programs to a disk. It is not good practice to save these programs to your working disk, because your working disk contains the Turbo Pascal Operating System, and you could accidentally erase or modify one of these programs. You need to take another disk and initialize it. (Refer to Appendix A if you're not sure how to do this). Using a felt tip pen, label the disk:

TURBO PASCAL
PROGRAMS
Disk 1
Your Name   Date

As you progress through this text, you will find that you may require more program disks to store all of your programs. So it's important to label and number them.

Two-Drive System

If you have a two-drive system, boot it up and insert your Turbo Pascal working disk in drive **A:**, and your program disk in drive **B:**. Get into Turbo from the DOS prompt by entering **TURBO.** Next get into the file menu (press RETURN/ENTER when **File** is highlighted). Note that the edit prompt line states: **A:NONAME.PAS.** Don't worry about what this means for now, just note that it's there.

For the two-drive system, you must tell the Turbo system that you will want your program stored on the disk in drive **B:** (where your program disk is) not in drive **A:** (where your Pascal working disk is). To do this, use the arrow keys to get down to the **Change dir** selection on the file menu, then press the RETURN/ENTER key. A window will appear stating that the current directory is **A:\.** You need to change this to **B:.** Do this from the keyboard by entering **B:** and then erasing the **A:** that has moved to the right by using the Del key on the bottom right of your keyboard. Once you have done this, press RETURN/ENTER and note that drive **B:** will turn on and the edit prompt line will change from **A:NONAME.PAS** to **B:NO NAME.PAS.** Now when you save or load programs to or from the disk, it will use the disk in drive **B:**, not drive **A:**.

One-Drive System

If you have a single-drive system, you will not change the drive from **A:** to **B:** as is necessary for a double-drive system. Instead, after getting into the Turbo Pascal environment, remove your Pascal working disk and insert your Pascal program disk in its place. Now any programs you save will automatically be placed on the program disk.

Checking the Directory

If you have been following along, step-by-step, you will still be in the file menu. Now you want to look at what is on your Pascal program disk. (If it is a newly formatted disk, there won't be any programs on it yet, but it's nice to know how to confirm this). Use the arrow keys to highlight **Directory** on the file menu, then press RETURN/ENTER. Another window will appear saying **Enter mask** and ***.*.** This simply means that all DOS files will be displayed. Press the RETURN/ENTER key again and a large window will appear displaying your disk files—if your disk has just been initialized, this window will be blank, of course, because there aren't any files on it. To get back to the file menu, press the Esc key (on the upper left of your keyboard).

Entering and Saving

Get into the editor (use F10 to get back to the main menu, use the arrow keys to highlight **Editor**, and press RETURN/ENTER). Once in the editor, note the effect the Caps Lock key has on the right side of the bottom line. Depressing it makes the word **CAPS** appear. Depressing it again will make the word disappear. This is simply an easy way to let you know if the Caps Lock key is on or off. If on, everything you type will be in caps; if off, you'll type in lowercase.

Now, enter the following program:

```
PROGRAM Test_It;
 BEGIN
  WRITELN('This will be saved to the disk!');
 END.
```

Now get back to the file by first getting to the main menu (use **F10**). Making sure **File** is highlighted, press RETURN/ENTER, and the file menu will appear. Using the arrow keys, highlight **Save** on the file menu and press RETURN/ENTER. Another window will appear that says **Rename NONAME.PAS.** If you don't give it another legal DOS name, it will be saved as **NONAME.PAS.** If you do the same thing with the next program you write, it will write over your first program; you should give each Pascal program you save a different legal DOS name. For this program, call it **MYFIRST.** Simply enter this into the window and then press RETURN/ENTER. In a two-drive system disk drive **B:** will be activated (disk drive **A:** in a single-drive system), and then stop.

Your program has now been saved to your program disk. Verify this by choosing the **Directory** option in the file menu; note the program there now:

```
MYFIRST.PAS
```

The **.PAS** extension was automatically put on your file by the Turbo system, to let you know that this is a Pascal source code program. You could, at this time, **Quit,** remove your disks, come back another time, and **Load** the same program back into the editor. But don't do that yet; let's investigate a few more capabilities first. Get out of the directory by pressing the Esc key, get to the main menu (F10), and from there, back to the editor.

Automatic Backup

Let's change your program just a bit.

```
PROGRAM Test_It;
 BEGIN
  WRITELN('This will be saved to the disk!');
  WRITELN('This program contains a modification.');
 END.
```

One way to make this modification is to place the cursor at the end of the first **WRITELN** statement and press RETURN/ENTER. This will automatically push down the **END.** statement to give you a blank line. Now you can enter the new line, and save the new program under the same name as your first program. (**F10** to main menu, **File** menu, then **Save).**

The disk drive that was activated before to save your program will again become active, saving your modified program. Not only is the new, modified program being saved, but your original unmodified program is also being saved as a **backup**, just in case you change your mind and want the old unmodified program. To verify this, get to the **Directory** in the file menu; now there are two programs on your Pascal program disk.

```
MYFIRST.PAS
MYFIRST.BAK
```

MYFIRST.PAS is the program with the latest change, and **MYFIRST.BAK** is the old program before it was changed.

Getting a Program from the Disk

Now you know how to save your program to the disk. To get a program from the disk, get to the file menu, use the arrow keys to get to **Load,** and press RETURN/ENTER. Another window will appear asking for the name of the program to load. Enter **MYFIRST.BAK.** The disk will activate for a while, and then you will see your old program appear in the editor screen. This should verify for you that your old program was saved as well as its most recent modification.

Making an Executable File

You know how to save and load your Pascal source code; you also need to know how to save your Pascal object code on the disk, so it can be executed directly from DOS. This is to enable others who may not be familiar with the Turbo Pascal environment or any programming at all to use the program.

Get to the main menu (F10), use the arrow keys to get to the **Compile** option, and press RETURN/ENTER. A pull-down window will appear showing the compile menu. Use the arrow keys to highlight the **Destination memory** option, and press RETURN/ENTER. You will see this option change to **Destination disk.** This means that the compiled object code will no longer be saved just to memory (which is lost when the computer is turned off), but now to the disk as well.

Return to the main menu, get into the **Run** option and execute the program. It will first compile automatically and then execute. Return back to the Pascal system with a keypress after viewing your message, get into the file menu, and look at the **Directory.** You should see three programs now.

```
MYFIRST.PAS
MYFIRST.BAK
MYFIRST.EXE
```

You already know that the first two are source code. The one with the **.EXE** extension is an executable Pascal program. To demonstrate this, leave the Turbo environment and return to DOS (use the **Quit** option of the file menu). You will be returned to DOS. All you need to do now is enter, at the DOS prompt, the name of your saved program without the extension

`MYFIRST`

and press RETURN/ENTER. The disk drive will activate and your message will appear on the screen!

Conclusion

This was a very important section; you learned how to save and load Pascal programs to a disk, how to use the Turbo Pascal environment to view the directory of your disk, and how to make an executable Pascal program that can even be executed by those not familiar with Pascal. Check your understanding of this section by trying the following section review.

2–7 Section Review

1 Explain why it is good practice to keep your Pascal programs on a separate disk from your Pascal working disk.
2 When you save your source code to the disk, what extension does the Turbo system give it?
3 What does the extension `.BAK` mean on a Pascal file?
4 Explain how to create an executable Pascal program using the Turbo system.

Summary

1 Computers are good at repetitive tasks, arithmetic, logical decisions, sorting information, and graphical representation.
2 Computers are not good at activities which require pattern recognition, judgement, adaptation, or which have incomplete information.
3 People are good at what computers do not do well.
4 An important and useful theorem is the Boehm and Jacopini theorem that says that any program logic, no matter how complex, can be resolved into action blocks, branch blocks, and loop blocks.
5 An action block is a straight sequence of computer instructions.
6 A branch block has the capability of causing a different sequence of computer instructions depending upon a predetermined condition.
7 A loop block has the capability of going back and repeating one or more programming instructions.
8 The difference between a branch block and a loop block is that a branch block will always move forward to new instructions, while an active loop block will go backward and repeat instructions.
9 There are two kinds of branch blocks: an open branch and a closed branch.
10 There are two kinds of loop blocks: a counting loop and a sentinel loop.
11 An open branch has one programming option, while a closed branch has two different programming options depending upon a previous condition.
12 A counting loop is used when the programmer knows how many times the loop will be

repeated; a sentry loop is used when the programmer does not know how many times a loop is to be repeated.

13   Pascal lends itself to block structure. Using block structure makes a program easier to read, understand, and modify.

14   Flowcharting is another method of representing a logical process.

15   Top-down design is a method of developing computer programs; the programmer starts with the most general concepts of the problem and saves the details of program coding as the very last step.

16   The Applied Programming Method (APM) represents a structured procedure for converting a problem into a computer program.

17   A procedure is a Pascal programming block used to perform a specific task.

18   The seven steps in APM are: 1] Understand the problem. 2] Solve the problem by hand. 3] Record all steps to solve the problem. 4] Decide which control methods will be required. 5] List and define all program constants and variables. 6] Write the program in block form using comments and just enough program code for compiling. 7] Enter the code and test the program.

19   An algorithm is a step-by-step set of instructions for exactly what the program will do. It is similar to a recipe for apple pie.

20   The Problem Statement Guide can be used as an aid in making sure you understand the problem—the first step in APM.

21   The Problem Solution Guide can be used as an aid in documenting all of the steps required to solve the problem by hand—the second step in APM.

22   The Algorithm Development Guide can be used as an aid in developing the algorithms for the program, to help develop the final structure of the program—the third step in APM.

23   The Control Blocks Guide can be used as an aid in developing all of the Pascal procedures, and analyze exactly what each Pascal procedure is to do—the fourth step in APM.

24   The Constants and Variables Guide can be used as an aid in defining all constants and variables you will use in the program; the type and its meaning are both defined here— the fifth step in APM.

25   The Pascal Design Guide is an outline that helps you enter the program in block form using only comments and those commands necessary to have the program compile. This is the sixth of the seven steps in APM.

26   Section 2–6 presents the step-by-step development of a Pascal program using APM.

Interactive Exercises

Directions

These exercises require that you have access to a computer and software that supports Pascal, specifically, Turbo Pascal Development Environment, version 3.0 (or higher), from Borland International. The exercises will give you valuable experience and, most importantly, immediate feedback on what the concepts and commands introduced in this chapter will do. They are also fun.

Exercises

1   What is displayed on the monitor when you enter, compile and run program 2–4?

2   Modify the last part of the above program as shown below, then compile and run the program. What is now displayed on the screen? How is this different from program 2–4? What caused this difference?

Program 2–4

```
PROGRAM See_What_Happens;

  PROCEDURE First_One;
    BEGIN
      WRITELN('This is the first procedure.');
    END;

  PROCEDURE Second_One;
    BEGIN
      WRITELN('This is the second procedure.');
    END;

  PROCEDURE Third_One;
    BEGIN
      WRITELN('This is the third procedure.');
    END;

BEGIN
    First_One;
    Second_One;
    Third_One;
END.
```

```
BEGIN
   Third_One;
   First_One;
   Second_One;
END.
```

3 What is displayed on the screen when program 2–5 is entered, compiled and executed? Explain why this happens. What kind of programming blocks are used in this program? What do you have to do to stop the program?

Program 2–5

```
PROGRAM What_Will_Happen_Here?;

  PROCEDURE Here_We_Go;
    BEGIN
      WRITELN('How many times do you see this?');
    END;

BEGIN
   WHILE (NOT KEYPRESSED) DO
     Here_We_Go;
END.
```

4   Enter and compile program 2–6. When you run the program what happens? Why does this happen? What kind of program blocks are used here?

Program 2–6

```
PROGRAM What_Will_Happen_Here?;

   VAR  Counter : INT;

   PROCEDURE Here_We_Go;
     BEGIN
       Counter := 0;
        REPEAT
          WRITELN('Now how many times do you see this?');
          Counter := Counter + 1;
        UNTIL Counter = 10 END;

   BEGIN
      Here_We_Go;
   END.
```

5   How many times does the above program repeat itself? Why? Modify the above program so it repeats itself half as many times. Modify it again so it repeats itself twice as many times. What did you have to do in each case?

6   Enter, compile, and run program 2–7. What happens when you enter a number smaller than 10? What happens when you enter a number larger than or equal to 10?

7   What kind of programming blocks does program 2–7 use? Modify program 2–7 so the decision is made if the number entered is less than 20. What did you do to make this happen?

8   Modify program 2–7 as shown below. What now happens when the program is executed? Why does this happen?

```
PROCEDURE Check_It_Out!;
 BEGIN
  IF User_Input < 10 THEN WRITELN('Small number!');
    ELSE WRITELN('That is no small number!');
 END;
```

9   What type of programming block is the modified program block in Exercise 8? What would you change in order to make the block display

```
It is less than 10!
```

if the number is less than 10, or

```
It is not less than 10 anymore!
```

if the number is ten or larger?

Program 2–7

```
PROGRAM This_is_Another_Exercise;

  VAR  User_Input : REAL;

    PROCEDURE Get_From_User;
      BEGIN
        WRITELN('Enter a number between 0 and 20.');
        WRITELN('Then press RETURN/ENTER.');
        READLN(User_Input);
      END;

    PROCEDURE Check_It_Out!;
      BEGIN
       IF User_Input < 10 THEN WRITELN('Small number!');
      END;

    PROCEDURE Ask_for_More;
      BEGIN
        WRITELN('Run the program again and try a different');
        WRITELN('number!');
      END;

  BEGIN
    Get_From_User;
    Check_It_Out!;
    Ask_for_More;
  END.
```

Pascal Commands

PROCEDURE Name; A program block in Pascal. Must contain a **BEGIN** and an **END;**. Note that the **END** is followed by a semicolon (**;**). The name given to the procedure may subsequently be used in the main programming sequence. Observe this example.

```
PROCEDURE Number_One;
  BEGIN
    WRITELN('This is procedure number one.');
  END;

PROCEDURE Number_Two;
  BEGIN
    WRITELN('This is the second procedure.');
  END;

BEGIN
 Number_One;
 Number_Two;
END.
```

Self-Test

Directions

Program 2–8 is a completed Pascal program containing all of the correct Pascal code needed to make the program operate. The program represents the resistor voltage problem which is similar to the resistor power problem presented in section 2–5. Answer all of the questions for this self-test by referring to program 2–8.

Questions

1 How many different procedures are contained in program 2–8?
2 List the name of each procedure.
3 Are there any procedures used inside another procedure (outside of the main programming sequence)? If so, which procedure uses other procedures? What are the name(s) of these other procedures?
4 How many action blocks are in the program? Which ones are they?
5 Are there any loop blocks in the program (counting the main programming sequence)? If there are, which ones are they and what kind of loops are used?
6 How many constants are used in the program? Name them and state their purposes.
7 Complete the Control Blocks Guide (Appendix E) for program 2–8.
8 Complete the Algorithm Guide (Appendix E) for program 2–8.
9 Complete the Problem Statement Guide (Appendix E) for program 2–8.
10 Complete the Constants and Variables Guide (Appendix E) for program 2–8.

Problems

General Concepts

1 State three activities that computers do well. [Section 2–1]
2 Name two activities that computers do not do well. [Section 2–1]
3 In your own words, state the completeness theorem. [Section 2–2]
4 Define an action block. Give an example. [Section 2–2]
5 What is a branch block? Give an example. [Section 2–2]
6 Explain what a loop block does, and give an example. [Section 2–2]
7 Name the two types of branch block. [Section 2–2]
8 Give the two types of loop blocks. [Section 2–2]
9 Identify the flowchart symbols in figure 2–13 (p. 88). [Section 2–3]
10 Sketch a flowchart of A) an open branch B) a counting loop C) a closed branch. [Section 2–3]
11 State the difference between a loop block and a branch block. [Section 2–3]
12 List three advantages of flowcharting. [Section 2–3]
13 What is the general idea of top-down design? [Section 2–4]
14 State the general format used in top-down design for technology applications. [Section 2–4]
15 Define a Pascal procedure. [Section 2–4]
16 List the seven steps in APM. [Section 2–5]
17 State which of the programming guides are used with the first six of the steps in APM. [Section 2–5]

Program 2–8

```
PROGRAM Self_Test_Chapter_2;

{**************************************************************
            Developed by:  Good· Student Programmer
                   Date:  October 15, 1991
 **************************************************************
     This program will solve for the voltage drop across a
given value of resistor for ten different values of current.
The program user must supply the value of the resistor in
ohms.  The values of current used are from 1 to 10 amps in
steps of 1 amp each and display the corresponding value of
voltage.
     The program will tell the user if the voltage drop across
the resistor ever becomes larger than 100 volts.
 **************************************************************
                     Constants used:
 --------------------------------------------------------------
        Starting_Current_Value = 1.0   :  REAL
        Increment_Current_Value = 1.0   :  REAL
         Ending_Current_Value = 10.0   :  REAL
              Maximum_Voltage = 100.0 :  REAL
 **************************************************************
                     Variables used:
 --------------------------------------------------------------
        Resistor_Value = Value of resistor        : REAL
        Resistor_Voltage = Voltage across resistor  : REAL
        Resistor_Current = Current in resistor      : REAL
        Program_Repeat = Does user want program
                         repeated?                : CHAR
 **************************************************************}
  CONST
      Starting_Current_Value = 1.0;
      Increment_Current_Value = 1.0;
       Ending_Current_Value = 10.0;
            Maximum_Voltage = 100.0;

  VAR
      Resistor_Value,
      Resistor_Voltage,
      Resistor_Current
      : REAL;

      Program_Repeat
      : CHAR;

  PROCEDURE Explain_Program_to_User;
 {--------------------------------------------------------------
     This procedure explains the purpose of the program to
the program user.
 [Action block]
 --------------------------------------------------------------}
    BEGIN
      WRITELN;
      WRITELN('This program will compute the voltage drop');
      WRITELN('across a resistor for ten different values');
      WRITELN('of current starting at 1 amp and going to');
```

Program 2–8 *continued*

```
      WRITELN('10 amps.');
      WRITELN;
      WRITELN('You must enter the value of the resistor ');
      WRITELN('in ohms.  The program will display the ');
      WRITELN('value of the voltage for each current and');
      WRITELN('let you know if the voltage drop ever is');
      WRITELN('greater than 100 volts.');
      WRITELN;
 {------------------------------------------------------------}
    END; {of procedure Explain_Program_to_User;}

  PROCEDURE Get_Resistor_Value;
 {-----------------------------------------------------------
      This procedure gets the value of the resistor from the
 program user.
 [Action block]
 -----------------------------------------------------------
      Variables used:  Resistor_Value}
    BEGIN
      WRITELN('Enter the value of the resistor in ohms:');
      READLN(Resistor_Value);
 {----------------------------------------------------------}
    END; {of procedure Get_Resistor_Value}

  PROCEDURE Check_It_Out;
 {-----------------------------------------------------------
      This procedure checks to see if the value of the voltage
 ever becomes larger than 100 volts.
 [Branch block]
 -----------------------------------------------------------
    Constants used:  Maximum_Voltage
    Variables used:  Resistor_Voltage }
    BEGIN
       IF Resistor_Voltage > 100 THEN
          WRITELN('Resistor voltage is larger than 100!')
       ELSE
          WRITELN('Resistor voltage is safe.');
 {----------------------------------------------------------}
    END; {of procedure Check_It_Out}

  PROCEDURE Display_the_Answer;
 {-----------------------------------------------------------
      This procedure displays the value of the voltage drop
  across the resistor.
  [Action block]
 -----------------------------------------------------------
    Variables used:  Resistor_Voltage }
    BEGIN
      WRITELN;
      WRITELN('Voltage drop is ',Resistor_Voltage,' volts.');
 {----------------------------------------------------------}
    END;   {of procedure Display_the_Answer}
```

Program 2–8 *continued*

```
  PROCEDURE Calculate_It;
{------------------------------------------------------------
     This procedure sets the initial value of the current, and
  calculates the voltage for ten different current values.
  [Loop block]
  ------------------------------------------------------------
   Constants used: Starting_Current_Value
                   Increment_Current_Value
                   Ending_Current_Value
   Variables used: Resistor_Value
                   Resistor_Voltage
                   Resistor_Current   }
    BEGIN
       Resistor_Current := Starting_Current_Value;
         REPEAT
            Resistor_Voltage := Resistor_Current * Resistor_Value;
              Display_the_Answer;
              Check_it_Out;
            Resistor_Current := Resistor_Current + 1;
         UNTIL Resistor_Current = 10;
{----------------------------------------------------------}
   END;  {of procedure Calculate_It}

  PROCEDURE Ask_for_Repeat;
{------------------------------------------------------------
     This procedure asks the program user if the program is
  to be repeated.
  [Action block]
  ------------------------------------------------------------
   Variables used:  Program_Repeat  }
    BEGIN
       WRITELN;
       WRITELN('Do you want the program repeated?');
       WRITELN('Enter Y or N and press RETURN/ENTER ');
       READLN(Program_Repeat);
{----------------------------------------------------------}
   END;  {of Ask_for_Repeat}

{ ‖ Main Programming Sequence  ‖
   ========================================================}
   BEGIN
      Explain_Program_to_User;
      WHILE (Program_Repeat <> 'N') DO
        BEGIN
          Get_Resistor_Value;
          Calculate_It;
          Ask_for_Repeat;
        END;
{==========================================================}
   END.  {of main programming sequence}
```

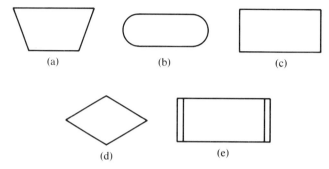

Figure 2–13   Flowchart Symbols Used for Problem 9

Program Analysis

18   Will program 2–9 compile? If not, why not?

Program 2–9

```
PROGRAM Problem_eighteen;
  BEGIN

  END.
```

19   What will be displayed on the screen when program 2–10 is executed?

Program 2–10

```
PROGRAM Another_Problem;

PROCEDURE This_is_a_Procedure;
   BEGIN
     WRITELN('This is from a procedure.');
   END;

BEGIN
   WRITELN('This is not from a procedure.');
   This_is_a_Procedure;
END.
```

20   What will program 2–11 display on the screen when executed?

Program 2–11

```
PROGRAM Another_Problem; PROCEDURE This_is_a_Procedure;
BEGIN  WRITELN('This is from a procedure.');  END;  BEGIN
WRITELN('This is not from a procedure.');
This_is_a_Procedure; END.
```

21 Correct any errors in program 2–12 so it will compile correctly.

Program 2–12

```
PROGRAM Check_This_Out;

   PROCEDURE First_One;
       WRITELN('This is the first procedure.');
   END;

   PROCEDURE Second_One;
       WRITELN('This is the second procedure.');
   END;

BEGIN
   Procedure First_One;
   Procedure Second_One;
END.
```

22 Will program 2–13 compile? If so, what will be displayed on the screen?

Program 2–13

```
PROGRAM Straighten_Me_Out; PROCEDURE Where_Are_We; BEGIN {
This procedure will be used for some calculations.} END;
PROCEDURE This_is_Another; BEGIN {This procedure will do
some screen displays.} END; BEGIN WRITELN('This is the
main programming block!'); END.
```

23 Make any necessary coding corrections in program 2–13 and rewrite it in block form so that it is easier to read.

Program Design

For the following programming problems, use the first six steps in the Applied Programming Method given in Appendix D. Use all of the Guides given in Appendix E. Each of the following problems will thus require a completed set of guides and a Pascal program using procedures, comments, and just enough code to make it compile (the code that has been presented up to this point in the text). Be sure to use the steps outlined in section 2–6 of this chapter.

Manufacturing Technology

24 Develop a Pascal program that will compute the number of items manufactured each 8-hour day, assuming that the same number of items are manufactured each hour. User input is the number of items made in one hour.

Drafting Technology

25 A Pascal program is required that will compute the area of a rectangle. User input is the height and width of the rectangle.

Computer Science

26 Create a Pascal program that will let the program user either double or triple a number. User input is the selection of either doubling or tripling and the value of the number.

Construction Technology

27 Develop a Pascal program that will compute the volume of a room. User inputs are the height, width, and length of the room.

Electronics Technology

28 Create a Pascal program that will give the power dissipation of a resistor for a range of current values and step increases selected by the program user. The program user is also to input the value of the voltage to be used. Program output is the resistor power. (Formula: $P = I \times E$ where P = power in watts, I = current in amps, and E = voltage in volts.)

Agriculture Technology

29 Develop a Pascal program that will compute acres of land. The program user is to input the width and length of the land in feet (assume a perfect rectangle). The program is to display the answer either in square feet or in acres as selected by the user. (Formula: 1 acre = 43,560 ft^2)

Health Technology

30 A Pascal program is required that will convert from degrees Fahrenheit to degrees centigrade over a range of temperatures and temperature increments selected by the program user. The output is to display both temperature scales. (Formula: $C = 5/9 \times (F - 32)$ where C = temperature in degrees centigrade and F is temperature in degrees Fahrenheit).

Business

31 Create a Pascal program that will compute the sales tax from the cost of three items entered by the program user. The output is the total cost of all the items including the sales tax. The program is to tell the user if the total will exceed $100.00.

3 Data, Expressions, and Assignments

Objectives

This chapter will give you the opportunity to learn:

1 Where semicolons are used.
2 The difference between a constant and a variable.
3 What identifiers are, and the rules for creating them.
4 Reserved words used in Pascal.
5 Pascal rules for assigning a constant and declaring a variable.
6 How to define the type of variable in Pascal.
7 Basic arithmetic operations used in Pascal.
8 The different types used in Pascal.
9 Defining your own types of variables.
10 Program development that will allow values to be entered by the program user.
11 Screen formatting that will give a professional touch to your programs.

Key Terms

Statement
Specific Operation
Semicolon
Compound Statement
Standard Pascal
Constant
Variable
Identifier
Reserved Words

Type
Declare
VAR
CONST
WRITELN
READLN
Assignment Operator
Operator
MOD

DIV	Exponential Notation
Carriage Return	LST
Compile-Time Error	

Outline

This is an important chapter. It presents the material Pascal is fussy about. Here you will learn how to create formulas in Pascal and use Pascal to solve them.

You had an introduction to this material in chapter two. The Constants and Variables Guide needed detailed information about the type of constant and variable to be used in your Pascal program; you undoubtedly had many questions about the use of this guide. This chapter will answer those questions for you.

You will also learn how to enable the user of your Pascal program to enter information into the program from the keyboard. This is a powerful feature that allows the computer to become a major problem-solving tool for technology.

3–1   Statements and Semicolons

Introduction

This section explains the use of those semicolons you've seen scattered around the Pascal programs you've been working with. You'll also learn here what statements are and what they do; this section sets the background for the rest of chapter three.

Statement

A Pascal **statement** is an instruction that tells the computer to perform a specific operation. You will learn ten statements in this book. Table 3–1 will give you an idea what these statements look like and what they do. Don't worry about understanding all of them now—that's what the rest of the book is about! Try to get a general idea of what a statement is and what it can do. Remember that the possibilities are limited; a statement can only loop, branch, or perform an action.

Statements, then, are the building blocks of any Pascal program. Now that you have an idea of what a statement in Pascal is, you're ready to see where to use those semicolons (;).

Table 3–1   Pascal Statements

Type	What it does	Example
Assignment Statement	Gives a value to a variable. In the example, the value of 5 will be assigned to A.	`A := 2 + 3;`
Procedure Call Statement	Activates a Pascal procedure. In the example, the procedure Explain__Program will become activated.	`Explain__Program;`

Compound Statement	Perhaps the most important statement in Pascal. Allows you to put many statements inside one statement, the **BEGIN . . . END;** statement.	`BEGIN`
		`  (Pascal statements);`
		`         .`
		`         .`
		`  (Pascal statements);`
		`END;`

IF-THEN or **IF-THEN-ELSE** Statement	Used for branching. In the example B is assigned a value of 4 **IF A = 3 .**	`IF A = 3 THEN B := 4;`

CASE Statement	Allows the program to select. In the example, the variable Resistor can have one of 3 values. The value it has will cause one of the three **WRITELN** statements to be executed.	`CASE Resistor OF`
		`  1 : WRITELN('10%');`
		`  2 : WRITELN('20%');`
		`  3 : WRITELN('30%');`
		`END;`

REPEAT-UNTIL Statement	Statements within the **REPEAT** are executed at least once and then repeated until a certain condition. In the example, the statements enclosed between the **REPEAT-UNTIL** will be done once and then repeated **UNTIL X = 5.**	`REPEAT`
		`    RT := R1 + R2;`
		`    X := X + 1;`
		`UNTIL X = 5;`

WHILE-DO Statement	No execution takes place if specific condition is not met. Otherwise compound statement is repeated. In example, compound statement is repeated as long as A = 'No'	`WHILE A = 'No' DO`
		`  BEGIN`
		`    WRITELN('Still No');`
		`    WRITELN('Try again.');`
		`    READLN(Answer);`
		`  END;`

FOR-TO-DO Statement	Creates a loop that begins and ends at a very specific count. In the example, the compound statement will be repeated exactly 5 times.	`FOR Count := 1 TO 5 DO`
		`  BEGIN`
		`    WRITELN('This is the');`
		`    WRITELN(Count);`
		`  END;`

WITH-DO Statement	A way of getting information in and out of items that are part of group of items. In the example, R-value, R__Watts and R__Toler are all items that are a part of Resistor__Record.	`WITH Resistor__Record DO`
		`  R__Value := 150;`
		`  R__Watts := 2;`
		`  R__Toler := 0.15;`
		`END;`

GOTO Statement	Causes the program to jump to another place in the program. It is considered very poor programming practice to use this statement.	`GOTO New__Spot;`

The Semicolon

In Pascal, **semicolons** are required between any two statements. They signify the end of one statement and the beginning of the other. Look at program 3–1.

Program 3–1

```
PROGRAM Using_Semicolons;

PROCEDURE Add_Them_Currents;
  BEGIN
   WRITELN('This is a statement.');
   Total_Current := I1 + I2 + I3;
  END;

BEGIN
  Add_Them_Currents;
END.
```

Note that each statement in program 3–1 is separated from the next statement by a semicolon. The only statement that doesn't have a semicolon is the last **compound statement** (**BEGIN/END**), which ends with a period because it is the main programming sequence.

In the original version of Pascal (called the **standard version**), the semicolon is not allowed to follow the last statement just before an **END,** because the **END** itself acts as a separator. Turbo Pascal doesn't care if you use a semicolon after the last statement just before an **END**, as illustrated in figure 3–1.

Conclusion

You have seen what Pascal statements look like and what they do. Pascal needs the semicolon in order to tell when one statement ends and another begins. You will find

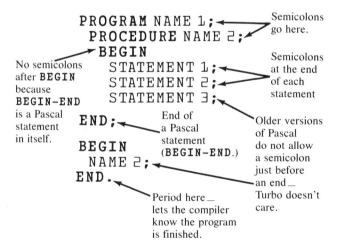

Figure 3–1   Use of the Semicolon in Pascal

that your biggest problem with errors in Pascal programming will be the omission of semicolons. Test your understanding of this section by trying the following section review.

3–1 Section Review

1   Explain what is meant by a Pascal statement.
2   How many different Pascal statements were listed in this section? Name three of them.
3   Where is a semicolon used in Pascal?
4   Explain the use of the semicolon in non-Turbo Pascal systems. Give the requirements of Turbo Pascal regarding semicolons.

3–2   Identifying Things

Introduction

This section shows you how to handle the information, called **data**, that you want your Pascal statements to manipulate. One of the fun things about Pascal is that you can create your own words to name the data or procedures. Here, you will learn some of the few rules you need to know to use these interesting and useful programming techniques.

Constants and Variables

The information you give the computer program to work with will consist of a constant or a variable. A **constant** is data (strings or numbers) that won't change when you execute the program. Here is an example.

```
WRITELN('This is a constant.');
```

The information inside the single quotation marks is a Pascal constant, because every time this **WRITELN** procedure is used it will cause the display of exactly the same thing.

A **variable** is a place in the computer's memory where information is kept. The value stored in this memory location can change during the execution of the program. Here is an example.

```
RT := R1 + R2;
```

The variable `RT` will have a value equal to the sum of whatever values are in variables `R1` and `R2`.

You need to know now how to tell Pascal that you want something to be a variable or a constant—how to **identify** the information you want the program to process.

Identifying Things

Each program or procedure used in Pascal has a name, or **identifier**. In Pascal an identifier is the name you, the programmer, give to data (constants or variables) and to major parts of your program (such as a procedure). Look at the program 3–2.

Program 3–2

```
PROGRAM Technician_Helper;

   PROCEDURE Resistance_Calculator;
     BEGIN
       Total_Resistance := R_1 + R_2;
     END;

   BEGIN
     Resistance_Calculator;
   END.
```

The identifiers in program 3–2 are
The name of the program: `Technician_Helper`
The name of the procedure: `Resistance_Calculator`
The variables: `Total_Resistance, R_1` and `R_2`
The procedure call statement: `Total_Resistance`

You will discover later in this chapter that you can also use identifiers to identify other parts of the Pascal program. One of the fun things about Pascal is that you can make up your own words as identifiers. The words you make up can say what the data or part of the program represents. There are a few rules you must use when creating identifiers. An identifier must begin with a letter of the alphabet or an underscore (`_`). This first character may be followed by any combination of letters, numbers or underscores, totaling one to 127 characters. No spaces are allowed. These, then, are all legal identifiers in Turbo Pascal.

```
R  R1  R_1  Resistor1  ResistorOne
This_is_the_value_of_resistor_number_one
Calculate_1_plus_23_plus_48
```

As you can see from the examples, the underscore is used in place of a space since spaces are not allowed in Pascal identifiers.

Pascal identifiers make no distinction between uppercase and lowercase letters. Thus, as far as Pascal is concerned, all of the following identifiers are exactly the same.

```
TRANSMITTER_POWER Transmitter_Power transmitter_power
tRANSMITTER_pOWER TrAnSmItTeR_pOwEr
```

Even though all of these identifiers are acceptable to Pascal, you should use the form that is easiest to read: `Transmitter_Power`.

Table 3–2 Pascal Reserved Words

AND	ARRAY	BEGIN	CASE	CONST
DIV	DO	DOWNTO	ELSE	END
FILE	FORWARD	FOR	FUNCTION	GOTO
IF	IN	LABEL	MOD	NIL
NOT	OF	OR	PACKED	PROCEDURE
PROGRAM	RECORD	REPEAT	SET	THEN
TO	TYPE	UNTIL	VAR	WHILE
WITH				

The following reserved words are additional features of Turbo Pascal.

ABSOLUTE	EXTERNAL	INLINE	OVERLAY
SHR	STRING	XOR	SHL

There is one more rule to keep in mind when making up your identifiers; you cannot use any of Pascal's **reserved words**. These words, listed in table 3–2, make up Pascal's instruction set.

Notice that you have already seen many of these reserved words in the sample programs. You are not allowed to use them as identifiers, as Pascal needs them to understand what you want it to do. But you can make up an identifier that uses a reserved word as a part of it.

```
Begin_This          End_Here          Goto_This_Procedure
```

These are all **legal** identifiers because the reserved word is not used by itself.

EXAMPLE 3–1

Which of the following are NOT valid identifiers? Why not?

```
Uncle_Pascal                Three-Fifty
One_Two_Three               GEORGEANDSAM
1_Two_Three                 Area of Circle
Resistor_1                  thisvalue
End                         ENDing
WOW!                        Forward
```

Solution
The identifiers that are not valid are

`1_Two_Three`	Starts with a number.
`Area of Circle`	Contains spaces.
`End`	Pascal reserved word.
`WOW!`	Contains a !
`Forward`	Pascal reserved word.
`Three-Fifty`	Uses a dash to separate words.

Conclusion

This section gave you an idea of the difference between a constant and a variable. You also saw how to identify things in Pascal and the rules relating to identifiers. Test your understanding of this section by trying the following section review.

3–2 Section Review

1 Explain the difference between data and statements.
2 Describe a constant.
3 Describe a variable.
4 What is a Pascal identifier?
5 State the rules for developing Pascal identifiers.
6 Explain what is meant by a Pascal reserved word. Name five Pascal reserved words.

3–3   Pascal Types

Introduction

This section will show you the different kinds of information Pascal can work with, and how to tell Pascal that you intend to use it. This is similar to the use of the Constants and Variables Guide, in which you indicated what kind of variable you were going to use.

Information Types

Remember that information used in a program is called data. In Pascal, all data must be classified into **types**, as seen in table 3–3.

Let's look again at the Constants and Variables Guide introduced in chapter two.

CONSTANTS AND VARIABLES GUIDE

Formula	Variable	Meaning	Type R I C B S	Block(s) Used

The column labeled "Type" in the Constants and Variables Guide is there to remind you that each variable you use in Pascal must have a type, but only one type. The initials in the smaller columns stand for: R = **REAL**, I = **INTEGER**, C = **CHAR**, B = **BOOLEAN**, and S = **STRING**. If you use the B column to mean the type **BYTE**, though, you must indicate this in the column headed "Meaning."

Table 3–3 Pascal Data Types

Data Type	Examples	Discussion
BOOLEAN	TRUE, FALSE	This type of data can have only two values. In spite of this apparent limitation, you will see that it is a very useful data type.
BYTE	2, 0, 255	This type of data can have any positive value from 0 to 255. This seems like a small range of values, but this data type has many useful applications.
CHAR	'B', '?', '3'	This means character and it can represent any of the keyboard characters. This data type is used for working with text information.
INTEGER	3, -3, 4500	This type of data is any positive or negative whole number from 32767 to -32768. You can think of these numbers as counting numbers. No decimal points are allowed and you cannot use a number larger than 32767 or smaller than -32768 for this data type.
REAL	1.2, -3.45E-02	This type is the one most frequently used in technical applications because it can represent any number from 1×10^{38} to 1×10^{-38} with eleven-digit accuracy. The **E** notation used in the example means the exponent of ten.
STRING[N]	'Hello there!' '6-+{[!*?/'	This means that you can have characters one after the other—strung together, hence the word STRING. The value N represents the length of the STRING. If the length is omitted the default value of 255 will be used.

*Note that Turbo Pascal contains other data types, not all of which are presented in this text. Some other data types will be introduced as needed, but for now, consider only those listed in this table.

EXAMPLE 3–2

Give the types of each of the following pieces of data.
    A 'H' B 5.06 C 'What is this?' D FALSE E -5 F 5

Solution
A 'H'   **CHAR**, because it contains a single character.
B 5.06   **REAL**, because the number includes a decimal point.
C 'What is this?'   **STRING**, because it contains a sequence of characters.
D FALSE   **BOOLEAN**, because it is one of two possible conditions (**TRUE** or **FALSE**).

E −5   **INTEGER**, because it is a whole number. An **INTEGER** can be negative, but this is not of type **BYTE**, because a **BYTE** cannot be a negative number.

F 5   Could be **INTEGER** or **BYTE**, because it is a positive whole number and less than 255 (if it were greater than 255, it could not be **BYTE**).

Declaring It

To use a variable in Pascal, you must first **declare** it. You do this by entering information about the variable at the very beginning of the Pascal program, after the reserved word **PROGRAM** and before any **PROCEDURE** or **BEGIN** statements, as shown in program 3–3.

Program 3–3

```
PROGRAM Declare_Your_Variables;

 VAR
   The_First : BOOLEAN;
   Second_Variable : BYTE;
   The_Third_One : CHAR;
   Here_is_4, And_Here_is_5 : INTEGER;
   Now_Six, Variable_7 : REAL;
   The_Last_One : STRING[10];

 PROCEDURE Lets_Use_Them;
   BEGIN
   { Body of this procedure }
   END;

 BEGIN   { Main programming sequence }
   Lets_Use_Them;
 END.
```

Observe that the reserved word **VAR** means that variables are to follow; you must always use **VAR** before you declare any Pascal variables.

Notice how the type **STRING** is declared. It is followed by brackets that contain a number. When you declare a string variable in Turbo Pascal, you must state the maximum number of string characters you intend your variable to have—the maximum possible is 255. The form of the Turbo **STRING** is:

STRING[N];

where N = a positive integer from 1 to 255

Remember that program 3–3 could also have been written as shown in program 3–4.

Program 3–4

```
    PROGRAM Declare_Your_Variables; VAR The_First : BOOLEAN;
Second_Variable : BYTE; The_Third_One : CHAR; Here_is_4 : INTEGER; Now_Five
: REAL; The_Last_One : STRING[10]; PROCEDURE Lets_Use_Them; BEGIN { Body of
this procedure } END; BEGIN   { Main programming sequence } Lets_Use_Them;
END.
```

Of course, the computer doesn't care which of the two forms you use; it will compile either one with equal ease. But the point of applying Pascal to technology problems is to make it as easy as possible to read and understand.

Figure 3–2 shows how variables are declared in a Pascal program.

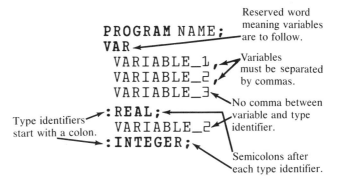

Figure 3–2   How Variables Are Declared

Commenting on Your Variables

In chapter two, you were introduced to a general Pascal structure (in Appendix G). Recall the part of that structure that suggested using a format listing all of the constants and variables used, with the type and an explanation of each. This part is shown below.

```
********************************************************
               Constants used:
--------------------------------------------------------
         Constant 1 = Value                  [Type]
         Constant 2 = Value                  [Type]
                .         .              .
                .         .              .
                .         .              .
         Constant N = Value                  [Type]
********************************************************
```

```
                    Variables used:
--------------------------------------------------------------
    Variable 1 = Explain what variable is.    [Type]
    Variable 2 = Explain what variable is.    [Type]
          .                      .                 .
          .                      .                 .
          .                      .                 .
    Variable N = Explain what variable is.    [Type]
    **********************************************************
```

This information in your Pascal program only consists of comments; you still need to declare your variables in the **VAR** section of your Pascal program. You can omit such variable comments if you carefully structure your variables when you declare them, as shown in program 3–5.

Program 3–5

```
PROGRAM Solve__It;

   {**************************************************************
            Developed By:  Electronics Student
               Date:  November 1991
    **************************************************************
      Program description:
          Description of program goes here.
    **************************************************************
                    Variables used:
   ------------------------------------------------------------}
   VAR

     First__Variable,        {Reason for this variable}
     Second__Variable        {Reason for this variable}
    :REAL;

     Third__Variable,        {Reason for this variable}
     Fourth__Variable        {Reason for this variable}
    :INTEGER;
   {************************************************************ }

   PROCEDURE This__One;
   {---------------------------------------------------------
     Purpose of this procedure is to:
   ---------------------------------------------------------}
     BEGIN

   {---------------------------------------------------------}
     END;   {of procedure This__One;}

   {|| Main Programming Sequence ||
   ==========================================================}
     BEGIN
     This__One;
   {==========================================================}
     END.   {of Main Programming Sequence.}
```

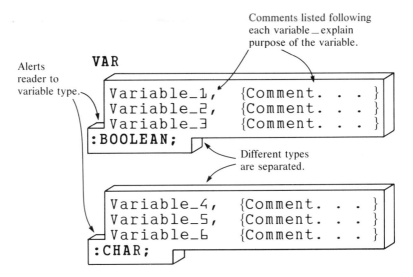

Alerts reader to variable type.

Comments listed following each variable — explain purpose of the variable.

```
VAR

   Variable_1,   {Comment. . . }
   Variable_2,   {Comment. . . }
   Variable_3    {Comment. . . }
  :BOOLEAN;

   Variable_4,   {Comment. . . }
   Variable_5,   {Comment. . . }
   Variable_6    {Comment. . . }
  :CHAR;
```

Different types are separated.

Figure 3–3 Advantages of VAR Structure

Figure 3–3 illustrates the advantages of the structure shown in program 3–5.

Assigning Constants

Pascal also allows you to **assign** values to constants. This is similar to the declaration of variables. All constant assignments must be done after the reserved word **PROGRAM** and before the variable declarations, as shown in program 3–6.

Note the use of the reserved word **CONST**. This means constants are to follow. The reserved word **CONST** must be used before any constants are assigned, to let Pascal know that you will be assigning constants. Figure 3–4 illustrates the structure.

If you add comments with your constants and variables, you don't need to use the "Constants used" and "Variables used" parts of the General Pascal Structure in Appendix G, unless a lengthy explanation is needed to adequately document the purpose of the constant and/or variable.

Conclusion

This section presented some of the most important requirements of the Pascal language: declaring variables and assigning constants. In the next section, you'll see how to use the data you have declared and assigned. Test your understanding of this section by trying the following section review.

Program 3–6

```
PROGRAM Name_It;

{******************************************************************
          Developed By:  Electronics Student
               Date:  November 1991
   ******************************************************************
       Program description:
   ******************************************************************
                   Constants used:
   ---------------------------------------------------------------}
   CONST

     This_Value = 123;          {Integer and comment}
     That_Value = 456.0;        {Real and comment}
     Another_Value = 'Hello'    {String and comment}
   {******************************************************************
                   Variables used:
   ---------------------------------------------------------------}
   VAR

     First_Variable,            {Reason for this variable}
     Second_Variable            {Reason for this variable}
   :REAL;

     Third_Variable,            {Reason for this variable}
     Fourth_Variable            {Reason for this variable}
   :INTEGER;
   {*****************************************************************}

   PROCEDURE This_One;
   {-----------------------------------------------------------------
     Purpose of this procedure is to:
   ---------------------------------------------------------------}
     BEGIN

   {---------------------------------------------------------------}
     END;  {of procedure This_One;}

   {|| Main Programming Sequence ||
     ===============================================================}
     BEGIN
     This_One;
   {===============================================================}
     END.   {of Main Programming Sequence.}
```

3–3 Section Review

1 Explain what is meant by a Pascal data type.
2 Name the data types presented in this section.
3 How many different values can a **BOOLEAN** type have? Name them.
4 State what is meant by assigning constants in Pascal. Where in the Pascal program must this be done?
5 State what is meant by declaring variables in Pascal. Where in the Pascal program must this be done?

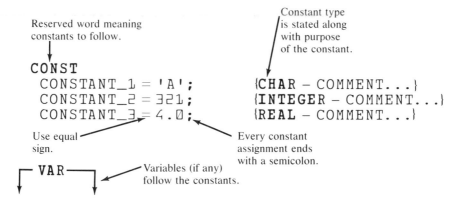

Reserved word meaning constants to follow.

Constant type is stated along with purpose of the constant.

```
CONST
  CONSTANT_1 = 'A';        {CHAR - COMMENT...}
  CONSTANT_2 = 321;        {INTEGER - COMMENT...}
  CONSTANT_3 = 4.0;        {REAL - COMMENT...}
```

Use equal sign.

Every constant assignment ends with a semicolon.

VAR

Variables (if any) follow the constants.

Figure 3–4 Structuring with Constants

3–4 Writing and Reading

Introduction

In this section you'll discover how to get information in your program displayed on the screen, and how to get information into your program while the program is being executed.

In the next section, you will learn about arithmetic operations so that you can begin using formulas in your Pascal programs. Armed with information on inputting, outputting, and arithmetic, you will be well on your way to the development of practical Pascal programs that will easily solve technology problems.

Writing to the Screen

You have seen the Pascal command **WRITELN** used in many of the previous programs. This is actually a built-in procedure in the Turbo Pascal system, prepared to make it easier for you to have information displayed on your screen or printer. The command is

```
WRITELN('This text appears on the screen.');
```

There are several things to note about this built-in procedure. First, the procedure word **WRITELN** simply means to write a line. Whatever is enclosed in the single quotes (**'**) (for instance, a string of characters) will appear on the screen when the program is executed. The last thing you see is the semicolon—remember that semicolons belong at the end of every Pascal statement. They are Pascal's way of signaling the end of a statement.

The **WRITELN** procedure can also be used to output the value of variables to the screen, using the format

```
WRITELN(Variable_Name);
```

When this is executed, it will display the value of `Variable_Name`, not the string `Variable_Name`; this is because the single quotation marks are omitted. Whenever Pascal fails to see the single quotation marks, whatever is inside the parentheses will be treated as data (which also means that it must have been declared in the **VAR** or **CONST** section of the program). Look at the example in program 3–7.

Program 3–7

```
PROGRAM Writeln_Demonstration;

CONST
        Number_One = 42;             {Integer}
        String_One =  'Hello';       {String}
        Number_Three = -55;          {Integer}

PROCEDURE Display_It;
{------------------------------------------------------------
      This procedure gives examples of the use of the
  WRITELN in displaying output to the screen.
  -------------------------------------------------------------}
    BEGIN
          WRITELN('Number_One');     {Displays the string.}
          WRITELN(Number_One);       {Displays the data.}
          WRITELN('String_One');     {Displays the string.}
          WRITELN(String_One);       {Displays the data.}
          WRITELN(Number_Three);     {Displays the data.}
          WRITELN('Number_Three');   {Displays the string.}
{-------------------------------------------------------------}
    END;    { of Procedure Display It;}

{ || Main Programming Sequence ||
========================================================}
 BEGIN
    Display_It;
{ ========================================================}
 END.    { of main programming sequence }
```

When program 3–7 is executed, the screen output will be

```
Number_One
42
String_One
Hello
-55
Number_Three
```

Reading From the Keyboard

You have seen how to use the **WRITELN** to display strings or the values assigned to identifiers used as constants. Here, you will learn how to construct a Pascal program so that the program user can enter a value. You will use a variable and will declare the identifier you create as such.

The Pascal command for reading data during program execution is

READLN(Variable);

where **Variable** is a previously declared variable. Program 3–8 illustrates the use of the **READLN**.

Program 3–8

```
PROGRAM Readln_Demonstration;

VAR

    First_Value,             { Variable used in demo. }
    Second_Value             { Another demo variable. }
:INTEGER;

PROCEDURE Get_Values;
{-------------------------------------------------------
        This procedure gets two values from the program
    user
    -------------------------------------------------}
    BEGIN
        WRITELN('Enter a whole number.');
        READLN(First_Value);
        WRITELN('Enter another whole number.');
        READLN(Second_Value);
   {----------------------------------------------------}
    END;    { of Procedure Get_Values;}

PROCEDURE Display_Values;
{-------------------------------------------------------
        This procedure displays the values entered by
    the program user.
    -------------------------------------------------}
    BEGIN
        WRITELN('The first value you entered was:');
        WRITELN(First_Value);
        WRITELN('The second value entered by you:');
        WRITELN(Second_Value);
{-------------------------------------------------------}
    END    { of Procedure Display_Values;}

{ || Main Programming Sequence ||
 ===================================================}
 BEGIN
    Get_Values;
    Display_Values;
{ ===================================================}
 END.    { of main programming sequence }
```

When program 3–8 is executed it will first ask the user for some values. Assume that the values entered by the program user are 10 and 5 (both whole numbers). The display will then become

```
Enter a whole number.
10
Enter another whole number.
5
The first value you entered was:
10
The second value entered by you:
5
```

You now know a way to allow the program user to enter data into your program. Notice that you had to declare the variables at the top of the Pascal program before you stated their type (in this case **INTEGER**) or wrote any procedures.

More Writing

There may be times when you wish to have the value of a declared variable or assigned constant appear on the same line as your string message. For example, in program 3–8, the last part of the display was

```
The first value you entered was:
10
The second value entered by you:
5
```

Suppose you wanted the output to appear as follows (values of the variables input by the program user are highlighted for emphasis):

```
You entered 10 as the first value,
and your second value was 5
```

This is how to modify the **WRITELN** statements in order to accomplish that output.

```
WRITELN('You entered ',First_Value,' as the first value,');
WRITELN('and your second value was ',Second_Value);
```

Observe that the strings are set off by single quotation marks and the declared variables are set off by commas.

EXAMPLE 3–3

Determine what the output of program 3–9 would be. Assume that the program user input is

```
12
Uncle Pascal
```

Program 3-9

```
PROGRAM Example_3_3;

CONST

    This_is_a_Constant = 100;        {Integer}

VAR

    Get_First               { Numerical value for demo.}
  :INTEGER;

    Get_Second              { String variable for demo. }
  :STRING[30];

PROCEDURE Ask_for_It;
{--------------------------------------------------------
        This procedure gets two values from the program
    user.
    ---------------------------------------------------}
    BEGIN
        WRITELN('Enter a whole number:');
        READLN(Get_First);
        WRITELN('Enter your name:');
        READLN(Get_Second);
{---------------------------------------------------}
    END;   { of Procedure Ask_for_it;}

PROCEDURE Show_What_Happens;
{--------------------------------------------------------
        This procedure displays an output using data
    entered by the program user.
    ---------------------------------------------------}
  BEGIN
    WRITELN(Get_First,' was the first number.');
    WRITELN('The second variable, Get_Second, was');
    WRITELN(Get_Second);
    WRITELN('Both of the variables were:');
    WRITELN(Get_First, Get_Second);
    WRITELN('Look at this ',This_is_a_Constant,' number.');
{---------------------------------------------------}
  END;   { of Procedure Show_What_Happens;}

{ || Main Programming Sequence ||
 =========================================================}
 BEGIN
    Ask_for_It;
    Show_What_Happens;
{ =========================================================}
 END. { of main programming sequence }
```

Solution

When program 3–9 is executed it will display

```
Enter a whole number:
12
Enter your name:
Uncle Pascal
12 was the first number.
The second variable, Get_Second, was
Uncle Pascal
Both of the variables were
12Uncle Pascal
Look at this 100 number.
```

Observe the output of the procedure `Show_What_Happens`. The first **WRITELN**,

```
WRITELN(Get_First,' was the first number.');
```

produces the **READLN** value of the variable `Get_First`, and then the string follows as you would expect.

The second **WRITELN**,

```
WRITELN('The second variable, Get_Second, was');
```

does not output any data that was input by the program user; `GetSecond` is treated as a string because it is inside single quotes.

The third **WRITELN**,

```
WRITELN(Get_Second);
```

outputs the **READLN** value of the variable `Get_Second`, because `Get_Second` is not enclosed by single quotes. Since it was declared as a variable it will display its assigned value.

The fourth **WRITELN**,

```
WRITELN('Both of the variables were:');
```

outputs the string enclosed by single quotations.

The fifth **WRITELN**,

```
WRITELN(Get_First, Get_Second);
```

causes the **READLN** values of both variables to be displayed, one after the other, because no single quotes were used. Note that the two identifiers were separated by a comma. This comma is necessary to let Pascal know where one variable identifier ends and the other begins.

The last **WRITELN**,

```
WRITELN('Look at this', This_is_a_Constant,' number.');
```

produces a string on either side of the constant identifier. Note that the left and right strings are both enclosed in single quotation marks and the constant identifier is separated from both by commas.

Conclusion

This section introduced two important Pascal skills: how to display strings and data on the monitor and how to get data into your program using Pascal. In the next section, you will see how to do arithmetic operations using the Pascal programming language. For now, test your understanding of this section by trying the following section review.

3−4 Section Review

1   Is **WRITELN** a reserved word in Pascal? Explain.
2   Explain what the difference will be when the following two **WRITELN**s are executed.
    ```
    WRITELN('This_One');
    WRITELN(This_One);
    ```
3   Give the Pascal command for allowing the program user to input data into an active Pascal program.
4   If the identifier `This_Value` were assigned a value of `'This string!'`, what would the following **WRITELN** display when activated?
    ```
    WRITELN('This_Value is ',This_Value,' This_String.');
    ```

3−5   Assigning Things

Introduction

This section shows you how to work with the information, or data, such as numbers and words, that will be used by your Pascal programs. Previous sections laid the groundwork for this section. You now have an idea of what a Pascal statement is, where to use semicolons, how to make up your own identifier, and how to assign constants and declare variables. In this section, you will see how to work with this new information.

Assigning Things

In Pascal, if you want to solve an equation, you must use the **assignment operator**. As an example, if you want to find the total voltage drop across two resistors, the formula is

$$V_T = V_{R1} + V_{R2}$$

where V_T = total voltage in volts

V_{R1}, V_{R2} = voltage drop across each resistor in volts

If you were to write this formula as a Pascal equation, you would have to use the Pascal assignment operator in place of the equal sign. The Pascal assignment operator is

```
:=
```

In a Pascal program the equation would thus be written

```
Total_Voltage := V_R1 + V_R2;
```

The assignment operator means: Solve the expression on the right of the `:=` and assign it to the memory location named on the left side of the `:=`.

Thus, if you have the Pascal statement

```
Variable_1 := Variable_1 + 1;
```

it does not mean `Variable_1` equals `Variable_1 + 1;`—it means: add one to the contents of the memory location identified as `Variable_1` and store the results in the same place (back in the same memory location called `Variable_1`). In other words, the above Pascal statement simply adds 1 to the number already stored in memory location `Variable_1`.

Program 3–10 illustrates the use of the Pascal assignment operator.

This is a lot of programming just to add two whole numbers! But program 3–10 does illustrate the first Pascal program that actually solves a technology problem. It is so well documented that anyone can understand what it is supposed to do.

When the program is executed, here is what will happen (assume the program user enters the values of 5 and 12 for the voltages across each resistor):

```
This program will compute the total voltage
across two resistors in a series circuit.
All you need to do is enter the voltage drop
across each resistor and the program will do
the rest.
Enter the voltage across R1:
5
Enter the voltage across R2:
12
The total voltage across two series
resistors when VR1 = 5 volts
and VR2 = 12 volts is equal to
17 volts.
```

Conclusion

This section presented the assignment operator. Here you saw how to develop a Pascal program that takes data from the program user, performs a computation with that data, and then displays the results. In the next section, you will see how to do other arithmetic operations using Pascal as well as how to write a program that does some decision making. For now, test your understanding of this section by trying the following section review.

Program 3–10 (pp. 113–114)

```
PROGRAM Total_Voltage;

 {****************************************************************
               Developed By:  Electronics Student
               Date: November 1991
      ****************************************************************
      Program description:  This program finds the total voltage
         drop across two resistors.
      ****************************************************************
                        Constants used:
      --------------------------------------------------------------}
                             None
      ****************************************************************
                        Variables used:
      --------------------------------------------------------------}
    VAR
      Total_Voltage,          {Total circuit voltage.}
      V_R1,                   {Voltage across first resistor}
      V_R2                    {Voltage across second resistor}
     :INTEGER;
   {****************************************************************}

PROCEDURE Explain_Program;
{--------------------------------------------------------------
   This procedure explains the purpose of the program to the
   program user.
   -----------------------------------------------------------}
   BEGIN
        WRITELN('This program will compute the total voltage');
        WRITELN('across two resistors in a series circuit.');
        WRITELN('All you need to do is enter the voltage drop');
        WRITELN('across each resistor and the program will do');
        WRITELN('the rest.');
{-----------------------------------------------------------}
   END;  {of procedure Explain_Program;}

 PROCEDURE Get_Values;
{--------------------------------------------------------------
   This procedure gets the value of each voltage drop from the
   program user.
   -----------------------------------------------------------}
   BEGIN
        WRITELN('Enter the voltage across R1:');
        READLN(V_R1);
        WRITELN('Enter the voltage across R2:');
        READLN(V_R2);
{-----------------------------------------------------------}
   END;  {of procedure Get_Values;}

 PROCEDURE Do_Computations;
{--------------------------------------------------------------
   This procedure computes the total voltage.
   -----------------------------------------------------------}
```

Program 3–10 *continued*

```
   BEGIN
            Total_Voltage := V_R1 + V_R2;
{-----------------------------------------------------------------}
   END;   {of procedure Do_Computations;}

 PROCEDURE Display_Answers;
{----------------------------------------------------------------
   This procedure displays the value of the total voltage.
-------------------------------------------------------------}
   BEGIN
      WRITELN('The total voltage across two series ');
      WRITELN('resistors when VR1 = ',V_R1,' volts');
      WRITELN('and VR2 = ',V_R2,' volts is equal to');
      WRITELN(Total_Voltage,' volts');
{--------------------------------------------------------------}
   END;   {of procedure Display_Answer;}

{|| Main Programming Sequence ||
 ================================================================}
   BEGIN
    Explain_Program;
    Get_Values;
    Do_Computations;
    Display_Answer;
{================================================================}
   END.   {of Main Programming Sequence.}
```

3–5 Section Review

1 Explain the purpose of the Pascal assignment operator.
2 Show how you would write a Pascal statement that would add three variables together (use X, Y, and Z as the variables and Sum as the sum).
3 State what the following equation means.

```
This_One := This_One + 2;
```

3–6   Expressing Things

Introduction

In this section you will see how to do arithmetic operations in Pascal, applying much of what you already know. When you complete this section, you will be ready to perform most of the math processes required of technology problems. This is an important section and very practical as well.

Table 3–4 Pascal Arithmetic Operators

Symbol	Operation	Example
*	Multiplication	3 * 4 = 12
/	Division with a REAL answer	9/2 = 4.5
DIV	Division of INTEGER with an INTEGER answer.	12 DIV 3 = 4
MOD	Test to see if one number is a factor of another—equals zero if a factor.	8 MOD 2 = 0 (2 is a factor of 8) 9 MOD 2 = 1 (2 is not a factor of 8)
+	Addition	5 + 3 = 8
−	Subtraction	5 − 3 = 2

Arithmetic Operators

Think of an **operator** as a symbol that asks Pascal to do something. The **arithmetic operators** used in Pascal are listed in Table 3–4.

You should see at least three arithmetic symbols familiar to you in table 3–4: division (/), addition (+), and subtraction (−). The multiplication symbol used is the asterisk (*). This is because standard keyboards do not have a separate symbol for multiplication; if you use the "X" key to try to multiply, Pascal will think you are trying to use an identifier rather than an operator.

You are probably wondering what the **MOD** and **DIV** really mean. You will see that you need these when dividing with INTEGER type numbers. When working with Pascal, you will be using two different types of numbers: REAL and INTEGER. Because of this, each of these arithmetic operators warrants further discussion.

The * Operator

All multiplication in Pascal is done with the * operator. You may freely mix REAL and INTEGER numbers in this operation. Table 3–5 summarizes the results you will get. (When you multiply two numbers, the multiplier is one of the numbers and the multiplicand is the other.)

Table 3–5 Multiplication of Data Types

Multiplicand Type	Multiplier Type	Result Type	Example
INTEGER	INTEGER	INTEGER	3 * 4 = 12
REAL	REAL	REAL	3.0 * 4.0 = 12.0
INTEGER	REAL	REAL	3 * 4.0 = 12.0
REAL	INTEGER	REAL	3.0 * 4 = 12.0

The combinations shown in table 3–5 are illustrated in example 3–4.

EXAMPLE 3–4

State the type of answer you would expect for the following multiplication problems.
A 5 * 8 =   B 12.0 * 0.2 =   C 6 * 0.01 =

Solution (Refer to table 3–5)
A   This is INTEGER * INTEGER = INTEGER
     Thus, 5 * 8 = 40
B   This is REAL * REAL = REAL
     Thus, 12.0 * 0.2 = 2.4
C   This is INTEGER * REAL = REAL
     Thus, 6 * 0.01 = 0.06

The / Operator

The / operator is used for division and always gives a REAL result, as shown in table 3–6. (In the division problem A/B, "A" is the dividend and "B" is the divisor).

Table 3–6   Division (/) of Data Types

Dividend Type	Divisor Type	Result Type	Example
INTEGER	INTEGER	REAL	12/3 = 4.0
REAL	REAL	REAL	12.0/3.0 = 4.0
REAL	INTEGER	REAL	12.0/3 = 4.0
INTEGER	REAL	REAL	12/3.0 = 4.0

The combinations shown in table 3–6 are illustrated in example 3–5.

EXAMPLE 3–5

State the type of answer you would expect for the following division problems.
A 8/2 =   B 7/2 =   C 6/12.0 =

Solution (Refer to table 3–6)
A   INTEGER/INTEGER = REAL
        8/2 = 4.0
B   INTEGER/INTEGER = REAL
        7/2 = 3.5
C   INTEGER/REAL = REAL
        6/12.0 = 0.5

The DIV Operator

If you need to divide by integers and you want the result to be INTEGER you are saying: divide 9 by 2 and get a whole number answer. This, you can see is a contradiction. 9 divided by 2 gives 4.5, by definition a REAL number. But Pascal has made provisions for this and when using DIV, will return the number on the left of the decimal point, leaving off the remainder and thus producing an INTEGER. For example, to divide 9 by 2 and get an INTEGER number, use this statement.

9 **DIV** 2 = 4 (The remainder is lost, but the answer is an INTEGER type.)

EXAMPLE 3–6

Give the results of the following INTEGER division problems.
A 10 DIV 2 =   B 5 DIV 2 =   C 3 DIV 10 =

Solution
A   First, do the actual division (10/2 = 5), then preserve everything to the left of the decimal point.
    10 DIV 2 = 5
B   Again, do the actual division (5/2 = 2.5), then preserve everything to the left of the decimal point.
    5 DIV 2 = 2
C   Again, the actual division (3/10 = 0.3), then preserve only that to the left of the decimal point.
    3 DIV 10 = 0

The MOD Operator

The **MOD** operator comes from the word modulus. This operator works in a way that is similar to the operation of a car odometer. If you had an odometer that reset itself to zero every 10 miles, at the end of 5 miles it would read 5, at the end of 9 miles it would read 9, and at the end of 10 miles it would read 0. This would be called modulus 10. You could also have an odometer that would reset itself to zero every 5 miles, or every 3 miles. This would represent modulus 5 or 3, as shown in table 3.7.
    Example 3–7 illustrates some applications of the Pascal MOD operator.

EXAMPLE 3–7

Give the results of the following INTEGER operations.
A 10 MOD 2 =   B 5 MOD 2 =   C 3 MOD 10 =

Solution
A   First, do the actual division (10/2 = 5). Since there is no remainder this means that 2 is a factor of 10 (the odometer would have turned back to 0).
    10 MOD 2 = 0

B   Again, do the actual division (5/2 = 2.5). Since there is a remainder, this means that 2 is not a factor of 5 (it also means that the odometer would not be zero, but 1 past zero).

5 MOD 2 = 1 (Note that the answer must be INTEGER.)

C   Repeat the division (3/10 = 0.3). (This means that the odometer would not yet have had a chance to get to 0.)

3 MOD 10 = 3 (Again the answer is of type INTEGER.)

The − Operator

As you would expect, the − operator in Pascal means subtraction. In the subtraction problem of A − B, "A" is the **minuend** and "B" is the **subtrahend**. Table 3−8 illustrates the resultant types in subtraction.

As you can see from Table 3−8, the only time the result of a subtraction is an INTEGER is when both minuend and subtrahend are INTEGERs. What applies to subtraction also applies to addition.

Table 3−7   Examples of Different Moduli

Number	MOD 10	MOD 5	MOD 3
0	0 MOD 10 = 0	0 MOD 5 = 0	0 MOD 3 = 0
1	1 MOD 10 = 1	1 MOD 5 = 1	1 MOD 3 = 1
2	2 MOD 10 = 2	2 MOD 5 = 2	2 MOD 3 = 2
3	3 MOD 10 = 3	3 MOD 5 = 3	3 MOD 3 = 0
4	4 MOD 10 = 4	4 MOD 5 = 4	4 MOD 3 = 1
5	5 MOD 10 = 5	5 MOD 5 = 0	5 MOD 3 = 2
6	6 MOD 10 = 6	6 MOD 5 = 1	6 MOD 3 = 0
7	7 MOD 10 = 7	7 MOD 5 = 2	7 MOD 3 = 1
8	8 MOD 10 = 8	8 MOD 5 = 3	8 MOD 3 = 2
9	9 MOD 10 = 9	9 MOD 5 = 4	9 MOD 3 = 0
10	10 MOD 10 = 0	10 MOD 5 = 0	10 MOD 3 = 1

Table 3−8   Subtraction (−) of Data Types

Minuend Type	Subtrahend Type	Result Type	Example
INTEGER	INTEGER	INTEGER	10 − 3 = 7
REAL	REAL	REAL	10.0 − 3.0 = 7.0
REAL	INTEGER	REAL	10.0 − 3 = 7.0
INTEGER	REAL	REAL	10 − 3.0 = 7.0

EXAMPLE 3–8

State the type of result you would expect to get from the following operations.
A 14 − 3 =   B 3 − 10 =   C 35− 6.2 =

Solution

A   This is INTEGER − INTEGER = INTEGER:
    14 − 3 = 11 (Answer is of type INTEGER.)
B   Here you have INTEGER − INTEGER = INTEGER:
    3 − 10 = −7 (Answer is of type INTEGER.)
C   This is INTEGER − REAL = REAL:
    35 − 6.2 = 28.8 (Answer is of type REAL.)

Precedence in Calculations

You can use parentheses to indicate the order you want your calculations performed. Pascal does all multiplication and division first, then addition and subtraction; otherwise, all operations are performed from left to right. When you use parentheses, all calculations inside them are done first. Look at this example and try to determine what operation was performed first.

```
8 + 5 * 3 = 23
```

In this calculation, Pascal's order of precedence caused the multiplication to be done first, followed by the addition. Now look at another example.

```
(8 + 3) * 3 = 39
```

Here, since parentheses are used, the addition was done first (Pascal always does calculations inside parentheses first), and then the multiplication was performed.

EXAMPLE 3–9

Determine the values of the following operations.
     A 4/2 − 1 =   B 4/(2 − 1) =   C (5 + 7)/6 − 3   D 5 + 7/(6 − 3)

Solution

A   4/2 − 1 = 2 − 1 = 1
B   4/(2−1) = 4/1 = 4
C   (5 + 7)/6 − 3 = 12/6 − 3 = 2 − 3 = −1
D   5 + 7/(6 − 3) = 5 + 7/3 = 5 + 2.333 = 7.333

Sample Program

Program 3–11 illustrates the interaction between INTEGER and REAL numbers. Read through the program first, then look at the sample output that follows.

Program 3–11

```
PROGRAM Arithmetic_Operations;

{**************************************************************
                Developed by:  The Author
                    Date: January, 1989
 **************************************************************
    This program demonstrates the interaction between INTEGER
and REAL variables.
 **************************************************************
                    Constants used:
 --------------------------------------------------------------
                    None
 **************************************************************
                    Variables used:
 -------------------------------------------------------------}
VAR
  First_Integer,        {Number entered by program user.}
  Second_Integer,       {A second number entered by program user}
  Integer_Answer        {Answer for arithmetic results}
 :INTEGER;
  First_Real,           {Number entered by program user.}
  Second_Real,          {A second number entered by program user}
  Real_Answer           {Answer for arithmetic results}
 :REAL;
{***************************************************************}

PROCEDURE Get_INTEGER_Numbers;
{--------------------------------------------------------------
    This procedure gets two integer numbers from the program user.
 -------------------------------------------------------------}
    BEGIN
      WRITELN('Give me an INTEGER:');
      READLN(First_Integer);
      WRITELN('Give me a second INTEGER:');
      READLN(Second_Integer);
{-------------------------------------------------------------}
    END;   {of procedure Get_INTEGER_Numbers;}

PROCEDURE Do_Integer_Operations;
{--------------------------------------------------------------
  This procedure does arithmetic operations on the two REAL
  numbers entered by the program user.
 -------------------------------------------------------------}
  BEGIN
    {Do INTEGER addition}
    Integer_Answer := First_Integer + Second_Integer;
    WRITELN('Adding the two numbers = ',Integer_Answer);

    {Do INTEGER subtraction}
    Integer_Answer := First_Integer - Second_Integer;
    WRITELN('Subtracting the two numbers = ',Integer_Answer);

    {Do INTEGER multiplication}
    Integer_Answer := First_Integer * Second_Integer;
    WRITELN('Multiplying first number by second = ',Integer_Answer);
```

Program 3-11 *continued*

```
     {Do INTEGER division}
     Integer_Answer := First_Integer DIV Second_Integer;
     WRITELN('The first number DIV second number = ',Integer_Answer);

     {Do INTEGER remainder}
     Integer_Answer := First_Integer MOD Second_Integer;
     WRITELN('The first number MOD second number = ',Integer_Answer);

     {Do real division using INTEGERS—Note answer must be REAL.}
     Real_Answer := First_Integer/Second_Integer;
     WRITELN('The first number / the second number = ',Real_Answer);
{-------------------------------------------------------------------}
  END;  {of procedure Do_Integer_Operations}

PROCEDURE Get_Real_Numbers;
{-------------------------------------------------------------------
  This procedure gets two REAL numbers from the program user.
  ----------------------------------------------------------------}
  BEGIN
     WRITELN('Give me a REAL number:');
     READLN(First_Real);
     WRITELN('Give me another REAL number:');
     READLN(Second_Real);
{-------------------------------------------------------------------}
   END;   {of procedure Get_Real_Numbers;}

 PROCEDURE Do_Real_Operations;
{-------------------------------------------------------------------
  This procedure preforms arithmetic operations on the REAL numbers
  entered by the program user.
  ----------------------------------------------------------------}
  BEGIN

   {Do REAL addition}
   Real_Answer := First_Real + Second_Real;
   WRITELN('The first number + the second number = ',Real_Answer);

   {Do REAL subtraction}
   Real_Answer := First_Real - Second_Real;
   WRITELN('The first number-the second number = ',Real_Answer);

   {Do REAL multiplication}
   Real_Answer := First_Real * Second_Real;
   WRITELN('The first number * the second number = ',Real_Answer);

   {Do REAL division}
   Real_Answer := First_Real / Second_Real;
   WRITELN('The first number / the second number = ',Real_Answer);
{-------------------------------------------------------------------}
   END;    {of procedure Do_Real_Operations;}

 PROCEDURE Do_Real_and_Integer;
{-------------------------------------------------------------------
  This procedure demonstrates arithmetic operations between REAL and
  INTEGER numbers.
  ----------------------------------------------------------------}
```

Program 3–11 *continued*

```
  BEGIN
    WRITELN;
    WRITELN('Operations with REAL and INTEGER numbers:');
    WRITELN;

    {Do REAL and INTEGER addition}
    Real_Answer := First_Real + First_Integer;
    WRITELN(First_Real,' + ',First_Integer,' = ',Real_Answer);

    {Do REAL and INTEGER subtraction}
    Real_Answer := First_Real - First_Integer;
    WRITELN(First_Real,' - ',First_Integer,' = ',Real_Answer);

    {Do REAL and INTEGER multiplication}
    Real_Answer := First_Real * First_Integer;
    WRITELN(First_Real,' * ',First_Integer,' = ',Real_Answer);

    {Do REAL and INTEGER division}
    Real_Answer := First_Real / First_Integer;
    WRITELN(First_Real,' / ',First_Integer,' = ',Real_Answer);
 {------------------------------------------------------------------}
   END;    {of procedure Do_Real_and_Integer;}
{|| Main Programming Sequence ||
 ==================================================================}
  BEGIN
     Get_Integer_Numbers;
     Do_Integer_Operations;
     Get_Real_Numbers;
     Do_Real_Operations;
     Do_Real_and_Integer;
{==================================================================}
   END.    {of Main Programming Sequence.}
```

Program 3–11 gives insights into Pascal arithmetic operations using REAL and INTEGER types of numbers. A sample execution of program 3–11 is now given.

```
Give me an INTEGER:
5
Give me a second integer:
8
Adding the two numbers = 13
Subtracting the two numbers = -3
Multiplying first number by second = 40
The first number DIV second number = 0
The first number MOD second number = 5
The first number / the second number = 6.2500000000E-01
Give me a REAL number:
12
Give me another real number:
```

```
0.34
The first number + the second number = 1.2340000000E+01
The first number - the second number = 1.1660000000E+01
The first number * the second number = 4.0800000000E+00
The first number / the second number = 3.5294117647E+01


Operations with REAL and INTEGER numbers:

1.2000000000E+01 + 5 = 1.7000000000E+01
1.2000000000E+01 - 5 = 7.0000000000E+00
1.2000000000E+01 * 5 = 6.0000000000E+01
1.2000000000E+01 / 5 = 2.4000000000E+00
```

Observe that the first REAL number entry request was answered by inputting an INTEGER, 12. This is permitted in Pascal because it will automatically be converted to a REAL number. However, the opposite is not true—you cannot input a REAL number for an INTEGER. Note also that for any mixed REAL and INTEGER operations, the answer must be of type REAL.

Conclusion

In this section you saw how to do arithmetic operations using Pascal. In the next section you will learn some important screen-formatting techniques that will give your program output real professional polish. For now, test your understanding of this section by trying the following section review.

3-6 Section Review

1 Name the six arithmetic operators used by Pascal.
2 State the two types of variables used in arithmetic operations.
3 Explain what happens when you do arithmetic operations that contain an INTEGER and a REAL.
4 What operator would you use to do INTEGER division and keep an INTEGER answer?

3-7 Screen Formatting

Introduction

Screen formatting can be thought of as how the information produced in your Pascal program will be displayed on the screen and printer. It is just as important to have user information displayed in an easy-to-read manner as it is to have your source code presented that way.

You have come a long way in your study of applying Pascal to the solution of technology problems. The next chapter will show you some powerful ideas inside the Pascal programming language.

More Writing

There are two forms of the procedure that allow the display of information on the monitor. They are

```
WRITELN('This is some text.');
```

and

```
WRITE('This is more text.');
```

WRITELN produces a **carriage return** at the end of the text, while **WRITE** does not.

```
WRITELN('This is the first statement.');
WRITELN('This is the second statement.');
```

when executed will produce

```
This is the first statement.
This is the second statement.
```

But, if the **WRITE** procedure is used,

```
WRITE('This is the first statement.');
WRITE('  This is the second statement.');
```

when executed will produce

```
This is the first statement.  This is the second statement.
```

and the cursor would be left at the right of the displayed line.

The **WRITE** and **WRITELN** options are useful when your program is asking for user input. To do this, programs up to this point were written

```
WRITELN('Input a number ==> ');
READLN('Value');
```

and when executed would display

```
Input a number ==>
□
```

The cursor would then prompt for user input on the left of the second line. However, if you used the **WRITE** statement,

```
WRITE('Input a number ==> ');
READLN('Value');
```

when executed it would display

```
Input a number ==> □
```

Now, the cursor is on the first line waiting for user input.

More Reading

There are also two forms of the **READ** procedure:

```
READLN(Value);
```

and

```
READ(Value);
```

The READLN procedure produces an automatic carriage return after the program user has entered a value; the READ procedure does not. Look at the following illustration.

```
WRITE('Input some data ==> ');
READLN(Value_1);
WRITE('Now, some more ==> ');
READLN(Value_2);
```

When executed, assuming the program user inputs 23 and 48, this example will produce

```
Input some data ==> 23
Now, some more ==> 48
□
```

If this were coded using **READ** procedures instead of **READLN**, no automatic carriage return would occur after the user entered the data.

```
WRITE('Input some data ==> ');
READ('Value_1);
WRITE('Now, some more ==> ');
READ(Value_2);
```

If the program user entered the same data, when executed, the above would produce

```
Input some data ==> 23 Now, some more ==> 48 □
```

There is thus a similarity you should note in the **READ** and **WRITE** procedures.

WRITELN and **READLN** <= = Automatic carriage return.
WRITE and **READ** <= = No carriage return.

More About the WRITELN

There is more you should know about the **WRITE** and **WRITELN** procedures that will be helpful to you in developing practical Pascal programs. The first question to be answered is how many characters you can put into a single **WRITE** or **WRITELN** procedure.

The Turbo Pascal editor will allow you to type in up to 125 characters on a single line. However, when this is displayed from a **WRITE** procedure, the string is automatically broken into the normal screen width of 80 characters. You can break a **WRITE** or **WRITELN** procedure with a carriage return if you treat it as a variable.

```
WRITELN('An example of using the ',
  ,'carriage return in a writeln.');
```

When executed this example would produce

```
An example of using the carriage return in a writeln.
```

Even though you can use this technique to allow more text in a single write procedure, it is not recommended because it makes the program more difficult to follow and the screen output harder to predict.

You may have also wondered how you could include a word with an apostrophe, such as "You'll," in a write procedure, since the ' is used to signify the end of a string. For instance, you would get a **compile-time error** if you tried

```
WRITE('Give me a number and I'll double it ==> ');
```

The Pascal compiler will think that the string ends at the apostrophe inside the contraction for "I will" and assume that the rest is an illegal variable (because `ll double it ==>` contains spaces.) To get around this, use a double apostrophe inside the word.

```
WRITE('Give me a number and I''ll double it ==> ');
```

Now you will get

```
Give me a number and I'll double it ==>
```

Displaying REAL Numbers

Turbo Pascal presents a REAL variable with an eleven-digit accuracy, as demonstrated by program 3–12.

Suppose that when program 3–12 was executed you entered the number 1. Pascal would take this number and convert it to a type REAL because the identifier `Value` was declared as REAL. Look at the resulting output.

```
Give me a number ==> 1
This is your number ==> 1.0000000000E+00
```

If you were to execute the program again but this time input the number 10,

```
Give me a number ==> 10
This is your number ==> 1.0000000000E+01
```

If you were to enter the number 0.1, the output would be

```
Give me a number ==> 0.1
This is your number ==> 1.0000000000E-01
```

As you can see, Turbo Pascal prints out REAL numbers with eleven-place accuracy and uses the **E** notation. Even though you did not enter the number with this accuracy,

Program 3–12

```
PROGRAM Simple;

  VAR

    Value : REAL;

  BEGIN
    WRITE('Give me a number ==> ');
    READLN(Value);
    WRITELN('This is your number ==> ',Value);

  END.
```

this is how Turbo handles REAL numbers. A review of E notation now follows; if this is the first time you have seen E notation, don't fret, as what follows will make you an expert.

E Notation

The term **E notation** comes from **exponential notation,** the same kind of notation that is used in your scientific calculator. It is a convenient way of representing numbers that are very large or very small. Here is the meaning of this notation.

$NE\pm P$

where N = the number
$\qquad E$ = the letter E to indicate E notation
$\qquad P$ = the power of ten
$\quad$ As an example,

$5E+2$

means

$5 \times 10^{+2}$

and 10^{+2} means $10 \times 10 = 100$. Therefore, $5 \times 10^{+2} = 5 \times 100 = 500$. So, $5E2 = 500$.

$\quad$ Look at another example.

$6.2E+4 = 6.2 \times 10^{+4} = 6.2 \times 10,000 = 62,000$.

$\quad$ Thus, in Turbo Pascal, the number 10 displayed as a REAL number would be $1.0000000000E+01$, which is the same as $1.0000000000 \times 10^{+01} = 10.000000000$.

$\quad$ A **negative exponent** indicates division by the power of ten. For example,

$5E-2$

Table 3–9   Examples of E Notation

E Notation	Power of Ten	Actual Value	Metric Prefix
E+06	10^{+6}	1,000,000	M (Mega)
E+05	10^{+5}	100,000	100 k (kilo)
E+04	10^{+4}	10,000	10 k (kilo)
E+03	10^{+3}	1,000	k (kilo)
E+02	10^{+2}	100	
E+01	10^{+1}	10	
E+00	10^{+0}	1	
E−01	10^{-1}	0.1	100 m (milli)
E−02	10^{-2}	0.01	10 m (milli)
E−03	10^{-3}	0.001	m (milli)
E−04	10^{-4}	0.0001	100 μ (micro)
E−05	10^{-5}	0.00001	10 μ (micro)
E−06	10^{-6}	0.000001	μ (micro)

means

$$5 \times 10^{-2}$$

and 10^{-2} means $1/(10^{+2}) = 1/(10 \times 10) = 1/100 = 0.01$
Therefore, $5E-2 = 5 \times 10^{-2} = 5 \times 0.01 = 0.05$

Table 3–9 lists examples of **E** notation you are most likely to encounter in technology problems along with the corresponding metric prefix.

EXAMPLE 3–10

Convert the following **E** notation numbers into regular notation.
A 3.3400000000E+02   B 5.2000000000E−03   C 2.9800000000E+00

Solution
A $3.3400000000E+02 = 3.34 \times 10^{+2} = 3.34 \times 100 = 334.0$
B $5.2000000000E-03 = 5.2 \times 10^{-3} = 5.2 \times 0.001 = 0.0052$
C $2.9800000000E+00 = 2.98 \times 10^{+0} = 2.98 \times 1 = 2.98$

EXAMPLE 3–11

Convert the following numbers into Turbo Pascal E notation.
A   234.50 B   0.00286 C   2.0

Solution
A $234.50 = 2.345 \times 100 = 2.345 \times 10^{+2} = 2.345E+02 = 2.3450000000E+02$
B $0.00286 = 2.86 \times 0.001 = 2.86 \times 10^{-3} = 2.86E-03 = 2.8600000000E-03$
C $2.0 = 2.0 \times 1 = 2.0 \times 10^{+00} = 2.0E+00 = 2.0000000000E+00$

Controlling the Output

You can suppress the E notation used in REAL numbers in Pascal.

WRITELN(Variable : field length : places after decimal);

The **field length** indicates how many spaces, including the decimal point, you want to reserve on the screen for the output of the variable. The second number, **places after decimal**, is optional and indicates how many places after the decimal point you want to be displayed. Look at the examples in program 3–13.

Program 3–13

```
PROGRAM Format_Example;

  VAR

    Value : REAL;

  BEGIN
    WRITE('Give me a value ==> ');
    READLN(Value);
    WRITELN;
    WRITELN(Value : 10 : 10);
    WRITELN(Value : 5 : 5);
    WRITELN(Value : 3 : 3);
    WRITELN(Value : 1 : 1);
    WRITELN(Value : 10 : 0);
    WRITELN(Value : 3 : 0);
    WRITELN(Value : 0 : 3);
  END.
```

The output when program 3–13 is executed will be

```
Give me a value ==> 1234.5678
1234.5678000000
1234.56780
1234.568
1234.6
      1234
1235
1234.568
Give me a value ==> 2
2.0000000000
2.00000
2.000
2.0
        2
  2
```

```
2.000
Give me a value ==> 0.0006
0.0006000000
0.00060
0.001
0.0
           0
   0
0.001
```

Observe that the value of the number left of the decimal point and the places after the decimal point are always displayed, regardless of the field length. The number of spaces reserved in the field is otherwise always observed; note that the number is rounded as the places after decimal are changed.

EXAMPLE 3–12

Write the display you would expect from the following **WRITELN** procedures. Assume that **Value** is a REAL number and that the number entered by the program user is 6.579.

```
A  WRITE(Value);   B WRITE(Value : 12 : 0);
C  WRITELN(Value : 0 : 0 );
D  WRITELN(Value : 8 : 2 , Value : 0 : 5);
```

Solution

A `6.5790000000E+00`   B `-----------7`

C `7`   D `----6.586.57900`

Units

Turbo Pascal has many built-in library functions, procedures, and other useful items. You will see more about what a library is in the next chapter, but basically a library contains what are called units. Turbo Pascal has several different units. In this text the units that will be used are: **CRT**, **PRINTER**, and **GRAPH**. The **CRT** unit contains several features such as screen displays, keyboard control, and sound. The **PRINTER** unit allows you to send information to the printer rather than the monitor, and the **GRAPH** unit is used for creating powerful graphics (the topic of Chapter 9).

When you wish to use any of these units, you must let Turbo Pascal know by including the **USES** statement, followed by the name of the unit, immediately after the program heading.

```
PROGRAM Demonstration;
  USES CRT, PRINTER;
```

The **CRT** and **PRINTER** units are contained in a file called

TURBO.TPL

which must be on your Pascal disk. As you will see in Chapter 9, the graph file called

GRAPH.TPU

must also be on your Pascal disk for the graphics commands to work.

Now let's see how to clear the screen, format the screen, and send information to the printer—a good introduction to the use of Turbo Pascal units.

Putting Things Where You Want Them

There is a convenient command in Turbo Pascal that allows you to clear the screen in order to present new information. There is also a Turbo procedure for placing information anywhere you want on the screen. Both are useful commands for displaying your data in a usable, easy to read manner.

The Turbo Pascal procedure for clearing the screen and bringing the cursor to the upper left hand corner of the screen is

CLRSCR;

Remember that Turbo Pascal makes no distinction between upper and lowercase letters; some programmers prefer to use **ClrScr**.

The Turbo procedure for placing information anywhere you want on the monitor screen is

GOTOXY(Xpos, Ypos**);**

where Xpos = the horizontal position on the monitor screen

Ypos = the vertical position on the monitor screen

The IBM monitor text screen contains 25 horizontal lines, each containing 80 characters. Examples of the **GOTOXY** command are illustrated in program 3–14. Both procedures require the **USES CRT;**

Program 3–14

```
PROGRAM GOTOXY_Example;
  USES CRT;

  VAR

    H_Pos, V_Pos
 :INTEGER;

BEGIN
  WRITE('Horizontal position ==> ');
  READ(H_Pos);
  WRITE(' Vertical position ==> ');
  READLN(V_Pos);
  CLRSCR;   {Clear the screen.}
  GOTOXY(H_Pos,V_Pos);
  WRITELN('This is the position of GOTOXY(',H_Pos,',',V_Pos,')');
  WRITELN('Here is more text.');
END.
```

Figure 3–5 illustrates the results of program 3–14 for several different input values. Note that the second **WRITELN** is left-justified.

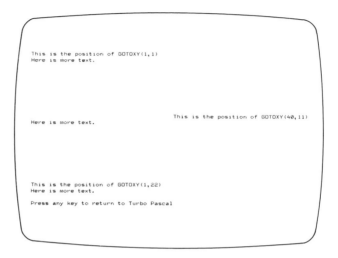

```
This is the position of GOTOXY(1,1)
Here is more text.




Here is more text.                        This is the position of GOTOXY(40,11)




This is the position of GOTOXY(1,22)
Here is more text.

Press any key to return to Turbo Pascal
```

Figure 3–5   Results of the GOTOXY Procedure

Getting Information to the Printer

Turbo Pascal allows you the option of outputting information with the **WRITE** or **WRITELN** procedure to either the monitor or your printer. As an example, when these two

```
WRITELN('This goes to the monitor.');
WRITELN(LST,'This goes to the printer.'0);
```

are executed, the first **WRITELN** will cause the words `This goes to the monitor.` to be displayed on the monitor screen. However, the second **WRI-TELN** will cause the words `This goes to the printer.` to appear on your printer and not the monitor, because of the **LST** command which precedes the string to be outputted. **LST** can also be used to output the values of variables to the printer.

```
WRITELN(LST,Value);
```

If the identifier `Value` had been assigned a number such as 24, when the above **WRITELN** was executed the output to the printer would have been `24`. It must be added that for **LST** to operate, **USES PRINTER;** must be included in the program following the program heading.

Conclusion

This section presented many useful programming skills that you can apply to technology problems, as well as various ways of controlling the placement of information on the screen of your computer. The interactive exercises present some exciting and, in many cases, surprising screen formatting exercises. Be sure to try them—this is the best way to learn what was presented in this chapter. Test your understanding of this section by trying the following section review.

3–7 Section Review

1   Explain the difference between **WRITELN** and **WRITE**.
2   State the difference between **READLN** and **READ**.
3   Show how Turbo Pascal would display the number 4 as a REAL number.
4   State how the following number would be displayed in Turbo Pascal (Assume that Value = 3.1496).
     WRITELN('Value : 8 : 2);
5   What statements would you use to cause the screen to be cleared and a string to appear starting in the center of the monitor screen?

3–8   Program Debugging and Implementation— Turbo System Features

Introduction

By now you know enough about the Turbo Pascal 4.0 environment to enter a program and save it on your disk. However, this is only a small fraction of the capability of the Turbo Pascal environment. In this section, you will learn about some of the many time-saving menu and editing features available with this powerful system.

Getting Help

Recall that one of the programs you saved on your Pascal working disk was **TURBO.HLP**. This program contains the Turbo **HELP** files. This can be accessed from anywhere in the Turbo system by simply pressing the F1 key (as indicated on the bottom line of the main menu). Pressing F1 will present a **HELP WINDOW** that tells you where you are in the system. For example, if you are at the main menu and **File** is highlighted, pressing F1 will give you information about this part of the file menu. This HELP screen will present its own menu selections that you may choose to get even more information about the different file menu items (such as **Directory**). To exit any HELP WINDOW simply press the Esc key on your keyboard. Some HELP WINDOWS contain more than one page of information. This will be indicated on the HELP WINDOW by a **PgDn** legend (indicating that you should use your PgDn key to get more information).

    You can get a menu of all 73 HELP screens by pressing F1 when you are in a HELP screen. When in a HELP screen you will note that the bottom line changes and pressing F1 again will bring up this menu.

Using Menus

Up to this point, you have been activating a menu item by first highlighting it and then pressing RETURN/ENTER. There are actually two other methods (both of them faster). The first method is to simply press the letter on your keyboard that is the same as the highlighted capital letter on any menu. For example, if you are in the main menu, you only need press the **F** key to get into the File menu. If you are in the file menu, you only need to press the **C** key in order to activate the Change directory.

Turbo Hot Keys

The Turbo Pascal environment has a built-in set of "hot keys" that give you another helpful short-cut for activating any part of the main menu. This hot key feature actually requires the depression of two keys at the same time. The six hot keys are shown in table 3–10.

You may use these hot keys from almost anywhere in the system. There are only two exceptions; the hot keys are disabled in an error box or a verify box. When these are encountered, you must use the key specified by the active box.

Changing the Bottom Line

You can see the effects of other Alt-key combinations by simply holding down the Alt key for a few seconds and observing the change in the bottom line. For example, Alt-F1 will bring up the last HELP screen you accessed.

Loading Files

To load a program from your program disk, first make sure you have your program disk in the active drive (**B:** for a two-drive system). From the main menu, press **F3** (Load) and a window will appear asking for the name of the file you wish to load. If you do not enter a file name, but erase everything in this window and press RETURN/ENTER, the directory will appear listing all of the files on your programming disk. These files will be in alphabetical order; to select the one you want to load, you

Table 3–10 Actions of the Six Hot Keys

Hot Key	Action
Alt-**F**	Activates the File menu.
Alt-**E**	Puts you into the Edit mode.
Alt-**R**	Runs your program (compiles it first if necessary).
Alt-**C**	Compiles your program.
Alt-**O**	Activates the Options menu.
Alt-**X**	Exits the Turbo Environment.

may highlight it by using the arrow keys or by pressing the letter that corresponds to the first letter of the file name. Once the file you want is highlighted, press RETURN/ENTER and the file will automatically be loaded into the editor.

If you allow Turbo to put the **.PAS** extension on all of your source files, when you use the F3 feature, just enter ***.PAS** in the file window that will appear and press RETURN/ENTER. Now the only files that will appear in the directory window will be your Pascal source files—none of the backup or **.EXE** files.

Saving Files

You can save your Pascal program easily by pressing **F2**. If you haven't given a file name to the program yet, a window will appear asking if you want to save it as **NONAME.PAS**. This gives you a chance to assign an appropriate legal DOS name and then press RETURN/ENTER. This is a handy feature to use while you are in the editor entering a program. It is a good rule to save your source code every 30 minutes of work; in case your computer goes off, you won't have lost more than 30 minutes of work.

Conclusion

This section presented some important short cuts that can be used in the Turbo environment. Be sure to try all methods presented and then select the one that works best for you. Check your understanding of this section by trying the following section review.

3–8 Section Review

1 How can you get help when working in the Turbo environment?
2 State the two methods of getting any menu item.
3 What is meant by a Turbo hot key?
4 How can you change the bottom line to see what effect an Alt key combination has on the system?
5 State a quick way of loading and saving your Pascal programs.

Summary

1 In Pascal, all statements must be separated by semicolons.
2 A Pascal statement tells the computer to do something.
3 There are ten different types of Pascal statements that will be presented in this book.
4 An identifier is a way of identifying constants, variables and Pascal procedures.
5 A constant is an identifier that will always contain the same information during the execution of the program.
6 A variable is an identifier that may contain different values during the execution of the program.
7 The specific identifier for a constant or a variable is a location in the computer's memory for holding the required data.

8   An identifier must begin with a letter of the alphabet (or an underscore) and may be followed by up to 255 letters, underscores, and numbers.

9   No distinction is made between upper- and lowercase letters for identifiers.

10   An identifier cannot be a Pascal reserved word.

11   A reserved word is an instruction reserved for Pascal.

12   There are five Turbo Pascal types explained in this text: BOOLEAN, BYTE, CHAR, INTEGER and REAL. There is a built-in Turbo procedure for a type STRING.

13   All variables must be declared in Pascal. This means the identifier you choose to represent the variable must be assigned a type by you before it can be used in the program.

14   Pascal declarations are done in the declaration block with constants assigned values under **CONST** and variables declared (given a type) under **VAR**.

15   The built-in procedure **WRITELN** can be used to display strings or the contents of identifiers used for constants or variables.

16   **WRITELN** produces an automatic carriage return, while **WRITE** does not.

17   **READLN** and **READ** are two ways of inputting information into an active Pascal program.

18   **READLN** produces an automatic carriage return while **READ** does not.

19   In Pascal **:** = is called the assignment operator. It evaluates any expression on the right and assigns the result to the identifier on the left.

20   There are six arithmetic operators in Pascal: ***** for multiplication; **/** for real division; **DIV** for integer division; **MOD** for integer modulus; **+** for addition; and **-** for subtraction.

21   Whenever a REAL and an INTEGER are combined in a legal arithmetic operation, the result is always of type REAL.

22   When solving expressions, Pascal does all multiplication and division first, then all addition and subtraction. Otherwise, all calculations take place from left to right.

23   Parentheses can be used to change the order of evaluation of an arithmetic expression. In Pascal, all expressions inside parentheses are evaluated first.

24   In Turbo Pascal REAL numbers are presented using E notation with eleven-place accuracy.

25   The output of a REAL number in Turbo Pascal can be controlled by the field specifier and the number of decimals to the left of the decimal point.

26   The built-in Turbo procedure **CLRSCR** causes the screen to be cleared and the cursor to be brought up to the upper left corner of the monitor screen.

27   The built-in procedure **GOTOXY** is used to locate information on the monitor screen.

28   In Turbo Pascal, the **LST** can be used in a **WRITE** or **WRITELN** to cause output to the printer rather than the monitor.

Interactive Exercises

Directions

These exercises require that you have access to a computer and software that supports Pascal, specifically the Turbo Pascal Development Environment version 4.0 (or higher), from Borland International. They are provided here to give you valuable experience and, most importantly, immediate feedback on what the concepts and commands introduced in this chapter will do. They are also fun.

Exercises

1   Turbo Pascal will accept the standard Pascal requirement that there not be any semicolons in the statement just before an **END** statement. Try program 3–15 and see what

happens. (Note that there are no semicolons separating the last statements from the **END** statement).

Program 3–15

```
PROCEDURE Not_Necessary;
  BEGIN
    WRITELN'Note the missing semicolon')
  END;

BEGIN
  Not_Necessary
END.
```

2 See if you can predict the output of program 3–16, then try the program. What Pascal type is each of the constants? Are the constants `Third` and `Fourth` the same type? If not, what are they?

Program 3–16

```
PROGRAM Constants;

CONST
    First = 'This is the first constant.';
    Second = 2.0;
    Third = 2;
    Fourth = '4';
    Fifth = TRUE;

BEGIN
    WRITELN(First);
    WRITELN(Second);
    WRITELN(Third);
    WRITELN(Fourth);
    WRITELN(Fifth);
END.
```

3 Program 3–17 (p. 138) illustrates the effect upper and lower case has on a Pascal identifier. See if you can predict the output of program 3–7, then actually try it to verify your predictions.

4 Program 3–18 (p. 138) illustrates the effect of the underscore in Turbo Pascal. Again, predict what program 3–18 will do and then try it.

5 Analyze program 3–19 (p. 138) and see if you can predict what the compiler will do before you compile it. Can you explain the results?

Program 3–17

```
PROGRAM Upper_And_Lower_Case;

 CONST
    constant_one = 'Case makes no difference.';

 BEGIN
   WRITELN(constant_one);
   WRITELN(Constant_One);
   WRITELN(CONSTANT_ONE);
   WRITELN(cONSTANT_oNE);
   WRITELN(CoNsTaNt_OnE);
 END.
```

Program 3–18

```
PROGRAM Some_Trouble;

 CONST
    _How_is_This = 1;
    How_is_This = 2;
    How_is_This_ = 3;

 BEGIN
   WRITELN(_How_is_This);
   WRITELN(How_is_This);
   WRITELN(How_is_This_);
 END.
```

Program 3–19

```
PROGRAM Identifiers;

 CONST
   BEGINNER = 'a';
   BEGIN_1 = 'b';
   BEGIN = 'c';

 BEGIN
   WRITELN(BEGINNER);
   WRITELN(BEGIN_1);
   WRITELN(BEGIN);
 END.
```

6   Enter and compile program 3–20. What happens? Why does this happen?

Program 3–20

```
PROGRAM Try_Some_Variables;
   VAR
      An_Integer
   :INTEGER;
      A_Real
   :REAL;
      A_Character
   :CHAR;
      A_String
   :STRING[5];
      A_Byte
   :BYTE;
      A_Boolean
   :BOOLEAN;
BEGIN
   WRITE('Enter an INTEGER type ==> ');
   READLN(An_Integer);
   WRITE('Enter a REAL type ==> ');
   READLN(A_Real);
   WRITE('Enter a CHARACTER type ==> ');
   READLN(A_Character);
   WRITE('Enter a STRING of five or less characters type ==> ');
   READLN(A_String);
   WRITE('Enter a BYTE type ==> ');
   READLN(A_Byte);
   WRITE('Enter a BOOLEAN type ==> ');
   READLN(A_Boolean);
{Display the results...}
   WRITELN;
   WRITELN('This is the INTEGER type ==> ',An_Integer);
   WRITELN('This is the REAL type ==> ',A_Real);
   WRITELN('This is the CHARACTER type ==> ',A_Character);
   WRITELN('This is the STRING type ==> ',A_String);
   WRITELN('This is the BYTE type ==> ',A_Byte);
   WRITELN('This is the BOOLEAN type ==> ',A_Boolean);
END.
```

7   Remove the statement

   READLN(A_Boolean);

   from program 3–20 and execute the program. Now, for each input, enter the number 2. Do all inputs accept this number? Why? Explain.

8   Run program 3–20 again, and this time try and enter the letter **A** for all inputs. What happens now? Why? Explain.

9   Run program 3–20 again; what happens if you enter a string that is longer than five characters for the **STRING** variable? Why does this happen? Explain.

10   What happens when you try to compile program 3–21? Explain.

Program 3–21

```
PROGRAM IE_7;
  BEGIN
    WRITELN('Here is an undeclared variable => ',Undeclared_Var);
  END.
```

11   Recall that there are ten Pascal statements. Now you can sample some of them, using the following series of mini-programs. You will see the **CASE** statement in program 3–22; it helps Pascal (and you) to create programs with decision-making capabilities. Read over program 3–22 to get the general idea. Then actually try it—nothing helps you learn what a program will do more than trying it. Input different values (such as 1, 2, or 3) and see if you can predict what will happen.

Program 3–22

```
PROGRAM Case_Statement;
VAR
  Selection
:INTEGER;
  PROCEDURE Get_Selection;
    BEGIN

    WRITE('Enter a number from 1 to 3 ==> ');
    READLN(Selection);
  END;

  PROCEDURE Case_Statement_Example;
    BEGIN
      CASE Selection OF
        1 : WRITELN('You entered a one.');
        2 : WRITELN('A two was entered by you.');
        3 : WRITELN('You selected a three!');
      END;   {Case statement}

    END;   {Procedure}

  BEGIN   {Main programming sequence}
    Get_Selection;
    Case_Statement_Example;
  END.
```

12   In program 3–23 you will find a decision-making block that uses the **IF–THEN–ELSE** statement. Read over the program and see if you can predict what it will do, then

actually try it. Pascal does make a distinction between uppercase and lowercase input strings (try inputting Left instead of LEFT).

Program 3–23

```
PROGRAM If_Then_Else_Example;
 VAR
   User_Input
 :STRING[10];

   PROCEDURE Get_User_Input;
    BEGIN
     WRITELN;
     WRITE('Enter the word LEFT or the word RIGHT ==> ');
     READLN(User_Input);
    END;

   PROCEDURE LEFT;
    BEGIN
      WRITELN;
      WRITELN('You selected the left side.');
      WRITELN('This information is from the PROCEDURE called');
      WRITELN('LEFT...');
    END;

  PROCEDURE RIGHT;
    BEGIN
      WRITELN;
      WRITELN('You selected the right side.');
      WRITELN('This information is from the PROCEDURE called');
      WRITELN('RIGHT...');
    END;

  PROCEDURE Far_Side;
    BEGIN
      WRITELN;
      WRITELN('You did not select either the RIGHT or the');
      WRITELN('LEFT.  Therefore you got the PROCEDURE called');
      WRITELN('the Far_Side.');
    END;

  PROCEDURE Make_Selection;
    BEGIN
     IF User_Input = 'LEFT' THEN LEFT
       ELSE IF User_Input = 'RIGHT' THEN RIGHT
         ELSE Far_Side;
    END;

BEGIN   {Main programming sequence}
  Get_User_Input;
  Make_Selection;
END.
```

13 Program 3–24 is an example of a **REPEAT–UNTIL** Pascal loop block. Read over the program to see if you can predict what it will do, then actually try it. Be creative and try many different values for input. Each time, see if you can predict the outcome.

Program 3–24

```
PROGRAM Repeat_Example;
  VAR
    Counter,
    Ending_Value,
    Starting_Value,
    Increment_Value
  :INTEGER;
  PROCEDURE Get_Information;
    BEGIN
      WRITE('Enter value you want loop to start with ==> ');
      READLN(Starting_Value);
      WRITE('Enter by how much you want to increment ==> ');
      READLN(Increment_Value);
      WRITE('Enter the value for the loop to end ==> ');
      READLN(Ending_Value);
    END;

  PROCEDURE Repeat_Example;
    BEGIN
      Counter := Starting_Value;
        REPEAT
          WRITE(Counter,' ');
          Counter := Counter + Increment_Value;
        UNTIL Counter >= Ending_Value;
    END;

  BEGIN   {Main programming loop}
    Get_Information; Repeat_Example;
  END.
```

14 Program 3–25 is an example of the Pascal **WHILE** loop. As before, read through the program and see if you can predict what will happen before actually trying the program.

15 Program 3–26 illustrates a third kind of Pascal loop block known as the **FOR–TO** loop. Read through it first, then try it.

16 Program 3–27 (p. 144) illustrates an important concept; it shows that a REAL number will convert an INTEGER into a REAL automatically. Try it and see what happens.

17 Look at program 3–28 on page 144, and see if you can predict what will happen, then try the program and see if your prediction is correct.

18 Program 3–29 (p. 144) demonstrates the difference between the **READ** and **READLN**. Predict what will happen, then try the program.

19 Program 3–30 (p. 145) gives you some valuable feedback concerning the field length specifier in Pascal. Try this program several times with different values for the field length.

Program 3–25

```
PROGRAM While_Example;

  VAR
    Check_This
  :CHAR;

    BEGIN
      WRITE('Input a character (enter R to repeat loop) ==> ');
      READLN(Check_This);

        WHILE Check_This = 'R' DO
          BEGIN
            WRITELN;
            WRITELN('You have entered an R so the loop is repeated.');
            WRITELN;
            WRITE('Input a character ==> ');
            READLN(Check_This);
        END;   {WHILE Loop}
        WRITELN('The loop is finished, and you did not enter an R.');
    END.
```

Program 3–26

```
PROGRAM For_To_Loop_Example;
  VAR
    Starting_Value,
    Ending_Value,
    Count
  :INTEGER;

    PROCEDURE Get_Information;
    BEGIN
      WRITELN;
      WRITE('Starting value of loop ==> ');
      READLN(Starting_Value);
      WRITE('Ending value of loop ==> ');
      READLN(Ending_Value);
    END;

  PROCEDURE Do_the_Loop;
    BEGIN
      FOR Count := Starting_Value TO Ending_Value DO
        BEGIN
          WRITELN;
          WRITELN('The count is = ',Count);
        END;    {of FOR/TO loop}
    END;   {of procedure}

  BEGIN    {Main programming sequence.}
    Get_Information;
    Do_the_Loop;
  END.
```

Program 3–27

```
PROGRAM Input_Integer;
  VAR
   Number
  :REAL;
    BEGIN
      WRITE('Input an integer = ');
      READLN(Number);
      WRITELN('The number is = ',Number);
    END.
```

Program 3–28

```
PROGRAM Write_and_Writeln;
  BEGIN
    WRITE('This information ');
    WRITE('will appear ');
    WRITE('on the same line.');
  END.
```

Program 3–29

```
PROGRAM Read_and_Readline;
  VAR
   First,
   Second,
   Third
  :CHAR;
    BEGIN
      WRITE('Give me a character = ');
      READ(First);
      WRITE(' and another = ');
      READ(Second);
      WRITE(' one more = ');
      READLN(Third);
      WRITE('Thank you!');
    END.
```

Program 3–30

```
PROGRAM Field_Length;
  VAR
    Value,
    Field_Length
  :INTEGER;
    BEGIN
      WRITE('Give me an integer number to display = ');
      READ(Value);
      WRITE(' and a field length = ');
      READLN(Field_Length);
      WRITELN;
      WRITELN('Number without the field specifier:');
      WRITELN(Value);
      WRITELN;
      WRITELN('Number with field specifier of ',Field_Length);
      WRITELN(Value : Field_Length);
      WRITELN;
      WRITELN('Value = ',Value : Field_Length);
    END.
```

20   Program 3.31 demonstrates the effect of the field length specifier with string variables. Try it for various field lengths.

Program 3–31

```
PROGRAM More_Field_Length;
  VAR
    Words
  :STRING[10];
    Field_Length
  :INTEGER;

    BEGIN
      WRITE('Give me ten characters or less => ');
      READLN(Words);
      WRITE('Give me a field length number = ');
      READLN(Field_Length);
      WRITELN;
      WRITELN('Without the field specifier:');
      WRITELN(Words,Words);
      WRITELN;
      WRITELN('With the field specifier:');
      WRITELN(Words : Field_Length, Words);
      WRITELN;
      WRITELN('With field specifier used twice:');
      WRITELN(Words : Field_Length, Words : Field_Length);
    END.
```

21   Program 3–32 demonstrates the effect of both the field length specifier and the places after decimal specifier. This is an important exercise; be sure to try this program for different values of both inputs.

22   Program 3–33 gives valuable feedback concerning the effect of the **GOTOXY** of Pascal. Try it for many different values of X and Y. Also note the effect of the Turbo Pascal **ClrScr** built-in procedure.

23   Program 3–34 demonstrates the output of information to the printer. You will need a printer connected to your system that is turned on. Note that when the program is executed, none of the information that appears on the printer will appear on your screen. When information appears on your screen it is called "echoing to the screen". If you want this echo to the screen you will need separate **WRITE** or **WRITELN** statements that do not contain the **LST** command.

Pascal Commands

Assignment Operator := Used in such Pascal equations as **A := B + C;** It means that the expression on the right will be evaluated and the resulting value assigned to the identifier on the left.

BOOLEAN, BYTE, CHAR, INTEGER, REAL   Pascal data types.

CLRSCR;   Built-in Turbo Pascal procedure for clearing the screen and bringing the cursor to the top left part of the monitor screen. Requires **USES CRT;** .

CONST   Used to indicate the beginning of the constant assignment block. If used, must appear before the VAR declaration block.

Program 3–32

```
PROGRAM Places_After_Decimal_and_Field_Length;
  VAR
    Number
 :REAL;
    Field_Length,
    Places_After_Decimal
 :INTEGER;

  BEGIN
    WRITE('Give me a number = ');
    READLN(Number);
    WRITE('Field length = ');
    READ(Field_Length);
    WRITE(' Places after decimal = ');
    READLN(Places_After_Decimal);
    WRITELN;
    WRITELN('This is the number with no field length or places:');
    WRITELN(Number);
    WRITELN;
    WRITELN('This is the number with field length only:');
    WRITELN(Number : Field_Length);
    WRITELN;
    WRITELN('This is the number with field length and places:');
    WRITELN(Number : Field_Length : Places_After_Decimal);
  END.
```

Program 3–33

```
PROGRAM Clearing_and_Moving;
  USES CRT;
  VAR
    Name
 :STRING[30];
    X_Pos,
    Y_Pos
 :INTEGER;
    BEGIN
      ClrScr;   {Clear the screen}
       WRITELN;
       WRITE('Enter your first name = ');
       READLN(Name);
       WRITELN;
       WRITE('Give X position = ');
       READ(X_Pos);
       WRITE(' and Y position = ');
       READLN(Y_Pos);
      ClrScr;
       GOTOXY(X_Pos, Y_Pos);
       WRITE(Name,' is at X = ',X_Pos,' Y = ',Y_Pos);
      END.
```

Program 3–34

```
PROGRAM Printer_Output;
  USES PRINTER;
  CONST
    Value_1 = 123;
    Word_1 = 'These words';
  VAR
    Value_One
  :INTEGER;

 BEGIN
      WRITE('Give me an integer number = ');
      READLN(Value_One);
      {Output to printer...}
      WRITELN(LST,'This line appears on your printer.');
      WRITELN(LST);
      WRITELN(LST);
      WRITELN(LST,Word_1,' also appear on your printer.);
      WRITE(LST,'And so do these numbers = ');
      WRITELN(LST, Value_1,'  ',Value_One);
 END.
```

DIV   The Pascal operator for INTEGER division.

GOTOXY(X__Val, Y__Val);   Built-in Turbo Pascal procedure for causing information to appear at a specific place on the monitor screen, where
X__Val = horizontal position (1 to 80)
Y__Val = vertical position (1 to 24). Requires **USES CRT;** .

MOD   The Pascal operator for modulus arithmetic.

READ(Value);   Same as **READLN** except no automatic carriage return is produced.

READLN(Value);   Allows data to be put into the identifier Value during program execution. Produces an automatic carriage return.

SEMICOLONS;   Must appear at the end of every Pascal statement. An exception to this is the last statement just before an **END;** or an **END.** Some forms of Pascal require that no semicolon be placed after this last statement. Turbo Pascal doesn't care.

STRING[N]   Built-in Turbo Pascal procedure that allows an identifier to be a string type. The number N can be from 1 to 255 and represents the maximum number of characters that the string variable can contain.

VAR   Used to indicate the beginning of the variable declaration block in Pascal.

WRITE   Same as **WRITELN** except an automatic carriage return is not produced.

WRITELN('String',Value);   A built-in Pascal procedure that outputs the data between the single quotes to the monitor and interprets information not enclosed by single quotes as a constant or variable and outputs its value to the monitor. Produces an automatic carriage return.

WRITELN(Value : Field__Length : Decimal__Places);   Used in Turbo Pascal to display output where
Field__Length = the number of spaces allowed for the display of Value right justified
Decimal__Places = the number of places to the right of the decimal point to be displayed
Regardless of the value of Field__Length, the digits to the left of the decimal point and the required number of decimal places to the right of the decimal point will always be displayed.

WRITELN(LST,Value); or **WRITE(LST,Value);**   Built into Turbo Pascal to cause the output of Value to appear on the printer, rather than on the monitor. Requires **USES PRINTER;** .

*   The Pascal operator for multiplication.

/   The Pascal operator for REAL division.

+ , −   The Pascal operators for addition and subtraction.

Self-Test

Directions

Program 3–35 is a completed Turbo Pascal program. It contains all the correct Pascal code needed to make the program operate. The program represents the solution to a parallel or a series circuit. It also attempts to include as many of the topics covered in this chapter as possible. For the purpose of the self-test, the program documentation is not complete. Answer all of the questions for this self-test by referring to program 3–35.

Program 3-35

```pascal
PROGRAM Self_Test_Chapter_3;
  USES CRT;

  CONST
     Prompt_User = 'Press RETURN/ENTER to continue...';
     X_Prompt = 15;     {Integer-Horizontal position for prompt.}
     Y_Prompt = 25;     {Integer-Vertical position for prompt.}

  VAR

     Series_Solution    {To indicate user's choice of circuit type.}
    :BOOLEAN;

     R_1,
     R_2,
     R_3,
     R_T
    :REAL;

     User_Selection
    :INTEGER;
     User_Name,
     Circuit_Type
    :STRING[20];

     Dummy,
     Program_Repeat
    :CHAR;

  PROCEDURE User_Prompt;
     BEGIN
       GOTOXY(X_Prompt, Y_Prompt);
       WRITE(Prompt_User);
       READ(Dummy);
     END;

  PROCEDURE Explain_Program;
    BEGIN
      CLRSCR;
      GOTOXY(5,5);
      WRITELN('This program will solve for the total resistance of',
      ' three resistors');
      WRITELN('in parallel or in series.');
      WRITELN;
      WRITELN('You must know the value of each resistor.');
      User_Prompt;
   END;

  PROCEDURE Get_User_Name;
    BEGIN
      GOTOXY(10,14);
      WRITE('Enter your first name ==> ');
      READ(User_Name);
      User_Prompt;
   END;
```

Program 3–35 *continued*

```
PROCEDURE Get_Selection;
  BEGIN
    GOTOXY(10,18);
    WRITE('Select by number: ');
    WRITELN('1] Series Circuit.  2] Parallel Circuit.');
    WRITELN;
    WRITE('Your selection ==> ');
    READLN(User_Selection);
      IF User_Selection = 1 THEN Series_Solution := TRUE
        ELSE Series_Solution := FALSE;
    User_Prompt;
  END;

PROCEDURE Get_Resistor_Values;
  BEGIN
    ClrScr;  {Clear the screen}
    GOTOXY(5,10);
    WRITE('Value of R1 = ');
    READ(R_1);
    WRITE(' of R2 = ');
    READ(R_2);
    WRITE(' of R3 = ');
    READ(R_3);
    GOTOXY(10,15);
    WRITELN('Thank you ',User_Name);
    User_Prompt;
  END;

PROCEDURE Calculate_Series;
  BEGIN
    R_T := R_1 + R_2 + R_3;
    Circuit_Type := 'series circuit';
  END;

PROCEDURE Calculate_Parallel;
  BEGIN
    R_T := 1/(1/R_1 + 1/R_2 + 1/R_3);
    Circuit_Type := 'parallel circuit';
  END;

PROCEDURE Do_Selection;
  BEGIN
    IF Series_Solution THEN Calculate_Series
    ELSE Calculate_Parallel;
  END;

PROCEDURE Display_Output;
  BEGIN
    CLRSCR;  {Clear the screen}
    GOTOXY(5,10);
    WRITELN('For a ',Circuit_Type,' where:');
    GOTOXY(5,15);
    WRITELN('R1 = ',R_1:5:2,' R2 = ',R_2:5:2,' and R3 = ',R_3:5:2,' ohms.');
    WRITELN;
    WRITELN('The total resistance is = ',R_T:4:3,' ohms.');
    User_Prompt;
  END;
```

Program 3–35 *continued*

```
BEGIN
  Explain_Program;
  Get_User_Name;
  Get_Selection;
  Get_Resistor_Values;
  Do_Selection;
  Display_Output;
END.
```

Questions

1 Describe, in your own words, what you think program 3–35 will do.
2 How many different data types are used in program 3–35? Which ones are they?
3 Could a type BYTE have been used for any of the variables? Explain.
4 How many different Pascal statements are used in program 3–35? What kind are they? Give an example of each from the program.
5 Are there any procedures that are called from another procedure? If so, state the case where this happens.
6 If the program user wished to solve for three resistors in series and the values of the resistors were 1, 2, and 3 ohms respectively, show how the final values would be displayed.
7 What would be displayed if the resistors were in parallel and each had a value of 9 ohms?
8 List the Pascal reserved words used in program 3–35.
9 What is the purpose of the procedure User_Prompt?
10 How does the program know which type of circuit the program user selected? Which procedure is involved and what kind of block is it?

Problems

General Concepts

Section 3–1

1 Explain what is meant by a Pascal statement.
2 How many different Pascal statements will be presented in this text?
3 State where semicolons are used. Are there any exceptions to this?
4 Give an example of three Pascal statements.

Section 3–2

5 Explain the difference between a constant and a variable.
6 What is meant by a Pascal identifier? Where are identifiers used? Give the rules for creating your own identifiers.
7 What are Pascal reserved words? Name five Pascal reserved words.
8 Which of the following are legal Pascal identifiers?
   A This_One;   B That_1   C That One   C Endthis   D Const   E 4_More

Section 3–3

9 Explain what is meant by a Pascal type.
10 How many different Pascal types are there? Name them.

11   Classify the following into correct Pascal types.
     A TRUE   B '6'   C 6   D 6.0
12   What is the maximum value that a Turbo Pascal INTEGER can have? What is its minimum value?
13   State the maximum value that a Turbo Pascal BYTE can have.
14   How are variables declared in a Pascal program?
15   How are constants assigned in a Pascal program?

Section 3–4
16   How would you develop a Pascal statement that would cause **This text.** to appear on the screen?
17   How would you cause the contents of a variable identified as **A__Variable** to appear on the screen?
18   Show a Pascal statement that would display text on either side of the contents of a constant called **This__Value**.
19   Write a Pascal statement that would allow the program user to enter a value during program execution.

Section 3–5
20   Write a Pascal assignment statement that computes the area of a rectangle.
21   Create a Pascal assignment statement that decreases the value of a memory location by 3.

Section 3–6
22   List the Pascal arithmetic operators and explain the purpose of each.
23   What results are achieved when two INTEGERs are multiplied? Added or subtracted? What results are achieved in these instances if the numbers are an INTEGER and a REAL?
24   State the result of adding an INTEGER and a REAL. What is the result if these same two numbers are multiplied? Divided using real division?
25   Give the difference between REAL and INTEGER division.
26   Explain the use of the MOD operator.
27   What is meant by precedence of operations? Explain the precedence of operations used by Pascal. How are parentheses treated in Pascal?
28   Evaluate the following expressions.
     A $2 + 6/2 =$ B $8 * 3 - 5 =$ C $(4 + 8)/6 + 2 =$ D $4 + 8/(6 + 2) =$

Section 3–7
29   Demonstrate the difference between **WRITE** and **WRITELN**.
30   Show the difference between **READ** and **READLN**.
31   Write a Pascal statement that allows a carriage return to appear in the middle of a string constant.
32   Write the number 3.479 as it would be displayed by Turbo Pascal from an active program.
33   Show how all REAL numbers are displayed by Turbo Pascal.
34   Convert the following numbers to standard notation.
     A $3.4000000000E+01$ B $5.72000000000E-03$
     C $4.2067000000E-6$ D $-7.38900000000E+6$
35   Develop a Turbo Pascal statement that would output a real number to the screen that will always have at least five spaces with a three decimal-point accuracy.
36   Explain what is meant by the term field length in a Pascal statement.
37   Give the Turbo Pascal procedure name for clearing the monitor screen.
38   Explain the Turbo Pascal built-in procedure for locating text anywhere on the monitor screen.
39   Write a Turbo Pascal statement that would output a string constant to the printer.

Program Analysis

40 Will program 3–36 compile with Turbo Pascal? If not, why not?

Program 3–36

```
BEGIN
    WRITELN('Will this program compile?')
END.
```

41 Check program 3–37 to see if you think it will compile. If you don't think it will compile state the reason why.

Program 3–37

```
PROGRAM Analysis;   BEGIN   WRITELN('What is this?');
END.
```

42 What will program 3–38 display on the screen when executed?

Program 3–38

```
PROGRAM A_Display;
   PROCEDURE Only_1;
      BEGIN
        WRITELN('These words...');
      END;
 BEGIN
     Only_1;
     Only_1;   Only_1;
     Only_1;   Only_1;   Only_1;
 END.
```

43 Will program 3–39 (p. 154) compile and execute? If not, why not?
44 Correct program 3–40 (p. 154) so that it will compile.
45 If you find any errors in program 3–41 (p. 154), correct them so the program will compile correctly.
46 Will program 3–42 (p. 154) compile correctly? If not, why not?
47 What will be displayed on the monitor when program 3–43 (p. 155) is executed?
48 Correct program 3–44 (p. 155) so it will compile without errors.

Program 3–39

```
PROGRAM Problem__43;
    PROCEDURE aSmallOne;
     BEGIN
      WRITELN('This is a small procedure.');
     END;
    BEGIN
      ASMALLONE;
    END.
```

Program 3–40

```
PROGRAM Begin;
    PROCEDURE Start Here;
        WRITELN('You'll find trouble here!');
    END;
 BEGIN
    Start Here;
 END.
```

Program 3–41

```
PROGRAM Problem__45;
    TYPE
        Hello = BOOLEAN;
    BEGIN
        Hello = TRUE;
    end.
```

Program 3–42

```
PROGRAM Another__problem;
    VAR
      Words, MoreWords : char;
    BEGIN
      READ(Words);
    END.
```

49   What will be displayed on the monitor when program 3–45 is executed?
50   If there are any errors in program 3–46 (p. 156), correct them.

Program 3-43

```
PROGRAM Problem_47;
    CONST
      Number = 33;
      Value = 47.2;
    BEGIN
      WRITE('Number = ',Number);
      WRITELN(Value,' is the Value');
    END.
```

Program 3-44

```
PROGRAM Problem_48;
  VAR
      Your_Name
   :STRING[25];
  CONST
      Her_Name = 'Joan';
  BEGIN
      WRITE('Enter your name => ');
      READ(Your_Name);
  END.
```

Program 3-45

```
PROGRAM Some_Calculations;
  CONST
    First_Number = 5;
    Second_Number = 2;
  VAR
    Answer
   :INTEGER;
    BEGIN
      Answer := First_Number + Second_Number;
      WRITE(Answer);
    END.
```

51 Determine what will be displayed when program 3-47 (p. 156) is executed.
52 Determine what will be displayed on the screen when programs 3-48 (p. 156) and 3-49 (p. 157) are executed.
53 Determine what will be displayed on the monitor when program 3-50 (p. 157) is executed.

Program 3–46

```
 PROGRAM Some_More_Calculations;
CONST
  First_Number = 5.0;
  Second_Number = 2;
VAR
  Answer
:INTEGER;
  BEGIN
    Answer := First_Number + Second_Number;
    WRITE(Answer);
  END.
```

Program 3–47

```
 PROGRAM Some_More_Calculations;
CONST
  First_Number = 5.0;
  Second_Number = 2;
VAR
  Answer
:REAL;
  BEGIN
    Answer := First_Number / Second_Number;
    WRITE(Answer:5:3);
  END.
```

Program 3–48

```
PROGRAM Some_More_Calculations;
CONST
  First = 8;
  Second = 12;
VAR
  Answer,
  Another_Answer
:INTEGER;
  BEGIN
    Answer := First(Second + First) - Second;
    Another_Answer := Second MOD First;
    WRITE(Answer:3:2);
    WRITE(Another_Answer);
  END.
```

Program 3–49

```
PROGRAM Some_More_Calculations;
   CONST
     First = 8.0;
     Second = 12.0;
   VAR
     Answer
   :REAL;
     BEGIN
       Answer := First*(Second + First) - Second;
       WRITE(Answer:3:2);
     END.
```

Program 3–50

```
PROGRAM Last_One;
    USES CRT;
    CONST
      Here = 10;
      There = 20;
  BEGIN
    CLRSCR;
      GOTOXY(Here, There);
        WRITE('Here I am...');
        WRITELN('Over there...');
        WRITE('and now here!');
    END.
```

Program Design

You now have enough information to actually develop all of the source code for the following programs. As with all Pascal programs you design, you are expected to use all of the steps in the Applied Programming Method given in Appendix A. Use all of the Guides given in Appendix B. Get in the habit of doing this now with these short programs; it will save you hours of work with longer and more complex programs. Be sure to completely document each program using the formats demonstrated in this chapter.

Electronics Technology

54 Develop a Pascal program that solves for the power dissipation of a resistor when the voltage across the resistor and the current in the resistor are known. The relationship for resistor power dissipation is

$$P = I \times E$$

where P = power dissipated in watts
I = resistor current in amps
E = resistor voltage in volts

55   Create a Pascal program to solve for the current in a series circuit consisting of three resistors and a voltage source. The program user must input the value of each resistor and the value of the voltage source. The relationship for total current is

$$I_t = V_t / (R_1 + R_2 + R_3)$$

where I_t = total circuit current in amps
V_t = voltage of voltage source in volts
R_1, R_2, R_3 = value of each resistor in ohms

56   Develop a Pascal program that will compute the inductive reactance for a particular frequency. The program user must input the value of the inductor and the frequency. The relationship for inductive reactance is

$$X_L = 2 \pi fL$$

where X_L = inductive reactance in ohms
f = frequency in hertz
L = value of inductor in henrys

57   Create a Pascal program that will solve for the capacitive reactance of a capacitor. The program user must input the value of the capacitor and the frequency. The relationship for capacitive reactance is

$$X_C = 1/(2\pi fC)$$

where X_C = capacitive reactance in ohms
f = frequency in hertz
C = value of capacitor in farads

58   Write a Pascal program that will find the total impedance of a series circuit consisting of a capacitor and an inductor. Program user is to input the value of the inductor, capacitor, and applied frequency. The relationship for series circuit impedance consisting of a capacitor and inductor is

$$Z = X_L - X_C$$

where Z = circuit impedance in ohms
X_L = inductive reactance in ohms
X_C = capacitive reactance in ohms

Manufacturing Technology

59   Create a Pascal program that will compute the number of items made in an eight-hour day assuming that the same number of items are made each hour. User input is the number of pieces manufactured in one hour.

Drafting Technology

60   Develop a Pascal program that will compute the area of a circle. User input is the radius of the circle.

Computer Science

61 Write a Pascal program that will allow the program user to format a REAL number from an active program. User input is the number to be formatted, field length, and number of places behind the decimal point.

Construction Technology

62 Make a Pascal program that will compute the volume of a room. User inputs are the height, length, and width of the room.

Agriculture Technology

63 Create a Pascal program that will compute the number of acres of land. Program user is to input the width and length (assume a perfect rectangle) of the land in feet. The program is to display the answer in acres, without E notation.
    (Formula: 1 acre $= 43,560$ ft^2)

Health Technology

64 Develop a Pascal program that will convert from degrees Fahrenheit to degrees centigrade. User input is the temperature in Fahrenheit. The relationship is
    $$C = 5/9(F - 32)$$
    where C = temperature in centigrade
          F = temperature in Fahrenheit

Business

65 Write a Pascal program that will compute a 6% sales tax for a purchase. The program user is to input the total amount of the purchase. The program is to return the original amount of purchase, the tax, and the total of the two.

4 Procedures and Functions

Objectives

This chapter will give you the opportunity to learn:

1 Where Pascal procedures can be used in a program.
2 How Pascal procedures can be used in different types of programming blocks.
3 The meaning of scope in terms of variables and constants.
4 The meaning of global and local constants and variables.
5 How to write a Pascal function and use it in your program.
6 When to use a **FORWARD** declaration and how this can be used in improved program documentation.
7 The use of a parameter list for procedures.
8 Various techniques for passing parameters between procedures.
9 The difference between a value parameter and a variable parameter.
10 The meaning of recursion.

Key Terms

Calling a Procedure   Declarations
Nested   Function
Recursion   Parameter List
Global   Value Parameter
Scope   Variable Parameter
Local   Forward Declarations

Outline

Procedures and functions are the basic building blocks of Pascal. You have already been using Pascal procedures; you have much more to learn about them. You will also be introduced to Pascal functions in this chapter.

You will first see how procedures can be used in a Pascal program—how a procedure can call other procedures or even call itself! You will find that a procedure can have its own set of constants and variables.

You will also see how to pass and receive values from procedures. This is a useful process used by advanced programmers and is presented in this chapter in an easy-to-understand format. Included here is a new, precise way of structuring a Pascal program that makes it easy to read, modify, and understand when you haven't looked at it for a long time.

You see that this is an important and useful chapter. The material covered here will be used in all the other chapters to follow. After completing this chapter, you will have added another necessary skill toward your goal of developing practical Pascal programs.

4–1 Review of Procedures

Introduction

Remember that a procedure can be thought of as a mini-program used inside your total Pascal program. As you have seen from the Pascal programs presented so far, a well-constructed Pascal program consists of a group of procedures. In this section, you will find out more about the versatility available in Pascal's procedures, and how procedures can be used in various interesting and useful ways.

Calling a Procedure

Remember that to **call a procedure** after first defining the procedure means to then write the identifier you used to name the procedure. The simple act of calling a procedure is shown in program 4–1.

When program 4–1 is executed, the displayed output will be

```
This is from a procedure!
```

Program 4-1

```
PROGRAM Call_It;

    PROCEDURE The_Procedure;
        BEGIN
           WRITELN('This is from a procedure!');
        END.

BEGIN    {Main programming sequence}
   The_Procedure;
END.
```

The important point to note is that the procedure was called in the main programming sequence block by just writing the name given to the procedure.

Multiple Calls

In Pascal, you can call a procedure as many times as you want, as shown in program 4-2.

Program 4-2

```
PROGRAM Call_It_Again_Sam;

    PROCEDURE The_Procedure;
        BEGIN
           WRITELN('This is from a procedure!');
        END.

BEGIN    {Main programming sequence}
   The_Procedure;
   The_Procedure;
   The_Procedure;
   The_Procedure;
END.
```

Execution of program 4-2 results in the following output.

```
This is from a procedure!
This is from a procedure!
This is from a procedure!
This is from a procedure!
```

Calling From a Procedure

You can call a procedure from within another procedure. All that is required is that you have already defined the procedure to be called, as illustrated in program 4–3.

Program 4–3

```
PROGRAM Calling_From_Another_Procedure;

   PROCEDURE First_Procedure;
       BEGIN
         WRITELN('This is from the First Procedure.');
       END;

   PROCEDURE Second_Procedure;
       BEGIN
         WRITELN('This came from the Second Procedure.');
         First_Procedure; {Calling a Procedure.}
       END;

BEGIN    {Main program sequence.}
   Second_Procedure;
END.
```

When program 4–3 is executed the output display will be

```
This came from the Second Procedure.
This is from the First Procedure.
```

You can call any previously defined procedure from another procedure, but you cannot do what you see in program 4–4.

Program 4–4 will not even compile, because when the compiler encounters the call for the second procedure found in the first procedure, it won't know what the second procedure is; it has not yet been defined. There is a method of getting around this, shown in the last section of this chapter. But what you see in program 4–4 is not allowed by the compiler.

Calling Order

A procedure may be called anywhere in the program or in another procedure, as long as it has been defined. You can think of the procedure identifier as a new word invented by you; the procedure itself then becomes the definition of that new word.

Procedures may be called in any order, as illustrated in program 4–5.

Program 4-4

```
PROGRAM This_is_not_Correct;

    PROCEDURE First_Procedure;
        BEGIN
          WRITELN('This is from the First Procedure.');
          Second_Procedure;  {This Procedure has not yet
                                       been defined!}

        END;

    PROCEDURE Second_Procedure;
        BEGIN
          WRITELN('This came from the Second Procedure.');
        END;

    BEGIN   {Main program sequence.}
       First_Procedure;
    END.
```

Program 4-5

```
PROGRAM Calling_Them_Again;

    PROCEDURE First_Procedure;
        BEGIN
          WRITELN('This is from the First Procedure.');
        END;

    PROCEDURE Second_Procedure;
        BEGIN
          WRITELN('This came from the Second Procedure.');
        END;

    BEGIN   {Main program sequence.}
       Second_Procedure;
       First_Procedure;
       First_Procedure;
       Second_Procedure;
    END.
```

Execution of program 4-5 results in the following output.

```
This came from the Second Procedure.
This is from the First Procedure.
This is from the First Procedure.
This came from the Second Procedure.
```

Calling Through a Procedure

You cannot call a nested procedure from outside its nest—doing so is equivalent to calling a procedure through another procedure, as shown in program 4–6.

Program 4–6

```
PROGRAM Calling_Through;    {This program will not compile!}

  PROCEDURE Outer_One;

    PROCEDURE Inside_Outer_One;
      BEGIN
        WRITELN('This is the inside Procedure.');
      END;  {of procedure Inside_Outer_One.}

  BEGIN
  END;   {of procedure Outer_One.}
BEGIN     {Main program block.}
  Inside_Outer_One;
END.
```

Program 4–6 will not compile, because the main program block will not know the identifier "Inside_Outer_One". This identifier is only known to **PROCEDURE Outer_One** and can be used by any other procedure nested within Outer_One, but nothing outside it.

Figure 4–1 summarizes the ways of structuring procedures in a Pascal program. Observe that a procedure may be **nested** inside another procedure. Actually, the Pascal compiler considers the main program block just to be another procedure and expects the same from it as other procedures.

Procedures in Decision-Making

Pascal procedures can be called from within a branch block, giving your programs powerful decision-making capabilities as shown in program 4–7 on page 168.

If the program user wanted the first procedure of program 4–7 displayed, program execution would result in

```
Select a procedure by number:
1] First 2] Second
Your selection ==> 1
This is from the first procedure.
```

If the program user selected the second procedure, the execution would result in

```
Select a procedure by number:
1] First 2] Second
```

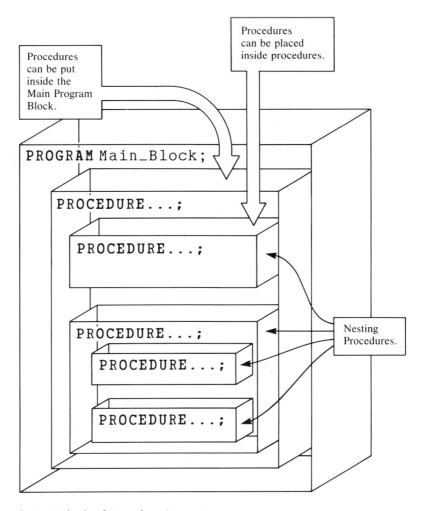

Figure 4–1   Methods of Procedure Structuring

```
Your selection ==> 2
This is from the second procedure.
```

Procedure calling from an open or closed branch is shown in figure 4–2 (p. 169).

Using Procedures in Loops

A Pascal procedure may also be used in loops, as illustrated in program 4–8.

Program 4–8 (p. 170) will keep repeating the procedure `Get_the_Guess` until the program user guesses the number 5. The following is a sample program execution of program 4–8. (Note that each time the computer screen is cleared and the message presented on line 10 at column 5.)

Program 4–7

```
PROGRAM Closed_Branch_Demo;

  VAR
    Selection          {User choice of Procedure.}
  :INTEGER;

    PROCEDURE First_Procedure;
        BEGIN
          WRITELN('This is from the first procedure.');
        END;

    PROCEDURE Second_Procedure;
        BEGIN
          WRITELN('This is from the second procedure.');
        END;

    PROCEDURE You_Select;
        BEGIN
          WRITELN('Select a procedure by number:');
          WRITELN('1] First   2] Second  ');
          WRITE('Your selection ==> ');
          READ(Selection);
        END;

    PROCEDURE Branch_Block;
        BEGIN
          IF Selection = 1 THEN First_Procedure
            ELSE Second_Procedure;
        END;
BEGIN     { Main programming sequence.}
    You_Select;
    Branch_Block;
END.
```


(Screen cleared)

Guess the number between 0 and 10 ==> 2

(Screen cleared)

Guess the number between 0 and 10 ==> 6

(Screen cleared)

Guess the number between 0 and 10 ==> 5
Congratulations, you got it!

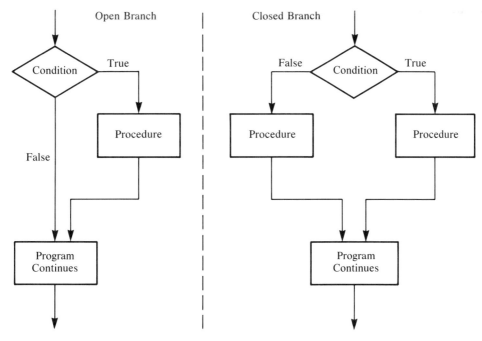

Figure 4–2 Calling a Procedure in Branch Blocks

Recursion

In terms of procedures, **Recursion** is the ability of a procedure to call itself, as illustrated in program 4–9 on page 170.

When program 4–9 is executed the resulting output is

1 2 3 4 5 6 7 8 9 10

The variable Count is first set to 0 (in the main programming sequence block). Next, one is added to Count, and as long as the variable Count is less than (<) 10, the procedure will call itself. As soon as Count equals 10, the procedure does not call itself.

Conclusion

This section presented the different ways a procedure could be called in a Pascal program. In the next section, you will see how a procedure can have its own set of constants and variables. Test your understanding of this section by trying the following section review.

Program 4-8

```
PROGRAM Loop_Example;
   USES Crt;
   CONST
      Secret_Number = 5;     {INTEGER—Number to guess.}

   VAR
     Response              {User guess of secret number.}
   :INTEGER;

    PROCEDURE Get_the_Guess;
       BEGIN
         ClrScr;   {Clear the screen.}
         GOTOXY(5,10);
         WRITE('Guess the number between 0 and 10 ==> ');
         READLN(Response);
       END;

    PROCEDURE Test_Answer;      {Loop block.}
       BEGIN
         REPEAT
           Get_the_Guess;
         UNTIL Response = Secret_Number;
       END;

    PROCEDURE Correct_Answer;
       BEGIN
         WRITELN('Congratulations, you got it!');
       END;

BEGIN    {Main programming sequence.}
   Test_Answer;
   Correct_Answer;
END.
```

Program 4-9

```
PROGRAM Recursion_Example;
   VAR
    Count
   :INTEGER;

   PROCEDURE Call_Yourself;
      BEGIN
        Count := Count + 1;
        WRITE(' ',Count);
        IF Count < 10 THEN Call_Yourself;
      END;

   BEGIN
     Count := 0;
     Call_Yourself;
   END.
```

4–1 Section Review

1 State the difference between defining and calling a procedure.
2 How many times may a procedure be called within a program?
3 Can a procedure be called from another procedure? Explain.
4 State when in a program a procedure may be called.
5 Explain what is meant by calling a procedure through a procedure. Will the Pascal compiler allow this? Explain.
6 Define recursion as it applies to a procedure.

4–2   Procedures, Constants, and Variables

Introduction

This section shows you how to make best use of your constants and variables in Pascal. It is considered good programming practice to assign constants and declare variables as locally as possible. You will discover what this means in this section.

Global Declarations

Up to this point, whenever you assigned a constant or declared a variable, it was done before any procedure was defined. Program 4–10 (p. 172) illustrates with a constant.

When program 4–10 is executed, the displayed output will be

```
This is Global
This is Global
```

The point is that a constant assigned before any procedure definitions is called **global**, because its value will be the same for all procedures (including nested procedures).

Execution of program 4–11 (p. 172) yields

```
This is global—from first procedure
This is global—from second procedure
This is global—from nested procedure
```

Figure 4–3 (p. 173) illustrates the concept of global declarations. What is true for constants is also true for variables. As you can see in figure 4–3, any **VAR** declaration made before the first procedure can be used by any of the procedures.

Program 4–10

```
PROGRAM Global_Demonstration;
 CONST
   Global_Constant = 'This is Global';

  PROCEDURE First_Procedure;
     BEGIN
      WRITELN(Global_Constant);
     END;

  PROCEDURE Second_Procedure;
     BEGIN
      WRITELN(Global_Constant);
     END;

 BEGIN
  First_Procedure;
  Second_Procedure;
 END.
```

Program 4–11

```
PROGRAM Another_Global_Demonstration;

 CONST
   Global_Constant = 'This is Global';

  PROCEDURE First_Procedure;
     BEGIN
      WRITELN(Global_Constant,' — from first procedure');
     END;

  PROCEDURE Second_Procedure;

      PROCEDURE First_Nested;
        BEGIN
           WRITELN(Global_Constant,'—from nested procedure');
        END;

    BEGIN          {procedure Second_Procedure}
       WRITELN(Global_Constant,' — from second procedure');
       First_Nested;  {Call the nested procedure.}
    END;    {of procedure Second_Procedure}

 BEGIN    {Main programming sequence.}
   First_Procedure;
   Second_Procedure;
 END.
```

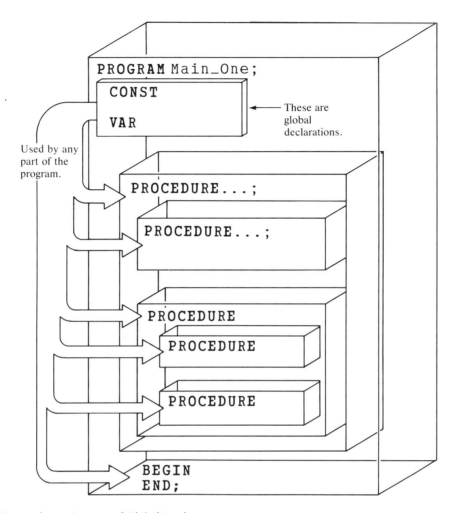

Figure 4–3   Concept of Global Declarations

Concept of Scope

Scope in Pascal refers to how far the constant assignment or variable declaration extends in the program. As you have just seen, the scope of a constant or variable declared before any procedure call extended to any procedure used in the program. Thus, the scope of a global constant or variable is the entire program.

As was mentioned before, it is not good programming practice to use only global constants and variables. If, for example, a constant or variable is used only by one procedure, that constant or variable should be **local** only to that procedure. The concept of a local constant is illustrated in program 4–12 on page 174.

Notice that there is an important difference in program 4–12. The constant assignment

Program 4–12

```
PROGRAM Local_Constant_Demonstration;

 PROCEDURE First_Procedure;
   CONST
     Local_Constant = 'This is local to first procedure.'

    BEGIN
      WRITELN(Local_Constant);
    END;

 PROCEDURE Second_Procedure;
    BEGIN
      WRITELN(Local_Constant);
    END;

 BEGIN        {Main programming sequence}
  First_Procedure;
  Second_Procedure;
 END.
```

```
Local_Constant = 'This is local to first procedure.'
```

is made inside the first procedure; it will thus be unknown to any other procedure except its own nested procedures. Because of this, program 4–12 will not compile because the second procedure tries to use Local_Constant again—but since it is not a global constant, the compiler will not accept it. This concept is illustrated in figure 4–4; an identifier has meaning only within the block in which it was declared and in procedures nested within that block.

Using the Constants and Variables Guide

Recall the Constants and Variables Guide presented in Chapter Two. It had a column that required you to identify the programming blocks (procedures) in which each constant or variable would be used.

CONSTANTS AND VARIABLES GUIDE

Formula	Variable	Meaning	Type R	I	C	B	S	Block(s) Used

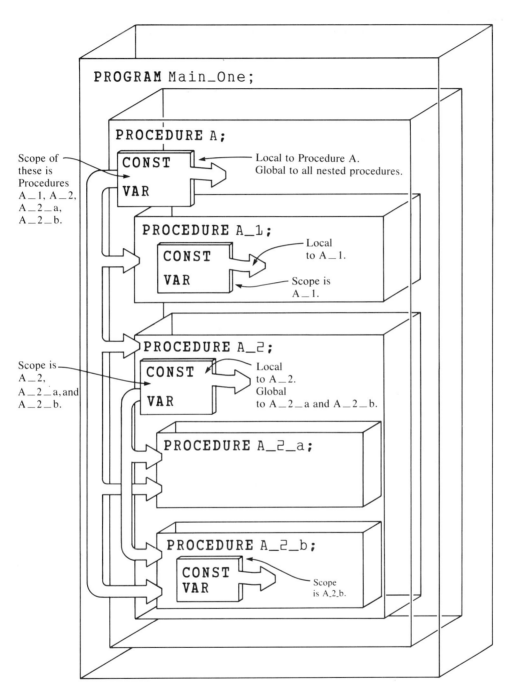

Figure 4-4 Scope of Local Declarations

The reason for the Block(s) Used column was to help you decide the scope of your constants and variables. As an example, suppose you wanted a program that would

1. Explain the program to the user.
2. Present a prompt at the bottom of the screen that states `Press RETURN/ ENTER to continue..` and clears the screen for the next presentation.
3. Get the value of two resistors from the user.
4. Compute the total resistance for the parallel combination of these two resistors.

For this program, the Constants and Variables Guide could appear as follows.

CONSTANTS AND VARIABLES GUIDE

Formula	Variable	Meaning	Type R	I	C	B	S	Block(s) used
RT = R1R2/(R1 + R2)	RT	Total R	X					`Do__Calculations;` `Display__Answer;`
	R1, R2	Value of each R	X					`Get__Values;` `Do__Calculations;`
(Dummy Variable)	Prompt	Wait for user input.			X			`User__Prompt;`

From the guide you know these facts about the variables that will be used in your program.

1. RT is of type REAL and is used in two procedures: `Do__Calculations;` `and Display__Answer;.`
2. R1 and R2 are of type REAL and are used in two procedures: `Get__ Values; and Do__Calculations;.`
3. Prompt is of type CHAR and is used in only one procedure: `User__ Prompt;.`

Variables used in more than one procedure (that are not nested in each other) can be considered to be global; variables used in only one procedure (and/or nested procedures) should be made local to that procedure. This is reflected in program 4–13 that uses the above Constants and Variables guide.

Note from program 4–13 the use of global and local variables.

Program 4–13

```
PROGRAM Two_Resistors_in_Parallel;
  USES Crt;
{***************************************************************************
              Developed By:  Electronics Student
                   Date:  November 1991
  ***************************************************************************
      This program will solve for the total resistance of two
  resistors in parallel.
  ***************************************************************************
                     Constants used:
  ------------------------------------------------------------------------
                          None.
  ***************************************************************************
                     Global Variables:
  ------------------------------------------------------------------------}
  VAR

    Total_Resistance,    {RT—Total resistance value.}
    Resistor_1,
    Resistor_2           {R1 and R2 values of each resistor}
  :REAL;
{***************************************************************************}
  PROCEDURE User_Prompt;
{------------------------------------------------------------------------
  Purpose of this procedure is to display a prompt message at
  the bottom of the computer screen and wait for user to press
  RETURN/ENTER. The screen is also cleared.
  ------------------------------------------------------------------------
                     Local Variables:
  ........................................................................}
    VAR
      Prompt            {Dummy variable for user response.}
    :CHAR;

  BEGIN
      GOTOXY(10,25);
      WRITE('Press RETURN/ENTER to continue...');
      READ(Prompt);
      ClrScr;       {Clear the screen.}
{------------------------------------------------------------------------}
  END;  {of procedure User_Prompt;}

  PROCEDURE Explain_Program;
{------------------------------------------------------------------------
  Purpose of this procedure is to explain the purpose of the
  program to the program user.
  ------------------------------------------------------------------------}
  BEGIN
      ClrScr;       {Clear the screen.}
      GOTOXY(2,5);
      WRITELN('This program will calculate the total resistance');
```

Program 4-13 *continued*

```
          WRITELN('of two resistors in parallel.');
          WRITELN;
          WRITELN('You must supply the value of each resistor in ohms.');
                User_Prompt;
 {------------------------------------------------------------------}
     END;  {of procedure Explain_Program;}

 PROCEDURE Get_Values;
 {------------------------------------------------------------------
   Purpose of this procedure is to get the values of the two
 resistors from the program user.
 ------------------------------------------------------------------}
    BEGIN
       GOTOXY(2,10);
       WRITE('Value of R1 ==> ');
       READ(Resistor_1);
       WRITE('Value of R2 ==> ');
       READ(Resistor_2);
          User_Prompt;
 {------------------------------------------------------------------}
    END;  {of procedure Get_Values;}

 PROCEDURE Do_Calculations;
 {------------------------------------------------------------------
   Purpose of this procedure is to calculate the total resistance
  of two resistors in parallel.
  ------------------------------------------------------------------}
 BEGIN
   Total_Resistance := Resistor_1*Resistor_2/(Resistor_1+Resistor_2);
 {------------------------------------------------------------------}
 END;  {of procedure Do_Calculations}

 PROCEDURE Display_Answer;
  {------------------------------------------------------------------
   Purpose of this procedure is to display the value of the total
   resistance of two resistors in parallel.
   ------------------------------------------------------------------}
    BEGIN
       GOTOXY(5,15);
       WRITE('The total resistance is ',Total_Resistance:5:3);
       WRITELN(' ohms.');
 {------------------------------------------------------------------}
    END;   {of procedure Display_Answer;}
{|| Main Programming Sequence ||
 ==================================================================}
    BEGIN
      Explain_Program;
      Get_Values;
      Do_Calculations;
      Display_Answer;
 {=================================================================}
    END.  {of Main Programming Sequence.}
```

When to Declare

It is good programming practice to keep your **declarations** as local as possible. There is a good reason for this which can be better understood if you see how Turbo Pascal allocates memory usage to your Pascal programs, illustrated in figure 4–5.

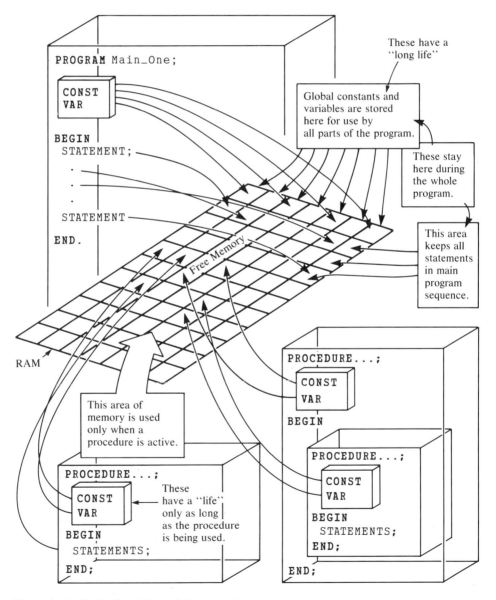

Figure 4–5   Turbo Pascal Use of Computer Memory

Figure 4–5 shows that global declarations reside in memory permanently, while procedures, along with their local declarations, are only in memory while the specific procedure is being used. It is not desirable to have your variables always accessible by any part of the program (as is the case with global declarations); they can too easily be changed by any part of the program. If you are not careful in your programming, you could leave a value in a global variable from one procedure and transfer that value to another procedure that uses the same variable—although this does have applications, it may not always be desirable. It is also possible that because of the way you write a particular procedure using global variables, you could erase an important value from a previous procedure.

If your constants and variables are all local, however, there is less chance of making a programming error with them. Programmers call keeping your declarations as local as possible protecting them. This is no "hard-and-fast" rule, but it is a style suggested by professionals—try to use it when you can.

Conclusion

This section presented the important concept of scope in Pascal, and you also saw the difference between a global and local identifier. You also had a sample program illustrating the applications of such information.

In the next section, you will learn how to use functions, another powerful Pascal feature. Functions will save you hours of programming time and will give your technical programs a real touch of professionalism. For now, test your understanding of this section by trying the following section review.

4–2 Section Review

1 What part of a Pascal program is the declaration?
2 Explain what is meant by a global variable.
3 What procedures can use global variables?
4 Explain what is meant by a local variable.
5 Define the term scope as used in Pascal.
6 What is considered good programming practice in terms of declarations?

4–3   Pascal Functions

Introduction

Most technology problems in all areas of technology contain formulas for their solution. This section presents a useful Pascal tool for the presentation of formulas. It is called the **function**, and like the Pascal procedure, once defined, it can be called to work for you.

Basic Idea

Suppose you developed a Pascal program that did nothing more than add two numbers together. This very simple task could be done as shown in program 4–14.

Program 4–14

```
PROGRAM Add_Them;
 VAR

    Answer                {Result of the addition.}
 :INTEGER;
  PROCEDURE Get_Sum;
    BEGIN
        Answer := 2 + 4;
    END;
  PROCEDURE Show_Answer;
    BEGIN
        WRITELN('The answer is ==> ',Answer);
    END;
BEGIN
  Get_Sum;
  Show_Answer;
END.
```

When program 4–14 is executed, the resulting display is

```
The answer is ==> 6
```

Your First Function

The same program will now use a Pascal function to add the two numbers instead of a Pascal procedure, as shown in program 4–15 on page 182.

Program 4–15 could also have been written as shown in program 4–16 (p. 182).

For both program 4–15 and program 4–16, the display when executed would be

```
The answer is ==> 6
```

This seems like a lot of work just to add two numbers, but it does serve to illustrate the basic idea of a Pascal function. A function will always return a value to the calling procedure; this is the main difference between a function and a procedure (a procedure does not need to return a value).

Note from figure 4–6 (p. 183) that the function itself must have a type, because it will return a value to the calling procedure. The function must also have a way of getting values from the calling procedure; it does this by using its **parameter list**. The

Program 4–15

```
PROGRAM Function_Demo;

    VAR
        Answer
    :INTEGER;

FUNCTION Sum_It(Value_1, Value_2 :INTEGER) :INTEGER;
    BEGIN
        Sum_It := Value_1 + Value_2;
    END;

PROCEDURE Show_Answer;
    BEGIN
        Answer := Sum_It(4,2);
        WRITELN('The answer is ==> ',Answer);
    END;

BEGIN
    Show_Answer;
END.
```

Program 4–16

```
PROGRAM Function_Demo_Again;

FUNCTION Sum_It(Value_1, Value_2 :INTEGER) :INTEGER;
    BEGIN
        Sum_It := Value_1 + Value_2;
    END;

PROCEDURE Show_Answer;
    BEGIN
        WRITELN('The answer is ==> ',Sum_It(2,4));
    END;

BEGIN
    Show_Answer;
END.
```

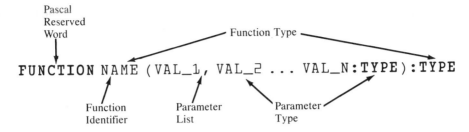

Figure 4–6   Construction of a Pascal Function

parameter list also contains numbers and must therefore have its own type. As indicated by figure 4–6, you can have different types within the parameter list of a function.

Figure 4–7 shows us how a function gets its parameters. The calling procedure is

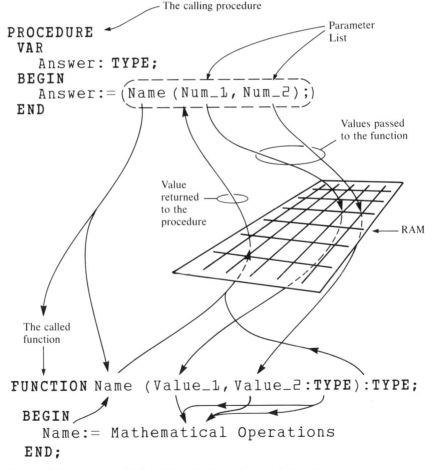

Figure 4–7   Passing of Values Between Procedure and Function

the procedure that is using the function, and the called function is the function that is used by the procedure.

Function Examples

Program 4–17 gives several examples of using Pascal functions.

Program 4–17

```
PROGRAM Several_Functions;

    VAR

        Num_1,
        Num_2
    :REAL;

  FUNCTION Multiply(Val_1, Val_2 :REAL) :REAL;
     BEGIN
        Multiply := Val_1 * Val_2;
     END;

  FUNCTION Divide(Val_1, Val_2 :REAL) :REAL;
     BEGIN
        Divide := Val_1 / Val_2;
     END;

  FUNCTION Square(Val_1 :REAL) :REAL;
     BEGIN
        Square := Val_1 * Val_1;
     END;

  FUNCTION Cube(Val_1 :REAL) :REAL;
     BEGIN
        Cube := Square(Val_1) * Val_1;
     END;

  FUNCTION Fourth_Power(Val_1 :REAL) :REAL;
     BEGIN
        Fourth_Power := Square(Val_1) * Square(Val_1);
     END;
```

Program 4–17   *continued*

```
PROCEDURE Get_Values;
    BEGIN
      WRITELN('Give me two numbers:');
      WRITELN;
      WRITE('First number ==> ');
      READ(Num_1);
      WRITE(' Second number ==> ');
      READLN(Num_2);
    END;

PROCEDURE Do_Things;
  BEGIN
  ClrScr     {Clear the screen}
  WRITELN('The product is ==> ',Multiply(Num_1,Num_2));
  WRITELN('First divided by the second ==> ',Divide(Num_1,Num_2));
  WRITELN('First number squared ==> ',Square(Num_1));
  WRITELN('First number cubed ==> ',Cube(Num_1));
  WRITELN('First number fourth power == ',Fourth(Num_1));
  END;

BEGIN    {Main programming sequence}
  Get_Values;
  Do_Things;
END.
```

If the program user entered the numbers 9 and 3 for the two numbers in program 4–17, the display after execution would be

```
Give me two numbers;
First number ==> 9 Second number ==> 3
The product is ==> 2.7000000000E+01
First divided by the second ==> 3.0000000000E+00
First number squared ==> 8.1000000000E+01
First number cubed ==> 7.2900000000E+02
First number fourth power ==> 6.5610000000E+03
```

There are some important points to note about program 4–17. The function that takes the cube calls the function that took the square of the number. This wasn't necessary; the number could have been multiplied by itself three times, but it does serve to illustrate a function calling another function. The function that took the fourth power called the function for squaring twice, which again wasn't necessary, but illustrated multiple calls of the same function from inside a function. The calling of a Pascal function is illustrated in figure 4–8 on page 186.

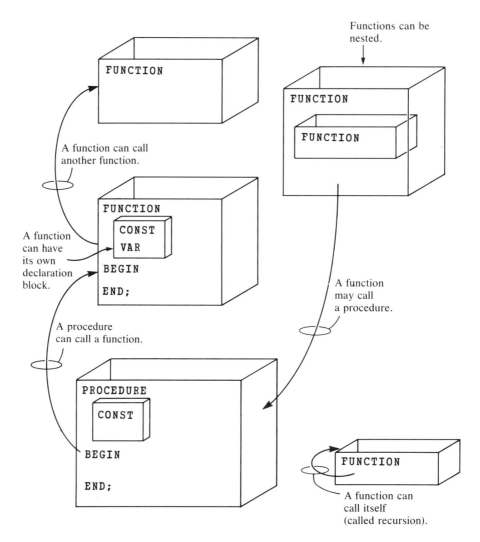

Figure 4–8 Calling Pascal Functions

Built-In Turbo Functions

Turbo Pascal has some useful functions built into it at the factory; they are listed in table 4–1.

Table 4-1   Built-In Turbo Functions (pp. 187–188)

ABS(Expression : **INTEGER** or **REAL**) : **INTEGER** or **REAL;**

Description

Returns the absolute value of Expression.

Example

Value := **ABS(**5 — 8**);**

Result

Value = 3.0000000000E+00

ARCTAN (Tangent :**INTEGER** or **REAL**) :**REAL;**

Description

Returns the angle in radians whose tangent is expressed in Tangent.

Example

Angle := **ARCTAN(**1**);**

Result

Angle = 7.8539816340E—01

COS (Radians :**INTEGER** or **REAL**) :**REAL;**

Description

Returns the cosine of the angle expressed in Radians.

Example

Cosine_of := **COS(**3.14159**);**

Result

Cosine_of = —9.9999999999E—01

FRAC (Number :**INTEGER** or **REAL**) :**REAL;**

Description

Returns the fractional portion of Number

Example

Fraction_Part := **FRAC(**3.14159**);**

Result

Fraction_Part = 1.4159000000E—01

INT (Number :**INTEGER** or **REAL**) :**REAL;**

Description

Returns the integer portion of Number

Example

Whole_Part := **INT(**3.14159**);**

Result

Whole_Part = 3.0000000000E+00

LN (Number :**INTEGER** or **REAL**) :**REAL;**

Description

Returns the natural logarithm of Number.

Example

Natural_Log := **LN(**25**);**

Result

Natural_Log = 3.218858249E+00

Table 4–1 *continued*

SIN (Angle :INTEGER or REAL) :REAL;
Descriptions
 Returns the sine of the angle in radians.
Example
 `Sine_of_Angle := SIN(3.14159);`
Result
 `Sine_of_Angle = 2.6535890356E-06`

SQR (Number :INTEGER or REAL) :REAL;
Description
 Returns the square of Number.
Example
 `Square_Value := SQR(7);`
Result
 `Square_Value = 4.9000000000E+01`

SQRT (Number :INTEGER or REAL) :REAL;
Description
 Returns the square root of Number
Example
 `Square_Root := SQRT(49);`
Result
 `Square_Root = 7.0000000000E+00`

ROUND (Number :INTEGER or REAL) :INTEGER;
Description
 Returns the value of Number rounded (0.5 or greater rounds up). This is a way of going from a REAL
type to an INTEGER.
Example
 `Integer_Number := ROUND(4.6);`
Result
 `Integer_Type = 5.0000000000E+00`

The last section of this chapter illustrates a practical applications program that utilizes some of the functions presented in table 4–1.

Conclusion

This section introduced the Pascal function. You saw that it was first used to develop a formula and could then be called from any part of the program from which a procedure could be called. In the next section, you will see how to pass values to a procedure just as you can to a FUNCTION. Test your understanding of this section by trying the following section review.

4–3 Section Review

1 State the use of a function in Pascal.
2 What are the similarities between a Pascal function and a Pascal procedure?
3 What are the differences between a Pascal function and a Pascal procedure?
4 Can a function call itself? What is this called?
5 What is a built-in Turbo Pascal function? Give an example.

4–4   More About Procedures

Introduction

You can pass values to a Pascal procedure just as you can to a Pascal function. But with a function, you always have to return a value; with a procedure, you may or may not return a value—the choice is yours. The ability to pass values to functions and procedures is one of Pascal's many features that make it a powerful programming language.

An Application

To explain why you might want to pass parameters to a procedure, suppose you needed a program that would construct a rectangle anywhere on the monitor screen. Actually, this is one of the important areas of computer programming for technology; that is how information is displayed. Spending some time and effort in this area can make your programs stand out from the others by increasing their readability; they will be more inviting and interesting.

If you are using an IBM PC (or a true compatible), there are some extra symbols that are available from the keyboard. These are shown in table 4–2.

The "Alt-" in table 4–2 means: Hold down the key marked "Alt" (on the bottom left side of the main keyboard) and at the same time type in the indicated numbers

Table 4–2 Extra Keyboard Symbols

Key Strokes	Resulting Symbol
Alt-218	⌐
Alt-192	∟
Alt-179	│
Alt-196	─
Alt-217	┘
Alt-191	┐

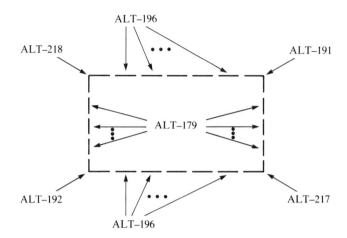

Figure 4–9   Developing a Rectangle

(using the number keypad on the right side of the keyboard). You can use this technique to develop a rectangle, as shown in figure 4–9.

For example, program 4–18 will create a rectangle in the center of the computer monitor and display a message inside it.

Program 4–18

```
PROGRAM Rectangle_1;
 USES Crt;
 PROCEDURE Draw_Rectangle;
 BEGIN
  ClrScr;   {Clear the screen}
  GOTOXY(28,12);
  WRITELN('┌────────────────────┐');
  GOTOXY(28,13);
  WRITELN('│       ELECTRONIC        │');
  GOTOXY(28,14);
  WRITELN('│         FORMULAS        │');
  GOTOXY(28,15);
  WRITELN('└────────────────────┘');
 END;

 BEGIN {Main programming sequence}
    Draw_Rectangle;
 END.
```

When executed, program 4–18 will display

```
┌─────────────────────────┐
│        ELECTRONIC       │
│        FORMULAS         │
└─────────────────────────┘
```

If you wanted to allow the program user to place this anywhere on the computer screen, you could write a procedure to make the rectangle as before, but have the parameters (values) of the X and Y position passed to it from another procedure. Program 4–19 illustrates how to do this.

Program 4–19

```
PROGRAM Rectangle__Again;
 USES Crt;
 PROCEDURE Draw__It (X__Val, Y__Val :INTEGER);
   BEGIN
    ClrScr;    {Clear the screen.}
    GOTOXY(X__Val, Y__Val);
    WRITELN('┌───────────────────┐');
    GOTOXY(X__Val, (Y__Val + 1));
    WRITELN('│    ELECTRONIC     │');
    GOTOXY(X__Val, (Y__Val + 2));
    WRITELN('│     FORMULAS      │');
    GOTOXY(X__Val, (Y__Val + 3));
    WRITELN('└───────────────────┘');
   END;

 PROCEDURE Select__Position;
   VAR
    X__Pos,
    Y__Pos
   :INTEGER;

   BEGIN
    WRITE('X position ==> ');
    READ(X__Pos);
    WRITE(' Y position ==> ');
    READ(Y__Pos);
    Draw__It(X__Pos, Y__Pos);
   END;

BEGIN {Main programming sequence}
     Select__Position;
END.
```

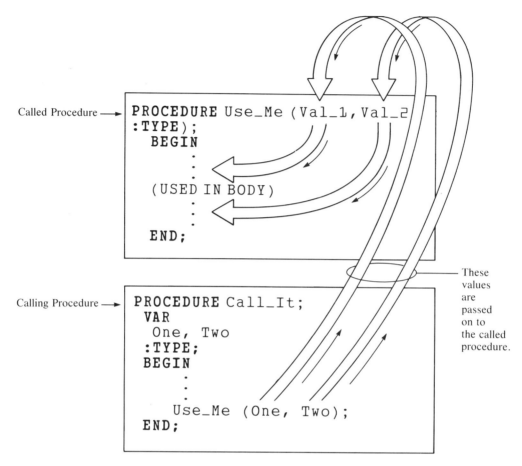

Called Procedure ➝

```
PROCEDURE Use_Me (Val_1, Val_2
:TYPE);
    BEGIN
        .
        .
        .
    (USED IN BODY)
        .
        .
        .
    END;
```

Calling Procedure ➝

```
PROCEDURE Call_It;
    VAR
      One, Two
    :TYPE;
    BEGIN
        .
        .
        .
        Use_Me (One, Two);
    END;
```

These values are passed on to the called procedure.

Figure 4–10   Procedure Passing Parameters

When program 4–19 is executed, the program user can now select the position of the box with the message. Note that the values of the position are passed to the procedure that constructs the message box in the same way values were passed to a Pascal function. The difference is that a value is not returned to the passing procedure. The construction of such a procedure is shown in figure 4–10.

Value and Variable Parameters

You can also use a procedure to return a value just as you do with a function. To illustrate the process, the simple example of adding two numbers will again be used in program 4–20.

Program 4–20

```
PROGRAM Passing_Parameters;

PROCEDURE Add_Them (Num_1, Num_2 :INTEGER; VAR Sum :INTEGER);
    BEGIN
      Sum := Num_1 + Num_2;
    END;

PROCEDURE Get_Values;
    VAR
     Answer
     :INTEGER;

    BEGIN
      Add_Them(3,4,Answer);
      WRITELN('The total is ==> ',Answer);
    END;

BEGIN {Main programming sequence}
  Get_Values;
END.
```

When program 4–20 is executed the resultant display will be

```
The total is ==> 7
```

What is happening in program 4–20 is illustrated in figure 4–11 on page 194.

You can see in program 4–20 that using a procedure to pass the value of a parameter is somewhat awkward compared to using a function to do the same thing. However, you will see that there are times when the ability to pass variables as well as values is quite useful.

Conclusion

This section demonstrated that values and variables could be passed between PROCEDURES. You will see many applications of this feature of Pascal as you develop more sophisticated technology programs. In the next section, you will learn about some useful built-in Turbo Pascal procedures. For now, test your understanding of this section by trying the following section review.

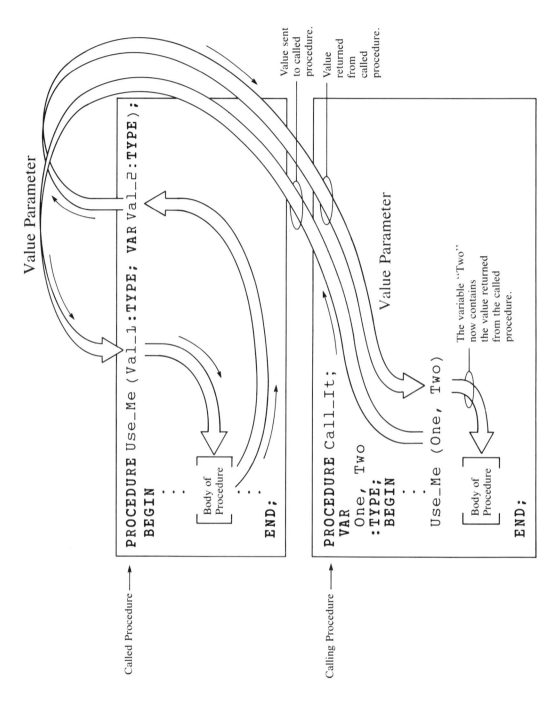

Figure 4–11 Using Value and Variable Parameters

194

4–4 Section Review

1 What is the difference between passing values to a procedure and to a function?
2 What is a value parameter?
3 What is a variable parameter?

4–5 Library Routines

Introduction

Library routines are procedures and functions that are routinely used in different programs. For example, the built-in procedure **WRITELN** is routinely used in almost all your Pascal programs. As you increase your programming skills in the area of technical programming, there will be many procedures and functions that you will want to save and use in other programs. You will be developing your own library of routines that you can simply copy and use in new programs.

What is a Good Routine?

To decide if a procedure or function is worth saving to use in other programs, use the following criteria.

1. Would this procedure or function find consistent usage?
   Is it so useful you would need to use it in most of your future programs?
2. Well-documented?
   Could you come back to it several months from now and understand exactly what each part of it is supposed to do, as well as how to modify it (if necessary) for a particular application?
3. Well-structured?
   Does it use local variables when possible? Can you easily identify the major parts—action, branch, or loop? Is there only one exit from it—at the end?
4. User-friendly (easily detects errors and can make its own corrections)?
   Is it clear to the program user exactly what is expected? If the program user makes an entry error can it easily recover?
5. No bad side effects?
   Could it potentially cause a problem in another part of the program? Could it develop characters that would lock out the keyboard or erase a disk?

Turbo Pascal has many useful procedures programmed into it at the factory. You have already used several of them, such as **WRITE, WRITELN, READ, READLN, CLRSCR,** and **GOTOXY**.

Useful Turbo Routines

In the last chapter you were introduced to Turbo Pascal UNITS, where Turbo's library routines are contained. Table 4–3 (pp. 196–197) contains a sampling of the many useful ones contained in the TURBO.TPL file. Remember, to activate any of these, you must include in your program

USES Crt;

Table 4–3   Representative Sample of Turbo Procedures

DELAY(Time :INTEGER);   **USES Crt;**

Description

Delays for the number of milliseconds indicated by the numerical value of the variable Time.

Example

DELAY(5000);

Result

A five second delay.

Use

Create a timed sequence of information presented to the user.

LOWVIDEO;   **USES Crt;**

Description

Dims the video display.

Example

LOWVIDEO

Result

All information displayed on the text screen is dimmed.

Use

Emphasize different parts of the screen display.

NORMVIDEO   **USES Crt;**

Description

Sets the video display back to normal intensity.

Example

NORMVIDEO;

Result

All information displayed on the text screen is at its normal intensity level.

Use

Can be used with **LOWVIDEO** to help emphasize different parts of the screen display.

WINDOW (Lft__Col, Top__Line, Rgt__Col, Bot__Line :INTEGER;
USES Crt;

Description

Redefines the size of your monitor screen to a smaller area. The parameters are

Lft__Col = left column of the window

Top__Line = top line of the window

Rgt__Col = right column of the window

Bot__Line = bottom line of the window

Example

WINDOW(5, 10, 30, 12);

Result

The text screen will now start at column 5, end at column 30 and have a height from row 10 to row 12.

Use

For multiple displays and "pull-down" windows. Allows you to clear the screen of selected windows while others are still being displayed.

Table 4–3 *continued*

SOUND(Frequency **:INTEGER);**   USES Crt;

Description

 Produces a sound using the computer's speaker. The frequency is determined by the value of Frequency. Sound lasts until the built-in procedure **NOSOUND** is used.

Example

 SOUND(261);

Result

 A tone near middle C.

Use

 To acknowledge an input or call attention to an output—adds another important dimension of information.

NOSOUND;   USES Crt;

Description

 Stops sound coming from the computer speaker.

Example

 NOSOUND;

Result

 No sound from the speaker.

Use

 Used in conjunction with **SOUND** and **DELAY** to produce different sounds from the computer speaker.

Conclusion

This section presented some useful built-in Turbo Pascal procedures that aid in the design of practical technology programs. Some are included in the interactive exercises for this chapter. Test your understanding of this section by trying the following section review.

4–5 Section Review

1 Describe a library routine.
2 What constitutes a "good" routine?
3 Name two of the most common built-in Pascal procedures you have used up to this point.
4 Give the built-in Turbo Pascal procedure for activating the speaker. Describe what it means.

4–6   Forwarding Things

Introduction

Recall that you can not call a function or procedure until you define it, to let the Pascal compiler know the meaning of your identifier. There are times when you need to call a function or procedure *before* it is defined. Pascal provides a method of doing just that, a method that also provides another option for structuring your Pascal programs.

When Needed

Program 4–21 is an example of a program that will not compile.

Program 4–21

```
PROGRAM Can_Not_Work;

  PROCEDURE First_Procedure;
     BEGIN
       WRITELN('This is the first procedure.');
       Second_Procedure;
     END;

  PROCEDURE Second_Procedure;
     BEGIN
        WRITELN('This is the second procedure.');
        First_Procedure;
     END;

BEGIN  {Main programming sequence}
   First_Procedure;
END.
```

Program 4–21 will not compile because the first procedure calls the second procedure before it has been defined. Rearranging the order would not work—the second procedure would then call the first procedure before it had been defined. Pascal does have a way out of this dilemma—it is called a **FORWARD** declaration. This is shown in program 4–22.

Program 4–22

```
PROGRAM Can_Work;

  PROCEDURE Second_Procedure;
  FORWARD;

  PROCEDURE First_Procedure;
     BEGIN
       WRITELN('This is the first procedure.');
       Second_Procedure;
     END;
```

Program 4–22 *continued*

```
   PROCEDURE Second_Procedure;
      BEGIN
         WRITELN('This is the second procedure.');
         First_Procedure;
      END;

BEGIN  {Main programming sequence}
   First_Procedure;
END.
```

Note that the addition in program 4–22 is the **FORWARD** declaration. The structure of this declaration is shown in figure 4–12.

Another Use

The **FORWARD** declaration statement allows you to structure your Pascal programs somewhat differently. One of the complaints about Pascal is that the main programming sequence is at the end of the program. This is necessary to have the program compile successfully. It would, however, make the program easier to read if the programming sequence could be put at the beginning of the program. Program 4–23 shows how this can be done using **FORWARD** declarations for all program procedures and functions.

Observe that in program 4–23 (pp. 200–201) the first defined procedure gives the main programming sequence. This regulates the normal main programming sequence (the last **BEGIN** and **END**.) to the rather insignificant, but important, role of simply calling the one procedure that calls all the others.

Also note that in all of the **FORWARD** declarations, the purpose of each procedure and function was stated. Now you can find all the procedures and functions defined in the body of the program. Moreover, even before you get there, you have a solid concept of what the program is going to do and how it is going to do it.

There is another important point; when a **FORWARD** declaration is used, the parameters are included. Notice, then, that you do not state the parameters again when the actual procedure or function is defined. This is evident in program 4–23.

```
PROCEDURE NAME (PARAMETER LIST);
FORWARD
```
                          or
```
FUNCTION NAME (PARAMETER LIST):TYPE;
FORWARD
```

Figure 4–12 Structure of the Forward Declaration

Program 4–23

```
PROGRAM New_Structure;

{*****************************************************************
                Developed By:  A. Pascal Student
                   Date:  November 1991
 *****************************************************************
        Program description:

 *****************************************************************
                   Global Constants used:
 ----------------------------------------------------------------}
  CONST

    This_Value = 123;         {Integer and comment}
    That_Value = 456.0;       {Real and comment}
 {*****************************************************************
                   Global Variables used:
 ----------------------------------------------------------------}
  VAR

    First_Variable,           {Reason for this variable}
    Second_Variable           {Reason for this variable}
  :REAL;

    Third_Variable,           {Reason for this variable}
    Fourth_Variable           {Reason for this variable}
   :INTEGER;
 {*****************************************************************}
    PROCEDURE First_One;
    FORWARD;   {Purpose of this procedure:                       }
 {---------------------------------------------------------------}
    PROCEDURE Second_One;
    FORWARD;   {Purpose of this procedure:                       }
 {---------------------------------------------------------------}
    FUNCTION First_Calculation(Value_1, Value_2 :REAL) :REAL;
    FORWARD;   {Purpose of this function:                        }
 {---------------------------------------------------------------}
    FUNCTION Second_Calculation(Value_1 :INTEGER) :INTEGER;
    FORWARD;   {Purpose of this function:                        }

  PROCEDURE Main_Sequence;
 {---------------------------------------------------------------
    The following is the sequencing of the entire program.
 ---------------------------------------------------------------}
    BEGIN
        First_One;
        Second_One;
 {---------------------------------------------------------------}
    END;   {of Main_Sequence;}

  PROCEDURE First_One;
 {---------------------------------------------------------------}
```

Program 4–23   *continued*

```
   BEGIN
         { Statements;}
{-----------------------------------------------------------------}
   END;   {of procedure First_One;}

 PROCEDURE Second_One;
{-----------------------------------------------------------------}
   BEGIN
         { Statements;}
{-----------------------------------------------------------------}
   END;   {of procedure Second_One;}

 FUNCTION First_Calculation;
{-----------------------------------------------------------------}
   BEGIN
         { Statements;}
{-----------------------------------------------------------------}
   END;   {of function First_Calculation;}

 FUNCTION Second_Calculation;
{-----------------------------------------------------------------}
   BEGIN
         { Statements;}
{-----------------------------------------------------------------}
   END;   {of function Second_Calculation;}

 BEGIN
   Main_Sequence;
 END.
```

Turbo Pascal already "knows" the parameter list and will not accept a redundant (with a chance of inconsistency) parameter list.

The Ultimate Documentation

Using the **FORWARD** declaration idea to develop a more easily-grasped program structure led to the development of the following style of structure. All of the "boxes" are available from the IBM keyboard (or a true compatible). Even though program 4–24 (pp. 202–204) is easy to read and understand, you do have to ask yourself if all of the extra typing is really worth it.

Conclusion

In this section you were introduced to the **FORWARD** declaration. You saw where such a declaration was needed and how to correct for it. Some new ideas were also presented in structuring Pascal programs. No matter what structure you decide to

Program 4–24

```
                              {
                         } PROGRAM Reactance; {

} USES Crt;                                                       {
                         Programmer's Block

    Developer's Name:

    Development Date:

    Program Description:

    Global Constants:

    Global Variables:


                      Program Procedures

} PROCEDURE Explain_Program;                                      {
} FORWARD; {This procedure explains the purpose of the
                program to the program user.

} PROCEDURE Select_Formula;                                       {
} FORWARD; {This procedure has the program user select
                the type of electronic formula needed.

} PROCEDURE Input_Values;                                         {
} FORWARD; {This procedure has the program user input
                the values needed to solve the formula.

} PROCEDURE Display_Answer;                                       {
} FORWARD; {This procedure displays the answer to the
                formula selected by the program user.


                      Program Functions

} FUNCTION X_L(Freq, Ind :REAL) :REAL;                            {
} FORWARD; {This function solves for the value of the
                inductive reactance of an inductor.

    Value Returned:  Inductive reactance in ohms.

    Parameters:
                Freq = applied frequency in hertz.
                Ind = value of inductor in henrys.
```

Program 4–24 *continued*

```
      Sample Call;
              X__L(1000,0.005);
              Solves for the inductive reactance of a
              5 mH inductor at a frequency of 1 kHz.
              X__L = 31.4 ohms.
```

```
} FUNCTION X__C(Freq, Cap : REAL) :REAL;                       {
} FORWARD;   {This function solves for the value of
                the capacitive reactance of a capacitor.

      Value Returned:  Capacitive reactance in ohms.

      Parameters:
              Freq = applied frequency in hertz.
              Ind = value of capacitor in farads.

      Sample Call:
              X__C(2000, 0.000010);
              Solves for the capacitive reactance of a
              10 micro farad capacitor at a frequency
              of 2 kHz.
              X__C = 7.96 ohms.
```

```
} PROCEDURE Main__Sequence;     {
```
```
} BEGIN                                                        {
}                                                              {
}     Explain__Program;                                        {
}     Select__Formula;                                         {
}     Input__Values;                                           {
}     Display__Answer;                                         {
}                                                              {
} END;                                                         {
```
```
  This completes PROCEDURE Main__Sequence.
```

```
} PROCEDURE Explain__Program; {
```
```
} VAR                                                          {
}   User__Response        {Pressing of the ENTER key.}         {
} :CHAR;                                                       {
```

Program 4–24 *continued*

```
} BEGIN                                                        {
}                                                              {
}    ClrScr;          {Clear the computer screen.}             {
}    GOTOXY(1,5);     {Starts message below top line.}         {
}                                                              {
}  WRITELN('This program will solve for the value');           {
}  WRITELN('of the inductive reactance of an');                {
}  WRITELN('inductor or the capacitive reactance ');           {
}  WRITELN('of a capacitor.');                                 {
}  WRITELN;  {Skip a space.}                                   {
}  WRITELN('You must select the formula and input');           {
}  WRITELN('the indicated values.');                           {
}                                                              {
}    GOTYXY(10,24);  {Message at screen bottom.}               {
}    WRITE('Press ENTER to continue...');                      {
}    READLN(User_Response);                                    {
}                                                              {
} END;                                                         {
```

This completes PROCEDURE Explain_Program.

```
          {Other procedures and functions go here.}
BEGIN
 Main_Sequence;
END.
```

make your own, select one that helps more than it hinders and then be consistent with its usage. Looking at the work of other programmers can be a valuable learning tool to see what kind of structure you may want to use (or, just as important, not want to use). Test your understanding of this section by trying the following section review.

4–6 Section Review

1 Normally, when may a procedure or function be called?
2 State the purpose of a **FORWARD** declaration.
3 Comment on the parameter list for a **FORWARD**-declared function or procedure.
4 is there any difference between **FORWARD** declarations for procedures and functions?

4–7 Program Debugging and Implementation— Features of the Turbo Editor

Discussion

You have already used the Turbo Pascal editor many times and the programs you are working with are much larger. At this stage you can appreciate some of the powerful features offered by the Turbo Pascal editor. This section will present several of the most commonly-used editor features.

Moving the Cursor

There are many ways to move the cursor where you want it on the editor screen. You have been using the arrow keys, backspace key, and the RETURN/ENTER key. Table 4–4 lists some of the other methods of moving the cursor while in the editor.

The listing in table 4–4 represents the most commonly-used cursor-movement methods. For in-depth information on other types of cursor movements, consult your Turbo Pascal Owner's Manual.

Inserting and Deleting

There are editor commands for inserting and deleting lines as well as for deleting single characters or words. These commands are listed in table 4–5 (p. 206) along with their intended actions.

Block Commands

Block commands are used to manipulate large sections of programming text. This usually means that you can move a whole Pascal procedure or function to another

Table 4–4   Cursor Movement

Key(s)	Action
Left Arrow or Ctrl-S	Moves left one character.
Right Arrow or Ctrl-D	Moves right one character.
Up Arrow or Ctrl-E	Moves up one line.
Down Arrow or Ctrl-X	Moves down one line.
Ctrl-A or Ctrl-Lft Arw	Moves one word to the left.
Ctrl-F or Ctrl-Rgt Arw	Moves one word to the right.
PgUp or Ctrl-R	Moves up one page.
PgDn or Ctrl-C	Moves down one page.
Ctrl-W	Scroll edit window up.
Ctrl-Z	Scroll edit window down.
Home or Ctrl-Q S	Moves to beginning of line.
End or Ctrl-Q D	Moves to end of line.
Ctrl-Home or Ctrl-Q E	Moves to top of edit window.
Ctrl-End or Ctrl-Q X	Moves to bottom of edit window.
Ctrl-PgUp or Ctrl-Q R	Moves to the beginning of program.
Ctrl-PgDn or Ctrl-Q C	Moves to the end of program.

Table 4–5 Inserting and Deleting

Key(s)	Action
Ins or Ctrl-V	Toggle between insert and overwrite.
Ctrl-N	Inserts a blank line between text.
Ctrl-Y	Deletes a line of text.
Ctrl-Q Y	Deletes from cursor to end of line.
Backspace or Ctrl-H	Deletes a character to cursor left.
Del or Ctrl-G	Deletes character under cursor.
Ctrl-T	Deletes character to cursor right.

place on the edit screen, or you can copy a block of text to another section of the edit screen or even to the printer. Another useful feature is the ability to save a part of your program (such as a particularly useful procedure or function) to a disk file and then call it back to any editor screen you wish.

The block commands available in the Turbo editor allow for great time savings and will add a degree of sophistication to your editing skills. They are summarized in table 4–6.

Other Important Commands

The Turbo editor has many other powerful commands to assist you in developing your Pascal program. Some that are most frequently used give you the ability to locate a word within your Pascal program. Another command gives you the option of finding a word and replacing it with another. Some of the other commands allow you to mark key places in long programs so that you may get easy access to them. Table 4–7 outlines these commands; a more complete presentation follows.

Table 4–6 Block Commands

Key(s)	Action
F7 or Ctrl-K B	Marks the beginning of a block.
F8 or Ctrl-K K	Marks the end of a block.
Ctrl-K P	Prints a block defined by the above.
Ctrl-K C	Copies a defined block to the current cursor position in the editor.
Ctrl-K Y	Deletes a defined block.
Ctrl-K V	Moves a defined block from its current position to the current cursor location.
Ctrl-K W	Writes a defined block to the disk under a name given by you.
Ctrl-K R	Reads a block you have saved to the disk.

Table 4–7 Other Commands

Key(s)	Action
Ctrl-U	Abort last operation.
Ctrl-O I or Ctrl-Q I	Turn autoindent on or off.
Ctrl-Q F	Find a word.
Ctrl-Q A	Find a word and replace it.
Ctrl-K n	Set a block marker where n = block marker number (0 to 3).
Ctrl-Q n	Bring cursor to block marker where n = block number (0 to 3).
Ctrl-Q L	Restore the line to previous condition.

A detailed summary of the commands in table 4–7 follows.

Abort last operation: Allows you to abort any command in progress; use this if you're in the middle of something that you can't get out of, or if you change your mind.

Turn autoindent on or off: This allows you to deactivate or reactivate the Turbo editor autoindent feature. When active, the cursor will be brought back to the position of the beginning line when RETURN/ENTER is pressed. When deactive, the cursor is brought back to the first column of the edit screen.

Find a word: Here you can look for any string sequence up to 30 characters in length. The 6 options listed here are available with this command. When evoked, a window at the top left of the screen will ask you for the string and one of these options. The options are indicated by entering one of the following letters or numbers.

    B: This causes a search for the string **B**ackwards from the current cursor position to the beginning of the text.

    G: Causes a **G**lobal search of the entire text regardless of the position of the cursor.

    L: This will search for the string occurrence only in the current marked block **L**ocally containing the cursor.

    n: A number that indicates the **n**th position to find the given string from the current cursor position.

    U: This does not make a distinction between **U**pper and lower case letters in the search string.

    W: Makes the string search look for **W**hole words only, such as apple if the search string is apple, rather than embedded words, such as the apple in applesauce.

Find a word and replace it: Works the same as finding a string, except that now you can replace the string you find with another string up to 30 characters long. This is very useful if you find you must change the name of a variable. The window will ask you for the replacement string. You simply type it in and then indicate one of the three following options.

    N: Will replace every found string automatically. Do not use this option until you are sure that you indeed want to change *every* occurrence of the string you find.

    n: Indicates the number of cases you want the new string to replace the found string; you enter the desired number.

    L: The only strings to be replaced will be those that are Local to the defined block where the cursor currently is found on the editor screen.

Set block number n: This allows you to mark up to four different places in your text (0–3). A useful feature for large programs that enables you to locate key procedures and functions.

Bring cursor to block number n: This finds the previously set block markers (from 0 to 3).

Restore line: If you haven't left the line it allows you to undo changes you may have made to it—useful if you change your mind about recent changes.

Conclusion

This section has presented some of the more common editor commands for the Turbo Pascal editor. Keep in mind that only some of the available commands have been presented here. Refer to your Turbo Pascal 4.0 Owner's Manual for a complete discussion of all of the available commands. Check your understanding of this section by trying the following section review.

4–7 Section Review

1   Name the major Turbo editor features presented in this section.
2   State the main purpose of cursor-movement commands as used in the Turbo editor.
3   Can the coding of a Pascal procedure be saved to the disk separately from the program in which it is contained? Explain.
4   Suppose you accidentally used a Pascal reserved word as an identifier throughout your program. How could you automatically correct this error?

Summary

1   A procedure can be thought of as a "mini-program" that may be used several times by the main program.
2   A procedure is defined by giving it a legal name, an assignment block (if necessary) and statements that go between the reserved words **BEGIN** and **END;**.
3   A procedure is called by using the name given to it when it was defined, followed by a semicolon (;).
4   A procedure may be called from the main programming block or from another procedure.
5   Procedures may be defined in a procedure; doing this is called nesting procedures.
6   A procedure cannot be called through another procedure.
7   Procedures already defined may be called in any order as long as they are at the same level of nesting.
8   Procedures can be called within a branch block.
9   Procedures may be used in loops.
10   The process of a procedure calling itself is called recursion.

11 A global declaration contains constants and variables that can be used anywhere in the program, by any procedure in the program.

12 A local declaration contains constants and variables that are declared within a procedure and may only be used by that procedure or procedures nested within it.

13 In Pascal, scope refers to how far the constant assignment or variable declaration extends into the program.

14 It is considered good programming practice to keep declarations as local as possible.

15 A Pascal function is similar to a Pascal procedure; both are defined similarly, and once defined, may be called.

16 A Pascal function also differs from a Pascal procedure; a function always returns a value to the calling procedure.

17 A function usually has a parameter list that defines the variable types that will be passed to the function.

18 All Pascal functions must have a type because they return a value.

19 Turbo Pascal has many built-in functions that can be used in the implementation of your programs.

20 The process of a function calling itself is also called recursion.

21 A procedure may also contain a parameter list just as a function does.

22 There are two types of parameters that can be used within the parameter list of a procedure: variable parameters and value parameters.

23 A value parameter allows a value to be passed to the called procedure.

24 A variable parameter allows a value to be returned back to the calling procedure, much as a value is returned back by a function.

25 A library routine is a procedure or function that is routinely used by different programs.

26 Turbo Pascal has many built-in library procedures that have wide application in the development and implementation of Pascal programs.

27 A procedure or function may be called before it is defined by using the **FORWARD** statement.

28 Being able to use the **FORWARD** technique with procedures and functions allows for many elaborate documentation and program-structuring schemes.

Interactive Exercises

Directions

These exercises require that you have access to a computer and software that supports Pascal, specifically the Turbo Pascal Development System, version 4.0 (or higher), from Borland International. They are provided here to give you valuable experience and immediate feedback on what the concepts and commands introduced in this chapter will do—they are also fun.

Exercises

1 Program 4–25 (p. 210) has nested procedures. Predict what the program will output to the screen, then enter, compile, and execute the program yourself.

2 Program 4–26 (p. 210) is identical to program 4–25 except that Procedure Three calls Procedure One. Will the program compile? If it does, what do you predict will be the output on the monitor when the program is executed?

Program 4–25

```
PROGRAM IE_1;
   PROCEDURE One;
      PROCEDURE Two;
         PROCEDURE Three;
            BEGIN    {Procedure Three.}
             WRITELN('This is from three.');
            END;
         BEGIN    {Procedure Two}
           WRITELN('This is from two.');
         END;
      BEGIN  {Procedure One;}
        WRITELN('This is from one.');
      END;   {Procedure One}

BEGIN
  One;
END.
```

Program 4–26

```
PROGRAM IE_2;
   PROCEDURE One;
      PROCEDURE Two;
         PROCEDURE Three;
            BEGIN    {Procedure Three.}
             WRITELN('This is from three.');
             One;
            END;
         BEGIN    {Procedure Two}
           WRITELN('This is from two.');
         END;
      BEGIN   {Procedure One;}
        WRITELN('This is from one.');
      END;   {Procedure One}

BEGIN
  One;
END.
```

3 Program 4–27 is similar to program 4–26, but this time the first procedure is calling the second, which calls the third, while the third procedure calls the first. Will the program compile? If not, why not? If it does, what do you predict will be the output to the monitor when it is executed?

4 Program 4–28 has three procedures at the same level. The main program block calls the third one, and the third one calls the other two. Predict what the output to the screen will be, then try it to check your answer.

Program 4–27

```
PROGRAM IE_3;
   PROCEDURE One;
      PROCEDURE Two;
         PROCEDURE Three;
            BEGIN    {Procedure Three.}
             WRITELN('This is from three.');
             One;
             END;
         BEGIN   {Procedure Two}
            Three;
            WRITELN('This is from two.');
         END;
      BEGIN  {Procedure One;}
        Two;
        WRITELN('This is from one.');
      END;  {Procedure One}

BEGIN
  One;
END.
```

Program 4–28

```
PROGRAM IE_4;
   PROCEDURE One;
     BEGIN
       WRITELN('This is from Procedure One.');
     END;

   PROCEDURE Two;
     BEGIN
       WRITELN('Here we are at Procedure Two.');
     END;

   PROCEDURE Three;
     BEGIN
       WRITELN('Look what''s from Procedure Three.');
       One;
       Two;
     END;

BEGIN
  Three;
END.
```

5 Program 4–29 illustrates the use of procedures in a branch block. You should be able to predict exactly what the program will do; confirm your predictions by actually trying the program. What happens if you enter a number other than 1 or 2? Why does this happen?

Program 4–29

```
PROGRAM IE_5;
 VAR
   Selection
 :INTEGER;

   PROCEDURE One;
     BEGIN
       WRITELN('This is from Procedure One.');
     END;

   PROCEDURE Two;
     BEGIN
       WRITELN('Here we are at Procedure Two.');
     END;

   PROCEDURE Three;
     BEGIN
       WRITE('Select 1 or 2 ==> ');
       READLN(Selection);
       IF Selection = 1 THEN One ELSE IF Selection = 2 THEN Two;
     END;

BEGIN
   Three;
END.
```

6 Program 4–30 gives you a chance to see recursion in action. Note that the procedure will call itself. What do you think the program will do? Try it and see.

Program 4–30

```
PROGRAM IE_6;
   VAR
     Counter
   :INTEGER;

    PROCEDURE One;
```

Program 4–30 *continued*

```
      BEGIN
        WRITELN('Number = ',Counter);
        Counter := Counter + 1;
        One;
      END;

  BEGIN
    Counter := 1;
    One;
  END.
```

7 Look at program 4–31 and see if you can predict the screen output. A global variable is used by a procedure.

Program 4–31

```
PROGRAM IE_7;
  VAR
    Global
  :INTEGER;

  PROCEDURE One;
    PROCEDURE Two;
      BEGIN
        Global := 2;
        WRITELN('Global = ',Global);
      END;
    BEGIN
      Global := 1;
      WRITELN('Global = ',Global);
    END;

BEGIN
  Global := 0;
  One;
END.
```

8 Program 4–32 (p. 214) uses a local as well as a global variable. See if you can predict what will happen when the program is executed. Check your prediction by trying it yourself.

Program 4–32

```
PROGRAM IE_8;
  VAR
    Global
  :INTEGER;

   PROCEDURE One;
    VAR
    Local_1
  :INTEGER;
      PROCEDURE Two;
        BEGIN
          Global := 2;
          Local_1 := 22;
          WRITELN('Global = ',Global);
          WRITELN('Local_1 = ',Local_1);
        END;
    BEGIN
     Local_1 := 11;
     Global := 1;
     WRITELN('Global = ',Global);
     WRITELN('Local_1 = ',Local_1);
    END;

BEGIN
  Global := 0;
  One;
END.
```

9 Program 4–33 demonstrates what is meant when programmers say that a global variable is not protected. The main programming block sets the value of the global variable to 0. The procedure then sets it to another value. Predict what the final output will be when the program is executed, then try it yourself.

Program 4–33

```
PROGRAM IE_9;
  VAR
    Global
  :INTEGER;
   PROCEDURE One;
    VAR
    Local_1
  :INTEGER;
```

Program 4–33   *continued*

```
       PROCEDURE Two;
         BEGIN
           Global := 2;
           Local_1 := 22;
           WRITELN('Global = ',Global);
           WRITELN('Local_1 = ',Local_1);
         END;
     BEGIN
      Local_1 := 11;
      Global := 1;
      WRITELN('Global = ',Global);
      WRITELN('Local_1 = ',Local_1);
     END;

 BEGIN
   Global := 0;
   One;
   WRITELN('Global from the main is = ',Global);
 END.
```

10 Program 4–34 demonstrates how local variables (and constants) are protected. Will the following program compile? If not, why not?

Program 4–34

```
PROGRAM IE_10;

    PROCEDURE One_1;
      CONST
          Constant_One = 'This definition';
      VAR
          Real_Number
       :REAL;

          BEGIN
            Real_Number := 3.5;
            WRITELN('The constant is ',Constant_One);
            WRITELN('The number is ',Real_Number);
          END;

BEGIN
  Real_Number := 5.8;
END.
```

11   Program 4–35 is similar to program 4–34, but now the local declarations made in the procedure are being made global as well. What do you predict will happen when you try to compile the following program? Why did this happen?

Program 4–35

```
PROGRAM IE_11;
 CONST
   Constant_One = 'A different definition';
 VAR
   Real_Number
  :REAL;

   PROCEDURE One_1;
     CONST
        Constant_One = 'This definition';
     VAR
        Real_Number
      :REAL;

         BEGIN
           Real_Number := 3.5;
           WRITELN('The constant is ',Constant_One);
           WRITELN('The number is ',Real_Number);
         END;

BEGIN
  Real_Number := 5.8;
END.
```

12   Program 4–36 demonstrates how local declarations are protected. Notice that the same identifiers are used in both procedures and each time with a different assignment and declaration. What do you predict will happen? Try it and see.

Program 4–36

```
PROGRAM IE_12;

   PROCEDURE One_1;
     CONST
        Constant_One = 'This definition';
     VAR
        Real_Number
      :REAL;
```

Program 4-36   *continued*

```
            BEGIN
              Real_Number := 3.5;
              WRITELN('The constant is ',Constant_One);
              WRITELN('The number is ',Real_Number);
            END;

      PROCEDURE Two_2;
        CONST
            Constant_One = 'A different definition.';
        VAR
            Real_Number
          :INTEGER;

            BEGIN
              Real_Number := 3.5;
              WRITELN('The constant is ',Constant_One);
              WRITELN('The number is ',Real_Number);
            END;

BEGIN
  One_1;
  Two_2;
END.
```

13   Program 4-37 (pp. 217–218) uses a function that does not have a parameter list. Will the program compile? If not, why not? If the program does compile, will it correctly compute the area of a circle (within the accuracy for the given value of pi)? Are any Turbo Pascal built-in functions used?

Program 4-37

```
PROGRAM IE_13;

    FUNCTION PI :REAL;
      BEGIN
        PI := 3.1415927;
      END;

    PROCEDURE Circle_Area(Radius :REAL);
      VAR
        Area
      :REAL;

      BEGIN
        Area := PI * SQR(Radius);
        WRITELN('Circle area = ',Area,' square units.');
      END;
```

Program 4–37   *continued*

```
    PROCEDURE Input__It;
       VAR
         User__Value
       :REAL;
         BEGIN
             WRITE('Value of radius = ');
             READLN(User__Value);
         END;

BEGIN
  Input__It;
END.
```

14   Program 4–38 gives a demonstration of the use of value parameters and variable parameters. See if you can predict what the program will do before you actually try to compile and execute it.

Program 4–38

```
PROGRAM IE__14;

    PROCEDURE Both__Ways(Input :INTEGER; VAR Output :INTEGER);
       BEGIN
          Output := 2 * Input;
       END;

    PROCEDURE Get__It;
      VAR
        User__Value,
        Answer
      :INTEGER;
        BEGIN
          WRITE('Give me a value = ');
          READLN(User__Value);
          Both__Ways(User__Value, Answer);
          WRITELN('The result is = ',Answer);
        END;

BEGIN
   Get__It;
END.
```

Pascal Commands

```
PROCEDURE Name (N₁, N₂,...Nₙ:TYPE₁; M₁,M₂,...Mₙ:TYPE₂,);
```

where $N_1, N_2, \ldots N_N$ = value parameter declarations of $TYPE_1$

$M_1, M_2, \ldots M_N$ = value parameter declarations of $TYPE_2$

A procedure may be followed by a parameter list consisting of an opening parenthesis, one or more parameter declarations separated by semicolons, and a closing parenthesis. This type of parameter is called a value parameter. A value parameter is used to pass values to the called procedure.

```
PROCEDURE Name (VAR N₁1,N₂,...Nₙ:TYPE₁;VAR M₁,M₂,...Mₙ:TYPE₂);
```

where $N_1, N_2 \ldots N_N$ = variable parameter declarations of $TYPE_1$

$M_1, M_2 \ldots M_N$ = variable parameter declarations of $TYPE_2$

When the parameter list contains the reserved word **VAR**, the parameters that follow are variable parameters. A variable parameter is used to return a value back to the calling procedure. Value parameters and variable parameters may be declared in the same parameter list.

```
PROCEDURE Name (VAR Yes :BOOLEAN; Number :INTEGER);
```

In this procedure, the parameter list has "Yes" as a BOOLEAN variable parameter and "Number" as a INTEGER value parameter.

```
FUNCTION Name (N₁,N₂,...Nₙ:TYPE₁;M₁,M₁,M₂,...Mₙ:TYPE₂) :TYPEf
```

where $N_1, N_2 \ldots N_N$ = value parameter declarations of $TYPE_1$

$M_1, M_2 \ldots M_N$ = value parameter declarations of $TYPE_2$

$TYPE_F$ = type of function

A function must have its own **type** because all functions return a value to the calling procedure.

Self-Test

Directions

Program 4–39 (pp. 221–234) is a complete Pascal program. It contains all the correct Pascal code to make the program operate. The program employs a unique structure that includes a thorough documentation of every procedure and function used in the program. All procedures and functions are declared **FORWARD** so that the first procedure defined in the program can illustrate the structure of the entire program. Hence, the first defined procedure is called the `Main_Programming_Sequence`. This method allows you to see the program structure before the program starts rather than looking for the structure in the end of the program as is necessary when a more traditional Pascal structure is used.

Figure 4–13 (p. 220) illustrates the structure of program 4–39. Observe that the parameter lists are repeated in the procedure and function definitions as comments. This is because Turbo Pascal allows you to define the parameter list only once for a given function or procedure; they are repeated as comments for easier program reading. Answer the self-test questions by referring to program 4–39.

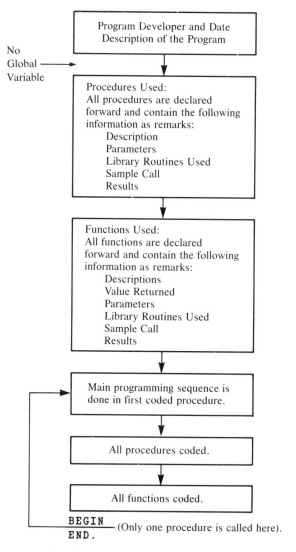

No
Global ———→
Variable

Figure 4–13   Structure of Self-Test Program

Questions

1   How many user-defined procedures and functions are used in program 4–39?
2   How many built-in Turbo Pascal procedures and functions are used? List them.
3   List all the global variables.
4   Where is it stated what the program does?
5   List the procedures that use value parameters. State what each of the values means. Where in the program do you look to find this out?
6   List the procedures that use variable parameters. State what each of these mean. Where in the program do you look to find them?

Program 4–39

```
  PROGRAM RLC_Circuit;
  USES Crt;
  {**************************************************************************
                   Developed By:  Electronics Student
                        Date:   November 1991
     **************************************************************************
         Program description:  This program will solve all of the
  electrical properties of a series RLC circuit.  The program user
  must input the value of the resistor, inductor and capacitor,
  the applied frequency, and voltage. The program will then compute:
      1.  Inductive and capacitive reactance.
      2.  Circuit impedance.
      3.  Circuit Q.
      4.  Circuit current.
      5.  Voltage drops across the inductor, capacitor, and
          resistor.
      6.  Phase relationship between circuit current and applied
          voltage.
      7.  Circuit resonant frequency.
     **************************************************************************}
  TYPE
     WORDS = STRING[79];
  {**************************************************************************
                   Procedures used:
  ----------------------------------------------------------------------}
  {PROCEDURE Main_Programming_Sequence;

      DESCRIPTION:
         This procedure calls all of the primary procedures used in this
         program.  It is the first procedure given in order to make the
         structure of the program easy to read and follow.

      PARAMETERS:
         None.

      SAMPLE CALL:
         BEGIN
           Main_Programming_Sequence;
         END.
  ----------------------------------------------------------------------}

  PROCEDURE Program_Title;
  FORWARD;
     {
      DESCRIPTION:
         This procedure displays the title:

            _____
           |      Series RLC Problem Solver      |
           |                by:                  |
           |             Your Name               |
           |_____|

PARAMETERS:
         None.

      SAMPLE CALL:
           Program_Title;
  ----------------------------------------------------------------------}
```

Program 4-39 *continued*

```
   PROCEDURE Explain_Program;
   FORWARD;
      {
         DESCRIPTION:
            This procedure explains the purpose of the program to the
            program user.  It also tells the program user what values are to
            be input and what the program will compute.

         PARAMETERS:
            None.

         SAMPLE CALL:
            Explain_Program;
- - - - - - - - - - - - - - - - - - - - - - - - - - - - - - - - - - - - - - - - - - - - - - - - - - - - - -}

   PROCEDURE Get_Values;
   FORWARD;
      {
         DESCRIPTION:
            This procedure gets the value of the resistor, inductor,
            capacitor, applied circuit frequency, and applied peak voltage
            from the program user.

         PARAMETERS:
            None.

         LIBRARY ROUTINES USED:
            Do_Calculations(R, L, C, F, VP :REAL) :REAL;

         SAMPLE CALL:
               Get_Values;

         RESULTS:
            Program user is prompted to input all of the above values and the
            procedure Do_Calculation is called with values passed to that
            procedure.
- - - - - - - - - - - - - - - - - - - - - - - - - - - - - - - - - - - - - - - - - - - - - - - - - - - - -}

   PROCEDURE Do_Calculations(R, L, C, F, VP :REAL);
   FORWARD;
      {
         DESCRIPTION:
            This procedure evokes the functions that do the actual
            computations of:
                Inductive Reactance
                Capacitive Reactance
                Circuit Impedance
                Circuit Current
                Phase angle of Total Voltage and Circuit Current
                Circuit Q
                Voltage across Resistor, Capacitor, and Inductor
                Circuit Resonant Frequency
```

Program 4-39 *continued*

```
    PARAMETERS:
        R [input]—Circuit resistance entered by user.
        L [input]—Circuit inductance entered by user.
        C [input]—Circuit capacitance entered by user.
        F [input]—Applied frequency entered by user.
        VP [input]—Applied peak voltage entered by user.

    LIBRARY ROUTINES USED:
        FUNCTION   X_L (Freq, Induct  : REAL) :REAL;
        FUNCTION   X_C (Freq, Capac  : REAL) :REAL;
        FUNCTION   Z (Ind_Rea, Cap_Rea, Resis  :REAL) :REAL;
        FUNCTION   Phase (Ind_Rea, Cap_Rea, Resis  :REAL) :REAL;
        FUNCTION   Circuit_Q (X_L, Res  :REAL) :REAL;
        FUNCTION   Circuit_Current(Peak_Volts, Circuit_Imp);
        FUNCTION   Resistor_Voltage(Circuit_Current, Res_Val);
        FUNCTION   Capacitor_Voltage(Circuit_Current, React_C);
        FUNCTION   Inductor_Voltage(Circuit_Current, React_L);
    PROCEDURE Imped_and_Phase (X_C, X_L, Res :REAL; VAR Z, Ph :REAL);
    PROCEDURE D_Out(X_Val,Y_Val:INTEGER;Show:WORDS;Value:REAL;Unit:WORDS);

    SAMPLE CALL:

        PROCEDURE  Do_Calculations (3.0, 0.5E-3, 0.2E-6, 15.0E3, 12.0);

    RESULTS:
        Inductive Reactance ==> 47.123890 ohms
        Capacitive Reactance ==> 53.051648 ohms
        Impedance ==> 6.643667 ohms
        Circuit Q ==> 15.707963 (no units)
        Phase angle ==> -1.102286 radians
        Voltage across R ==> 5.418694 volts
        Voltage across L ==> 85.116647 volts
        Voltage across C ==> 95.823550 volts
        Resonant frequency ==> 15915.494309 Hz
----------------------------------------------------------------------}

  PROCEDURE Imped_and_Phase(X_C, X_L, Res :REAL; VAR Z_P, Ph :REAL);
  FORWARD;
    {
      DESCRIPTION:
          This procedure uses the functions "Impedance" and "Phase" to
          compute the circuit impedance and phase angle of the circuit.

      PARAMETERS:
          X_C [input]—Capacitive reactance.
          X_L [input]—Inductive reactance.
          Res [input]—Value of circuit resistor.
          Z_P [output]—Circuit impedance.
          Ph [output]—Phase angle of applied voltage to circuit current.

      SAMPLE CALL:
            Imped_and_Phase(Cap_Rea, Ind_Rea, Tot_Z, Angle);

      RESULTS:
          Will return the circuit impedance (Tot_Z) and the phase angle
          (Angle) in radians.
    ----------------------------------------------------------------}
```

Program 4–39 *continued*

```
PROCEDURE D_Out(X_Val,Y_Val:INTEGER;Show:WORDS;Value:REAL; Unit:WORDS);
FORWARD;
   {
      DESCRIPTION:
           This procedure causes data to be displayed on the monitor
           screen.  It will display numeric values mixed with strings at a
           location selected by the program user.

      PARAMETERS:
           X_Val [input]—Starting column for output.
           Y_Val [input]—Starting row for output.
           Show [input]—String to be displayed.
           Value [input]—Real number to be displayed in form of 3:6.
           Unit [input]—String to be displayed.

      TYPES REQUIRED:
           WORDS = STRING[79];

      SAMPLE CALL:
              D_Out(1,2,'Inductive Reactance',18,' ohms');

      RESULTS:
              Inductive Reactance ==> 18 ohms
------------------------------------------------------------------------}

   PROCEDURE Note(Pitch, Duration :INTEGER);
   FORWARD;
      {
        DESCRIPTION:
          This procedure plays a tone on the computer speaker at a
          frequency and duration determined by the programmer.

        PARAMETERS:
          Pitch [input]—Frequency of the tone.
          Duration [input]—Length of the tone in milliseconds.

        SAMPLE CALL:
             Note(262,1000);

        RESULTS:
          Will play a note of about middle "C" for 1 second from the
          computer speaker.
------------------------------------------------------------------------}

   PROCEDURE Thank_You;
   FORWARD;
      {
        DESCRIPTION:
          This procedure makes a distinct sound to
          let the program user know that input has been accepted.

        PARAMETERS:
          None.

        LIBRARY ROUTINES USED:
          Note(Frequency, Duration :INTEGER);
```

Program 4–39   *continued*

```
            SAMPLE CALL:
                Thank_You;

            RESULTS:
                Pleasant set of notes.
-------------------------------------------------------------------------}

    PROCEDURE User_Prompt;
    FORWARD;
        {
            DESCRIPTION:
                This procedure displays the message

                    Press RETURN/ENTER to continue...

                at the bottom of the computer screen and waits for the program
                user to depress the RETURN/ENTER key.  The program gives the
                "Thank_You" sound.

            PARAMETERS:
                None.

            LIBRARY ROUTINES USED:
                Thank_You;

            SAMPLE CALL:
                    User_Prompt;

            RESULT:
                (Prompting sound)
                    Press RETURN/ENTER to continue..
                (Acknowledgement sound)

****************************************************************************
                    Functions used:
 -------------------------------------------------------------------------}

    FUNCTION X_L(Freq, Induct : REAL) :REAL;
    FORWARD;
        {
        DESCRIPTION:
            This function computes the value of the inductive reactance
            for a given frequency and value of inductor.

        VALUE RETURNED:
            Inductive reactance in ohms.

        PARAMETERS:
            Freq [input]—Applied frequency in Hz.
            Induct [input]—Inductor value in henrys.

        SAMPLE CALL:
                Result := X_L(15.0E3, 0.5E-3);

        RESULT:
                Result = 4.7123889804E+01
 -------------------------------------------------------------------------}
```

Program 4–39 *continued*

```
FUNCTION X_C(Freq, Capac : REAL) :REAL;
FORWARD;
  {
  DESCRIPTION:
      This function computes the value of the capacitive reactance
      for a given frequency and value of capacitor.

  VALUE RETURNED:
      Capacitive reactance in ohms.

  PARAMETERS:
      Freq [input]—Applied frequency in Hz.
      Capac [input]—Capacitor value in farads.

  SAMPLE CALL:
      Result := X_C(15.0E3, 0.2E-6);

  RESULT:
      Result = 5.3051647697E+01
-------------------------------------------------------------------------}

  FUNCTION Z(Ind_Rea, Cap_Rea, Resis :REAL) :REAL;
  FORWARD;
  {
  DESCRIPTION:
      This function computes the circuit impedance of a series RLC
      circuit when the inductive and capacitive reactances are known.

  VALUE RETURNED:
      Impedance in ohms.

  PARAMETERS:
      Ind_Rea [input]—Inductive reactance in ohms.
      Cap_Rea [input]—Capactive reactance in ohms.
      Resis [input]—Resistance in ohms.

  SAMPLE CALL:
      Result := Z(47.0, 53.0, 3);

  RESULT:
      Result = 6.6436671836E+00
-------------------------------------------------------------------------}

  FUNCTION Phase_Ang(Ind_Rea, Cap_Rea, Resis :REAL) :REAL;
  FORWARD;
  {
  DESCRIPTION:
      This function computes the phase angle in radians between the
      applied voltage and circuit current for a series RLC circuit.

  VALUE RETURNED:
      Phase angle in radians.

  PARAMETERS:
      Ind_Rea [input]—Inductive reactance in ohms.
      Cap_Rea [input]—Capacitive reactance in ohms.
      Resis [input]—Circuit resistance in ohms.
```

Program 4–39 *continued*

```
          SAMPLE CALL:
               Result := Phase_Ang(47.1, 53.0, 3);

          RESULT:
               Result = -1.1022857741E+00
   ----------------------------------------------------------------------}

     FUNCTION Circuit_Q(X_L, Res :REAL) :REAL;
     FORWARD;
          {
          DESCRIPTION:
               This function computes the Q of a series RL circuit.

          VALUE RETURNED:
               Circuit Q (no units).

          PARAMETERS:
               X_L [input]—Inductive reactance in ohms.
               Res [input]—Circuit resistance in ohms.

          SAMPLE CALL:
               Result := Circuit_Q(47.1, 3);

          RESULT:
               Result = 1.5707963268E+01
   ----------------------------------------------------------------------}

     FUNCTION Resonance(Capacitance, Inductance :REAL) :REAL;
     FORWARD;
          {
          DESCRIPTION:
               This function computes the resonant frequency of an LC
               circuit.

          VALUE RETURNED:
               Frequency in Hz.

          PARAMETERS:
               Capacitance [input]—Circuit capacitance in farads.
               Inductance [input]—Circuit inductance in henrys.

          SAMPLE CALL:
               Result := Resonance(0.2E-6, 0.5E-3);

          RESULT:
               Result = 1.5915494309E+04
   ----------------------------------------------------------------------}

FUNCTION Total_Current(Peak_Volts, Circuit_Imp :REAL) :REAL;
FORWARD;
     {
          DESCRIPTION:
               This function calculates the total current when the impedance
               and applied circuit voltage are known.

          VALUE RETURNED:
               Current in amps.
```

Program 4-39 *continued*

```
        PARAMETERS:
            Peak_Volts [input]—Applied peak voltage in volts.
            Circuit_Imp [input]—Circuit impedance in ohms.

        SAMPLE CALL:
            Result := Total_Current(12.0, 6.64);

        RESULT:
            Result = 1.8062313561E+00
    -------------------------------------------------------------------------}

FUNCTION Resistor_Voltage(Circuit_Current, Res_Val :REAL) :REAL;
FORWARD;
    {
    DESCRIPTION:
        This function computes the voltage drop across a resistor if the
        resistor current and resistor value are known.

    VALUE RETURNED:
        Voltage in volts.

    PARAMETERS:
        Circuit_Current [input]—Current in resistor in amps.
        Res_Val [input]—Value of resistor in ohms.

    SAMPLE CALL:
        Result := Resistor_Voltage(3,20);

    RESULT:
        Result = 6.0000000000E+01
    -------------------------------------------------------------------------}

FUNCTION Inductor_Voltage(Circuit_Current, React_L :REAL) :REAL;
FORWARD;
    {
    DESCRIPTION:
        This function computes the voltage across an inductor when the
        inductor current and reactance are known.

    VALUE RETURNED:
        Inductive reactance in ohms.

    PARAMETERS:
        Circuit_Current [input]—Current in the inductor in amps.
        React_L [input]—Inductor reactance in ohms.

    SAMPLE CALL:
        Result := Inductor_Voltage(3.0, 12.0);

    RESULT:
        Result = 3.6000000000E+01
    -------------------------------------------------------------------------}
```

Program 4-39 *continued*

```
FUNCTION Capacitor_Voltage(Circuit_Current, React_C :REAL) :REAL;
FORWARD;
    {
      DESCRIPTION:
          This function computes the voltage across a capacitor when the
          capacitor current and reactance are known.

      VALUE RETURNED:
          Capacitive reactance in ohms.

      PARAMETERS:
          Circuit_Current [input]—Capacitor current in amps.
          React_C [input]—Reactance of capacitor in ohms.

      SAMPLE CALL:
          Result := Capacitor_Voltage(2.0, 12.0);

      RESULT:
          Result = 2.4000000000E+01
  ------------------------------------------------------------------------}

{*************************************************************************}

  PROCEDURE Main_Programming_Sequence;
  {=======================================================================}

    BEGIN

      Program_Title;      {Displays the title of program in a box.}
      Explain_Program;    {Explain program to user.}
      Get_Values;         {Get circuit values from program user.}

  {=======================================================================}
    END;  {of procedure Main_Programming_Sequence}


  PROCEDURE Program_Title;
  {-----------------------------------------------------------------------}

      BEGIN

          ClrScr;  {Clear the screen.}
          WINDOW(18,10,65,15);   {Creates a window in center of screen.   }

          WRITELN('┌─────────────────────────────────┐');
          WRITELN('│      Series RLC Problem Solver   │');
          WRITELN('│                by:               │');
          WRITELN('│             Your Name            │');
          WRITELN('└─────────────────────────────────┘');

          User_Prompt;    {Display prompt message.}

  {-----------------------------------------------------------------------}
      END;   {of procedure Program_Title}

  PROCEDURE Explain_Program;
  {-----------------------------------------------------------------------}
```

Program 4–39 *continued*

```
    BEGIN

        ClrScr;  {Clear the screen.}
        WRITELN;
        WRITELN('This program will compute the circuit values of an RLC');
        WRITELN('circuit.');
        WRITELN;
        WRITELN('You must enter the value of the inductor, capacitor, ');
        WRITELN('circuit resistance, applied frequency ');
        WRITELN('and peak voltage.');
        WRITELN;
        WRITELN('The program will then compute the following:');
        WRITELN;
        WRITELN('1] Inductive Reactance in Ohms.');
        WRITELN('2] Capacitive Reactance in Ohms.');
        WRITELN('3] Circuit Impedance in Ohms.');
        WRITELN('4] Circuit Current in Amps.');
        WRITELN('5] Phase angle of Total Voltage and Circuit Current.');
        WRITELN('6] Circuit Q.');
        WRITELN('7] Voltage across —');
        WRITELN('    * Resistor');
        WRITELN('    * Capacitor');
        WRITELN('    * Inductor');
        WRITELN('8] Circuit Resonant Frequency.');

            User_Prompt;   {Display prompt at bottom of screen.}

{-------------------------------------------------------------}
    END;  {of procedure Explain_Program}

PROCEDURE Get_Values;
  {-------------------------------------------------------------}

    VAR

        Circuit_R,      {Circuit resistance in ohms.}
        Circuit_L,      {Circuit inductance in henrys.}
        Circuit_C,      {Circuit capacitance in farads.}
        Circuit_F,      {Applied frequency in Hertz.}
        Circuit_VP      {Applied peak voltage in volts.}
      :REAL;

    BEGIN

        ClrScr;     {Clear the screen.}
        GOTOXY(5,2);
        WRITELN('Enter the following values:');
        WRITELN;
        WRITE(' Circuit resistance [from 1 to 1,000 ohms] ==> ');
        READLN(Circuit_R);
        WRITELN;
        WRITE(' Circuit inductance in henrys ==> ');
        READLN(Circuit_L);
        WRITELN;
        WRITE('Circuit capacitance in farads ==> ');
        READLN(Circuit_C);
```

Program 4–39   *continued*

```
        WRITELN;
        WRITE(' Applied frequency in Hertz ==> ');
        READLN(Circuit_F);
        WRITELN;
        WRITE(' Applied peak voltage in volts ==> ');
        READLN(Circuit_VP);

  Do_Calculations(Circuit_R, Circuit_L, Circuit_C, Circuit_F, Circuit_VP);

    {--------------------------------------------------------------}
      END;  {of procedure Get_Values;}

    PROCEDURE Do_Calculations{(R, L, C, F, VP :REAL)};
    {--------------------------------------------------------------}

        VAR

            Inductive_Reactance,   {Inductive reactance of inductor.}
            Capacitive_Reactance,  {Capacitive reactance of capacitor.}
            Resonant_Frequency,    {Frequency of circuit resonance.}
            Quality_Q,             {Quality of circuit.}
            Imped_Z,               {Circuit impedance.}
            Phase,                 {Phase angle in radians.}
            I_T,                   {Total circuit current.}
            V_R,                   {Voltage across circuit resistor.}
            V_C,                   {Voltage across circuit capacitor.}
            V_L                    {Voltage across circuit inductor.}
          :REAL;

        BEGIN

            Inductive_Reactance := X_L(F,L);
            Capacitive_Reactance := X_C(F,C);
            Resonant_Frequency := Resonance(C, L);
            Quality_Q := Circuit_Q(Inductive_Reactance, R);

  Imped_and_Phase(Capacitive_Reactance,Inductive_Reactance,R,Imped_Z,Phase);

            I_T := Total_Current(VP, Imped_Z);
            V_R := Resistor_Voltage(I_T, R);
            V_L := Inductor_Voltage(I_T, Inductive_Reactance);
            V_C := Capacitor_Voltage(I_T, Capacitive_Reactance);

          D_Out(2,14, 'Inductive Reactance', Inductive_Reactance,' ohms');
          D_Out(2,15, 'Capacitive Reactance', Capacitive_Reactance,' ohms');
          D_Out(2,16, 'Impedance', Imped_Z,' ohms');
          D_Out(2,17, 'Circuit Q', Quality_Q,' (no units)');
          D_Out(2,18, 'Circuit Current',I_T,' amps');
          D_Out(2,29, 'Phase angle', Phase,' radians');
          D_Out(2,20, 'Voltage across R', V_R,' volts');
          D_Out(2,21, 'Voltage across L', V_L,' volts');
          D_Out(2,22, 'Voltage across C', V_C,' volts');
          D_Out(2,23, 'Resonant frequency', Resonant_Frequency,' Hz');

    {--------------------------------------------------------------}
      END;   {of procedure Do_Calculations}
```

Program 4–39 *continued*

```
PROCEDURE Imped_and_Phase{(X_C, X_L, Res:REAL;VAR Z_P, Ph:REAL)};
{-----------------------------------------------------------------------}

      BEGIN

         Z_P := Z(X_L, X_C, Res);
          Ph := Phase_Ang(X_L, X_C, Res);

{-----------------------------------------------------------------------}

      END;   {of procedure Imped_and_Phase}

PROCEDURE D_Out{(X_Val,Y_Val:INTEGER;Show:WORDS;Value:REAL; Unit:WORDS)};
{-----------------------------------------------------------------------}

      BEGIN

        GOTOXY(X_VAl, Y_Val);
        WRITELN(Show,' ==> ',Value:3:6,Unit);
{-----------------------------------------------------------------------}

      END;   {of procedure D_Out;}

PROCEDURE Note{(Pitch, Duration :INTEGER)};
{-----------------------------------------------------------------------}

        BEGIN

          SOUND(Pitch);   {Causes tone on computer speaker.}
          DELAY(Duration);{Allows sound to be heard for a time.}
          NOSOUND;            {Turns the sound off.}

{-----------------------------------------------------------------------}

      END; {of procedure Note;}

PROCEDURE Thank_You;
{-----------------------------------------------------------------------}

        BEGIN

          Note(300,200);   {Uses the "Note" procedure.}
          Note(500,100);

{-----------------------------------------------------------------------}

      END; {of procedure Thank_You;}

PROCEDURE User_Prompt;
{-----------------------------------------------------------------------}

      VAR

        Response       {Variable to receive user input of }
      :CHAR;               {RETURN/ENTER key.}
```

Program 4-39   *continued*

```
        BEGIN

            LOWVIDEO;   {Dim the output display.}
            WINDOW(5,23,75,25);  {Set display window for screen bottom}
            Note(600,100);       {Prompt program user with sound.}
            WRITE('     Press RETURN/ENTER to continue..');
            READ(Response);
            NORMVIDEO;  {Return video to normal display.}
            Thank_You;  {Thank you note.}
            WINDOW(1,1,80,25);   {Return window to full display.}
            ClrScr;   {Clear the entire screen.}

{------------------------------------------------------------------}
        END;   {of procedure User_Prompt;}

 FUNCTION X_L{(Freq, Induct : REAL) :REAL};
{------------------------------------------------------------------}
    BEGIN
            X_L := 2 * PI * Freq * Induct;
{------------------------------------------------------------------}
    END;  {of function X_L}

 FUNCTION X_C{(Freq, Capac : REAL) :REAL};
{------------------------------------------------------------------}
    BEGIN
            X_C := 1/(2 * PI * Freq * Capac);
{------------------------------------------------------------------}
    END;   {of function X_C;}

 FUNCTION Z{(Ind_Rea, Cap_Rea, Resis :REAL) :REAL};
{------------------------------------------------------------------}
    BEGIN
            Z := SQRT (SQR(Ind_Rea - Cap_Rea) + SQR(Resis));
{------------------------------------------------------------------}
    END;   {of function Z}

 FUNCTION Phase_Ang{(Ind_Rea, Cap_Rea, Resis :REAL) :REAL};
{------------------------------------------------------------------}
    BEGIN
            Phase_Ang := ARCTAN((Ind_Rea - Cap_Rea)/Resis);
{------------------------------------------------------------------}
    END;   {of function Phase;}

 FUNCTION Circuit_Q{(X_L, Res :REAL) :REAL};
{------------------------------------------------------------------}
    BEGIN
            Circuit_Q := X_L/Res;
{------------------------------------------------------------------}
    END;   {of function Circuit_Q;}

 FUNCTION Resonance{(Capacitance, Inductance :REAL) :REAL};
{------------------------------------------------------------------}
    BEGIN
            Resonance := 1/(2 * PI * SQRT(Capacitance * Inductance));
{------------------------------------------------------------------}
    END;   {of function Resonance;}
```

Program 4–39 *continued*

```
FUNCTION Total_Current{(Peak_Volts, Circuit_Imp :REAL) :REAL};
{----------------------------------------------------------------------}
     BEGIN
          Total_Current := Peak_Volts/Circuit_Imp;
{----------------------------------------------------------------------}
     END;    {of function Total_Current;}

FUNCTION Resistor_Voltage{(Circuit_Current, Res_Val :REAL) :REAL};
{----------------------------------------------------------------------}
     BEGIN
          Resistor_Voltage := Circuit_Current * Res_Val;
{----------------------------------------------------------------------}
     END;    {of function Resistor_Voltage;}

FUNCTION Inductor_Voltage{(Circuit_Current, React_L :REAL) :REAL};
{----------------------------------------------------------------------}
     BEGIN
          Inductor_Voltage := Circuit_Current * React_L;
{----------------------------------------------------------------------}
     END;    {of function Inductor_Voltage;}

FUNCTION Capacitor_Voltage{(Circuit_Current, React_C :REAL) :REAL};
{----------------------------------------------------------------------}
     BEGIN
          Capacitor_Voltage := Circuit_Current * React_C;
{----------------------------------------------------------------------}
     END;    {of function Capacitor_Voltage;}

BEGIN
Main_Programming_Sequence;
END.
```

Questions *continued*

7 Describe the main sequence of the program. Where do you look to find this?

8 For this program, does it make any difference in what order the procedures or functions are defined? Give the reason for your answer.

9 State how the procedure `Do_Calculations` gets the values of the circuit impedance and circuit phase angle.

10 State the purpose of the procedure `User_Prompt`.

Problems

General Concepts

Section 4–1

1 Describe a procedure.

2 Explain the difference between calling and defining a procedure.

3 State how many times a procedure may be called.

4 Give an example of a procedure calling another procedure.
5 Explain when it is not legal (program will not compile) for one procedure to call another.
6 Describe what is meant by calling through a procedure. What happens when this is attempted?
7 Show, by an example, what is meant by nesting procedures.
8 Give an example of procedures used in a branch block, and in a loop block.

Section 4–2

9 Define the term global as it is used in Pascal. Give an example.
10 Which procedures can use a global variable? A global constant?
11 Explain the meaning of scope as it is used in Pascal.
12 Define the term local as it is used in Pascal. Give an example.
13 Give an example of nested procedures using a local constant.
14 Is it better programming practice to have global variables and constants or local variables and constants? Explain.

Section 4–3

15 Define what a Pascal function does. How does this differ from a Pascal procedure?
16 Describe the similarities between a Pascal function and procedure.
17 Can a Pascal function call itself? What is this process called?
18 Develop a Pascal function that will return the square root of the sum of the squares of two numbers.
19 Describe a built-in Turbo Pascal function. Give an example of one.

Section 4–4

20 Can a value be returned with a Pascal procedure? How is this different from a Pascal function?
21 Explain what is meant by passing values to a procedure.
22 Define a value parameter. Give an example.
23 Define a variable parameter. Give an example.
24 Discuss the difference between a variable parameter and a value parameter.

Section 4–5

25 Explain what is meant by a library routine.
26 What is the built-in Turbo Pascal procedure for clearing the screen? For creating a window?
27 State some of the most common built-in Turbo Pascal procedures you have used to this point.

Section 4–6

28 Explain when a **FORWARD** declaration is used in Pascal.
29 State how Turbo Pascal deals with the parameter list of a forwarded procedure or function.
30 Give an example of how using **FORWARD** declarations can help in the structure of a Pascal program.

Program Analysis

31 Will program 4–40 (p. 236) compile? If not, why not?
32 How many nested procedures are there in program 4–41 (p. 236)? Will the program compile? If not, why not?

Program 4-40

```
PROGRAM Problem_31;
  BEGIN
    PROCEDURE First_One;
      WRITELN('This is the first procedure.');
    END;
  END.
```

Program 4-41

```
PROGRAM Problem_32;
  PROCEDURE The_First;
    BEGIN
      PROCEDURE The_Second;
        WRITELN('This is the second procedure');
      END;
      WRITELN('This is the first procedure');
    END;

BEGIN  {Main programming sequence.}
    The_First;
END.
```

33   Program 4-42 will not compile. What is the problem? What would you do to correct it?

Program 4-42

```
PROGRAM Problem_33;
  PROCEDURE First_One;
    PROCEDURE Second_One;
      BEGIN
        WRITELN('This is the second procedure.');
      END;
    BEGIN
        WRITELN('This is procedure one.');
    END;

BEGIN  {Main programming sequence}
  Second_One;
END.
```

34 Will program 4–43 compile? If it does, what output do you predict it will have?

Program 4–43

```
PROGRAM Problem_34;

   PROCEDURE Outer_One;
      PROCEDURE Inner_One;
         PROCEDURE Inside_Inner_One;
            BEGIN
               Inner_One;
            END;
         BEGIN
            Outer_One;
         END;
         WRITELN('Where did this come from?');
      END;

BEGIN    {Main programming sequence}
   Outer_One;
END.
```

35 Analyze program 4–44 and predict what it will do. If it won't compile, why do you think it won't? What change(s) would you make in order to make it compile?

Program 4–44

```
PROGRAM Problem_35;
  VAR
     Main_One
 :INTEGER;

   PROCEDURE Calculator;
     VAR
        Main_One
     INTEGER;
        BEGIN
           Main_One := 3 * 4;
        END;

BEGIN    {Main programming sequence}
   Calculator;
END.
```

36 Analyze program 4–45 (p. 238) to see if it will compile. If not, why not?

Program 4–45

```
PROGRAM Problem_36;

  PROCEDURE Calculator_1;
    VAR
      Main_One
    :INTEGER;
        BEGIN
          Main_One := 3 * 4;
        END;

  PROCEDURE Calculator_2;
    VAR
      Main_One
    :INTEGER;
        BEGIN
          Main_One := 8 * 2;
        END;

BEGIN  {Main programming sequence}
  Calculator_1;
  Calculator_2;
  WRITELN(Main_One);
END.
```

37 If program 4–45 does compile, what will be displayed on the monitor when the program is executed?

38 Analyze program 4–46. Determine if it will compile and if not, why not?

Program 4–46

```
PROGRAM Problem_38;
  VAR
    Results
  :INTEGER;
  FUNCTION First_One;
  FORWARD;

  FUNCTION Second_one (A_Variable :INTEGER);
  FORWARD;

  PROCEDURE Do_Some_Calculations(A_Number :INTEGER);
  FORWARD;

  PROCEDURE Get_the_Number;
    VAR
      Another_Variable
```

Program 4–46   *continued*

```
        :INTEGER;
          BEGIN
             WRITELN('Give me a number => ');
             READLN(Another_One);
             Do_Some_Calculations(Another_One);
          END;

      PROCEDURE Do_Some_Calculations(A_Number :INTEGER);
         VAR
            Results
         :INTEGER;

         BEGIN
             Results := First_One  +  Second_One(A_Number);
         END;

      FUNCTION First_One;
         BEGIN
             First_One := 3.14159;
         END;

      FUNCTION Second_One;{(A_Variable :INTEGER)}
         BEGIN
             Second_One := A_Variable * A_Variable;
         END;
  BEGIN  {Main programming sequence}
    Get_The_Number;
    WRITELN('The answer is => ',Results);
  END.
```

39 Correct program 4–46 so it will compile. When the program is executed, are the correct results displayed on the screen? If not, why not?

40 If program 4–47 (pp. 239–240) will compile, predict what it will display when executed. If it will not compile, make the necessary corrections, and state what they are.

Program 4–47

```
PROGRAM Problem_40;

   PROCEDURE First_One(One :INTEGER; VAR Two :REAL);
   FORWARD;

   PROCEDURE Do_It;

      CONST
         One = 5;
         Two = 8;
```

Program 4–47   *continued*

```
        VAR
          Result
        REAL;

        BEGIN
          First_One(One, Result);
          Display_It(Result);
          First_One(Two, Result);
          Display_It(Result);
        END;

      PROCEDURE Display_It (Output_Value);
        BEGIN
          WRITELN('This is the result => ',Output_Value);
        END;
  BEGIN  {Main programming sequence}
    Do_It;
  END.
```

Program Design

Use the structure developed in the Pascal program in the self-test for the following programs. This means:

1. All procedures are declared **FORWARD** and documented.
2. All functions are declared **FORWARD** and documented.
3. The first defined procedure presents the program sequence.
4. The traditional main programming sequence (between the final **BEGIN** and **END.**) calls only one procedure (the procedure, explained in 3 above).

Electronics Technology

41 Using Pascal functions, develop a program that will calculate the total current in a series circuit consisting of three resistors. Program user is to input the value of each resistor and the value of the voltage source.

   The relationship of the resistor and voltage values to the total circuit current is

$$I_T = V/R_T$$

where I_T = total circuit current in amps
   V = voltage of the voltage source in volts
   R_T = total circuit resistance in ohms
and

$$R_T = R_1 + R_2 + R_3$$

where R_T = total circuit resistance in ohms

R_1, R_2, R_3 = value of each resistor in ohms

42　Expand the program you wrote for problem 41, using function(s) where necessary so that the program will calculate the voltage drop across each resistor and its power dissipation. The relationships are

$$V_{RN} = I_T \times R_N$$

where V_{RN} = voltage drop across indicated resistor in ohms

$\quad I_T$ = total circuit current in amps

$\quad R_N$ = value of indicated resistor in ohms

and

$$P_{RN} = I_T \times V_{RN}$$

where P_{RN} = power dissipation of indicated resistor in watts

$\quad I_T$ = total circuit current in amps

$\quad V_{RN}$ = voltage drop across indicated resistor in ohms

43　If you haven't already done so, redo the program for problem 42 so there are no global variables.

44　Redo the program for problem 43 so that all functions are replaced by procedures. Recall that you must use variable parameters as well as value parameters to accomplish this.

Manufacturing Technology

45　Create a Pascal program that will compute the cost of producing a solid wire cable. Assume that the wire cable is a perfect cylinder. Program user inputs are

1. Density of the material.
2. Radius of the cable.
3. Length of the cable.
4. Cost of material per unit weight.
5. Cost to produce the material per unit length.

　　The program output will be

1. Weight of material required.
2. Cost of material.
3. Cost of production.
4. Total cost of material and production.

Drafting Technology

46　Design a Pascal program that will compute the following for a rectangular solid.

1. Surface area.
2. Volume.
3. Weight.

Program user input is

1. Dimensions of height, width, and thickness.
2. Density of material.

Computer Science

47 Develop a Pascal program that illustrates the use of global and local variables and demonstrates the meaning of scope. The program should also demonstrate how local variables are protected, and the difference between a value parameter and a variable parameter.

Construction Technology

48 Design a Pascal program that will compute

1. Floor area in square feet.
2. Roof area in square feet.
3. Exterior wall area in square feet. (including space used by interior walls)

Program user input is

1. Height, width, and length of the house.
2. Pitch of the roof (assume no overhang).
3. Total exterior door and window area.

Agriculture Technology

49 Create a Pascal program that will compute the weight of a cylindrical shipping container for farm produce. The user input is the length and radius of the shipping container and the density of the produce.

Health Technology

50 Develop a Pascal program that will compute the total bill of a hospital patient. The user inputs are

1. Number of days in hospital.
2. Surgery cost.
3. Medication cost.
4. Miscellaneous cost.
5. Cost per day.
6. Insurance deductible.

The program must also compute for insurance purposes

1. Total cost.
2. Total cost less insurance deductible.
3. Total cost less cost of medication and deductible.

Business

51 Design a Pascal program that will balance two different accounts consisting of five items each. The user inputs are

1. Number of widgets ordered.
2. Number of widgets shipped.
3. Shipping cost per widget.
4. Cost of each widget.
5. Handling charge per widget.

The program output is to display

1. Number of widgets on back order.
2. Total cost for widgets shipped.
3. Total cost when all widgets are shipped.

5 Branching and Strings

Objectives

This chapter will give you the opportunity to learn:

1 The use of relational operators in the Pascal language.
2 The meaning of the logic expressions: AND, OR, NOT and XOR.
3 How to use Boolean operators and expressions.
4 The order of precedence of Boolean operators.
5 How to use Boolean operators and expressions with the **IF—THEN** structure in Pascal.
6 The use of the Pascal **CASE** statement.
7 The meaning of string variables and constants.
8 How to use the Pascal built-in string operators, functions, and procedures.
9 How to create library units using Turbo Pascal.

Key Terms

Relational Operators	Nested
Boolean Algebra	Character Strings
AND	Null String
OR	Concatenation
NOT	Library
XOR	UNIT

Outline

In this chapter you will see how to make programs that make the computer look smart. You will also learn to work with words as well as numbers. With this combination, you'll be able to develop programs that will allow the program user to "talk" to the computer using words as well as numbers.

The chapter starts with some necessities of basic computer logic. You will see that this is a very simple form of logic (there are only two conditions allowed), yet you will learn to use Pascal to develop technical programs that will help other technicians troubleshoot robot systems.

If you just learn these basic concepts and practice them in the interactive exercises and problems, you will be able to create Pascal programs with useful applications in the world of technology.

5–1   Logic Decisions

Introduction

In this section, you will see how to make the computer look "smart" using Pascal. There are two basic ways of doing this: by using relational operators and by using logical operators. In this section, you will see practical examples of both types of operators. You will then be well on your way to creating Pascal programs that will perform decision-making tasks.

Relational Operators

Relational operators are symbols that indicate a relation between two quantities which may be either variables or constants. The important point about these relations is that they are either TRUE or FALSE; there is nothing in between. The relational operators used by Turbo Pascal are shown in table 5–1.

The applications of the relational operators shown in table 5–1 will be clearer in program examples. One of the major uses of relational operators is in branch blocks. There are several Pascal commands that can cause a branch.

The IF-THEN

In Chapter 2 you saw that the computer is good at making a decision based on a YES or NO answer. This is the same as saying that a computer can make a decision

Table 5–1   Relational Operators

Symbol	Meaning	TRUE Examples	FALSE Examples
>	Greater than	5 > 3	3 > 5
		C > A	f > j
>=	Greater than or equal to	6 >= 6	18 >= 25
		E >= B	D >= X
<	Less than	3 < 5	5 < 3
		A < C	Y < M
<=	Less than or equal to	10 <= 10	110 <= 55
		C <= C	g <= b
=	Equal to	35 = 35	35 = 50
		t = t	T = t
IN	Is a member of	3 IN [1..4]	5 IN [20..100]
		e IN [a..z]	F IN [A..E]

based on a TRUE or FALSE answer. Recall that this was used in describing how one of the three programming blocks, a branch block, worked. The relational operators are used in both open and closed branch blocks. The basic construction of the **IF-THEN** statement is

IF (Expression) **THEN** (Statement)

This means: Anytime (Expression) is TRUE, then (Statement) will be executed. The following examples illustrate applications of the relational operators using the **IF-THEN** statement.

Program 5–1

```
PROGRAM Relational_Operators;
 VAR
   User_Input        {Number input by program user}
 : INTEGER;

BEGIN
  WRITE('Give me a whole number from 1 to 9 => ');
  READLN(User_Input);

    IF User_Input = 5 THEN WRITELN ('The number is 5!');
    IF User_Input < 5 THEN WRITELN('Number less than 5.');
    IF User_Input > 5 THEN WRITELN('Number is larger than 5.');
    IF User_Input <> 5 THEN WRITELN('Number is not 5.');
    IF User_Input IN [1..5] THEN WRITELN('Number from 1 to 5.');
    IF User_Input IN [6..9] THEN WRITELN('Number from 6 to 9.');
  END.
```

Program 5–1 (p. 245) asks the program user to input a number from 1 to 9 and uses several decision-making statements that depend upon the relations between two conditions: the number input by the user and the number 5. Several examples of program execution are illustrated.

If the program user entered 3

```
Give me a whole number from 1 to 9 => 3
```

the resultant output would be

```
Number less than 5.
Number is not 5.
Number from 1 to 5.
```

If the program user entered 5

```
Give me a whole number from 1 to 9 => 5
```

the resultant output would be

```
The number is 5.
Number from 1 to 5.
```

As you saw in table 5–1, characters can also be compared. Recall that characters are stored in their ASCII code as numbers. Essentially, the computer makes relational comparisons of their ASCII number values; the letters of the alphabet have a number code in ascending order with 'A' at 65 and Z at 90. Lowercase letters also have ASCII number codes from 97 for "a" to 122 for "z". This was done to make it easy to create alphabetizing programs; the computer thus sorts letters in numerical order. A program for CHAR variables using relational operators is shown in program 5–2.

Program 5–2

```
PROGRAM Relational_Operators_Char;
VAR
 User_Input
:CHAR;

BEGIN
  WRITE('Give me a character => ');
  READLN(User_Input);
    IF User_Input IN ['a'..'m']
      THEN WRITELN('Lowercase letter from ''a'' to ''m''.');
    IF User_Input IN ['1'..'9']
      THEN WRITELN('Number character!');
    IF User_Input IN ['M'..'Z']
      THEN WRITELN('Uppercase letter from ''M'' to ''Z''.');
END.
```

Program 5–2 asks the user to input a character. Relational comparisons are then made according to the character that has been input. It's important to note that when a number symbol is input as a type CHAR, it is converted to its ASCII code and not treated as the value of the original number—only as its ASCII value (the ASCII value for 0 is 48, for 1 is 49 and for 9 is 57). The number would be now the character "0" to "9" and wouldn't be treated as a number anymore than "A" or "B" is treated as a number. Some examples of program execution are now illustrated.

If the program user entered a "k"

```
Give me a character => k
```

the resultant program output would be

```
Lowercase letter from 'a' to 'm'.
```

If the program user entered a 7

```
Give me a character => 7
```

the resultant program output would be

```
Number character!
```

Boolean Algebra

There was a mathematician named George Boole who published a book in 1854 that described a system of logic that had only two conditions: TRUE or FALSE. Of course George didn't know anything about computers, nor did he suspect how important his system of logic would become in computer programming, but this system of analysis is named in honor of him, and called **Boolean algebra**.

You will see that Boolean algebra consists of a few very simple rules and basic relationships. You will be able to use them to give your Pascal programs powerful decision-making capabilities.

Boolean Operators

A **Boolean operator** is used with a **Boolean expression**. This is an expression that has only a TRUE or FALSE condition and uses specific comparisons called Boolean operators.

The Boolean operators recognized by Turbo Pascal are
The **AND**:

```
(Expression₁) AND (Expression₂)
```

The entire statement will be TRUE if both Expression$_1$ and Expression$_2$ are TRUE. Otherwise, the entire statement is FALSE.

Table 5–2   Boolean Operators

Boolean operator	Example	Meaning
NOT	Y = NOT B	Y is the opposite of B; thus, if B is TRUE, Y is FALSE.
AND	Y = A AND B	Y is TRUE only if both A and B are TRUE. Otherwise Y is FALSE.
OR	Y = A OR B	Y is TRUE if A is TRUE or B is TRUE or if A and B are TRUE. Otherwise, Y is FALSE.
XOR	Y = A XOR B	Y is TRUE if A is TRUE or B is TRUE. Otherwise, Y is FALSE.

The **OR**:

```
(Expression₁) OR (Expression₂)
```

For the entire statement to be TRUE, either Expression$_1$ or Expression$_2$ or both Expression$_1$ and Expression$_2$ must be TRUE. The only time the entire statement is FALSE is when both Expression$_1$ and Expression$_2$ are FALSE.

The **NOT**:

```
NOT (Expression)
```

The statement will be the opposite of the expression. If the expression is FALSE, the entire statement will be TRUE. If the expression is TRUE, the entire statement will be FALSE.

The **XOR** (Exclusive **OR**):

```
(Expression₁) XOR (Expression₂)
```

The entire statement will be TRUE only when Expression$_1$ or Expression$_2$ are TRUE. The entire statement will be FALSE if both Expression$_1$ and Expression$_2$ are FALSE, or if both Expression$_1$ and Expression$_2$ are TRUE. These boolean operators are summarized in table 5–2.

Program 5–3 illustrates an application of the Boolean operators with the **IF–THEN** statement.

Program 5–3 uses all of the Boolean operators available in Turbo Pascal. Let us consider each one separately.

The **AND**:

```
IF (Response_1 = 'Y') AND (Response_2 = 'Y') THEN
WRITELN('Go home!');
```

Program 5–3

```
PROGRAM BOOLEAN_Operation;
  VAR
    Response_1,
    Response_2
  :CHAR;
```

Program 5–3 *continued*

```
BEGIN
  WRITE('Is it after 5 O''clock (type Y or N) => ');
  READLN(Response_1);
  WRITE('Is the system shut down (type Y or N) => ');
  READLN(Response_2);
   IF (Response_1 = 'Y') AND (Response_2 = 'Y') THEN
   WRITELN('Go home!');
   IF (Response_1 = 'Y') XOR (Response_2 = 'Y')
   THEN WRITELN('You must stay; only one is true!');
   IF (Response_1 = 'Y') OR (Response_2 = 'Y') THEN
   WRITELN('At least one is true, maybe both!');
   IF NOT((Response_1 = 'Y') OR (Response_2 = 'Y'))
   THEN WRITELN('Neither one is true; you must stay!');
END.
```

The expression

```
(Response_1 = 'Y') AND (Response_2 = 'Y')
```

will be TRUE only if both conditions are TRUE. Only then will the statement

```
WRITELN('Go home!');
```

be executed.
The **XOR**:

```
IF (Response_1 = 'Y') XOR (Response_2 = 'Y')
THEN WRITELN('You must stay; only one is true!');
```

The expression

```
(Response_1 = 'Y') XOR (Response_2 = 'Y')
```

will be TRUE only if either one or the other is TRUE, but not both. Only then will the statement

```
THEN WRITELN('You must stay; only one is true!');
```

be executed.
The **OR**:

```
IF (Response_1 = 'Y') OR (Response_2 = 'Y') THEN
WRITELN('At least one is true, maybe both!');
```

The only time the expression

```
(Response_1 = 'Y') OR (Response_2 = 'Y')
```

is FALSE is when both parts are FALSE. The expression will be TRUE if either one or both are TRUE. When TRUE the following statement is executed.

```
WRITELN('At least one is true, maybe both!');
```

The **NOT**:

Table 5–3   Summary of Program 5–3

Response__1	Response__2	Resulting output
FALSE	FALSE	Neither is true; you must stay.
FALSE	TRUE	You must stay; only one is true! At least one is true, maybe both.
TRUE	FALSE	You must stay; only one is true! At least one is true, maybe both.
TRUE	TRUE	Go home! At least one is true, maybe both.

```
IF NOT((Response__1 = 'Y') OR (Response__2 = 'Y'))
THEN WRITELN('Neither one is true; you must stay!');
```

The only time the expression

```
NOT((Response__1 = 'Y') OR (Response__2 = 'Y'))
```

is TRUE is when the inner expression

```
(Response__1 = 'Y') OR (Response__2 = 'Y')
```

is FALSE. This will happen only when both conditions are FALSE; then, the following statement will be executed.

```
THEN WRITELN('Neither one is true; you must stay!');
```

Table 5–3 summarizes the output of program 5–3 for all possible input conditions.

Combining Relational and Boolean Operators

You can combine relational and boolean operators; this is a very useful technique in programming. Program 5–4 illustrates an application.

Program 5–4

```
PROGRAM Both__Together;
   VAR
      User__Input
   :INTEGER;

BEGIN
   WRITE('Input a number between 8 and 10 => ');
   READLN(User__Input);
   {Check user input}
      IF (User__Input >= 8) AND (User__Input <= 10) THEN
      WRITELN('Input is OK!');
END.
```

Program 5–4 contains a Pascal statement that includes relational and boolean operators.

```
IF (User_Input >= 8) AND (User_Input <= 10) THEN
```

The entire statement will be TRUE only when both expressions are TRUE. Thus the statement

```
WRITELN('Input is OK!');
```

will be executed when the value of the user input is an integer from 8 to 10.

Conclusion

This section introduced you to two very important decision-making tools used in programming: relational and boolean operators. You will use these for all of your branching blocks. The next section presents some important details of the **IF-THEN** statement to bring you even more programming power. Test your understanding of this section by trying the following section review.

5–1 Section Review

1 State the difference between a relational operator and a Boolean operator.
2 Give the relational operators and explain their meanings.
3 Give the Boolean operators used in Turbo Pascal and explain their meanings.
4 Give an example of a Pascal statement that uses both the relational and Boolean operators.

5–2   The IF-THEN-ELSE

Introduction

You have already worked with the **IF-THEN** statement in Pascal. It is an example of an open branch. In this section, you will learn how to use a closed branch and design a program so that it will choose between two or more alternatives.

Basic Idea

Figure 5–1 (p. 252) reviews open and closed branches. Note the difference. In one, you may choose to do or not do one thing and then go on to something else (the open branch). In the other, you will do one thing or the other and then continue with something else (the closed branch).

Recall that in Pascal, the **IF-THEN** statement is

```
IF (Expression) THEN (Statement)
```

For the closed branch, the **IF-THEN-ELSE** statement is

```
IF (Expression) THEN (Statement₁) ELSE (Statement₂).
```

This is illustrated in program 5–5 on page 252.

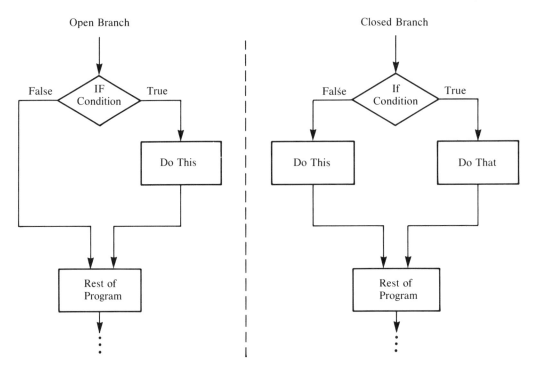

Figure 5-1 Review of Open and Closed Branch

Program 5-5

```
PROGRAM If_Then_Else_1;
   VAR
      User_Input
   :INTEGER;

   BEGIN
      WRITE('Enter a whole number from 1 to 9 => ');
      READLN(User_Input);

         IF (User_Input < 5 THEN WRITELN('Less than five.')
         ELSE WRITELN('It''s five or more.');

      WRITELN('That concludes this program.');

   END.
```

Observe in program 5–5 that there are no semicolons within the **IF–THEN–ELSE** statements, because the **THEN** and **ELSE** serve as statement terminators. Execution of program 5–5 yields

```
Enter a whole number from 1 to 9 => 3
Less than five.
```

```
That concludes this program.
```
or
```
Enter a whole number from 1 to 9 => ?
It's five or more.
That concludes this program.
```

Do you see why it's a closed branch? It's because the number you enter will cause either one statement or the other to be executed (never both) and the program will continue onward.

Sequencing IF-THEN-ELSE Statements

You can also do **IF-THEN-ELSE** statements, giving you several choices depending upon previous choices. The idea is shown in figure 5–2.

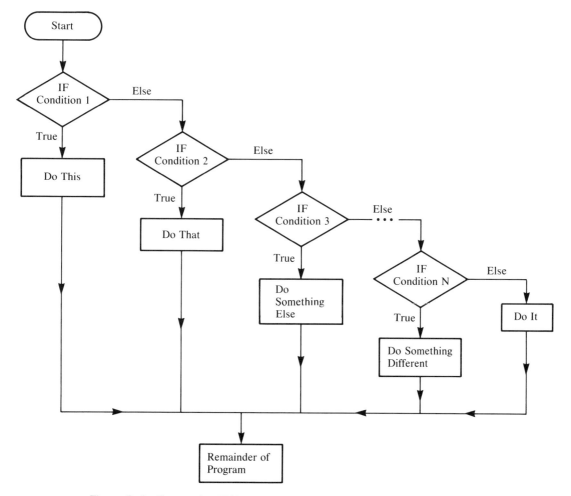

Figure 5–2   Sequencing IF-THEN-ELSE Statements

Program 5–6 gives an example of how you would do this using compound statements.

Program 5–6

```
PROGRAM If_Then_Else_2;
  VAR
   User_Input
 :CHAR;

   Line,
   Area
 :REAL;
  BEGIN
    WRITE('Select by letter:  C)ircle or S)quare => ');
    READLN(User_Input);

    IF User_Input = 'C' THEN
     BEGIN
       WRITE('Give me the value of the radius => ');
       READLN(Line);
       Area := 3.14159 * Line * Line;
       WRITELN('The area is = ',Area,' square units.');
     END

   ELSE

     IF User_Input = 'S' THEN
      BEGIN
       WRITE('Give me the value of one side => ');
       READLN(Line);
       Area := Line * Line;
       WRITELN('The area is = ',Area, 'square units.');
     END

   ELSE

    BEGIN
      WRITELN('I don''t understand what you want.');
      WRITELN('You must run the program again and press');
      WRITELN('either C or S.');
    END;

  END.
```

Note again in program 5–6 that there are no semicolons after the **END** statements within the **THEN–ELSE** sequence. However, within the compound statements, there are semicolons (for example, after each **WRITELN** statement).

A more elegant way of programming the same problem is to use procedures within the **IF–THEN–ELSE** statements. This is illustrated by program 5–7.

Program 5–7

```
PROGRAM If_Then_Else_Procedures;

  PROCEDURE Select_Circle;
 {------------------------------------------------------------}
  VAR
   Radius,
   Area
 :REAL;

    BEGIN
      WRITE('Give me the value of the radius => ');
      READLN(Radius);
      Area := 3.14159 * Radius * Radius;
      WRITELN('The area is = ',Area,' square units.');
 {------------------------------------------------------------}
  END;   {of procedure Select_Circle.}

  PROCEDURE Select_Square;
 {------------------------------------------------------------}
  VAR
   Side,
   Area
 :REAL;

    BEGIN
      WRITE('Give me the value of one side => ');
      READLN(Side);
      Area := Side * Side;
      WRITELN('The area is =',Area,' square units.');
 {------------------------------------------------------------}
  END;   {of procedure Select_Square.}

  PROCEDURE Try_Again;
 {------------------------------------------------------------}
    BEGIN
      WRITELN('I don''t understand what you want.');
      WRITELN('You must run the program again and press');
      WRITELN('either C or S.');
 {------------------------------------------------------------}
   END;   {of procedure Try_Again.}

  PROCEDURE Select_Figure;
 {------------------------------------------------------------}
    VAR
     User_Input
    :CHAR;

      BEGIN
        WRITE('Select by letter:  C)ircle or S)quare => ');
        READLN(User_Input);
```

Program 5–7 *continued*

```
        IF User_Input = 'C' THEN
            Select_Circle

        ELSE

        IF User_Input = 'S' THEN
            Select_Square

        ELSE

            Try_Again;
    {----------------------------------------------------------}
    END;   {of procedure Select_Figure.}
{MAIN PROGRAMMING BLOCK
    ==========================================================}
    BEGIN
    Select_Figure;
    END.
{==========================================================}
```

Note that the **IF–THEN–ELSE** statements of program 5–7 call procedures. Programs 5–6 and 5–7 both do the same thing, but it is considered better programming practice to use procedures with local variables. Therefore, program 5–7, even though it is longer, uses the preferred programming style for Pascal.

More Sequencing

Another example of sequential **IF–THEN–ELSE** statements is illustrated in program 5–8.

Program 5–8

```
PROCEDURE Multiple_Selection;
  VAR
    User_Selection
  :INTEGER;

  BEGIN

    WRITELN('Select which formula by number:');
    WRITELN('1] V = IR  2] I = V/R  3] R = V/I');
    WRITE('Your selection => ');
    READLN(User_Selection);
```

Program 5–8 *continued*

```
        IF User_Selection = 1  THEN Voltage_Formula

        ELSE

        IF User_Selection = 2 THEN Current_Formula

        ELSE

        IF User_Selection = 3 THEN Resistance_Formula

        ELSE

        Quit_Procedure;

  END;   {of procedure Multiple_Selection.}
```

The procedure in program 5–8 is an example of another type of **IF–THEN–ELSE** sequencing. In fact, this is such a common type of programming requirement that it has its own unique Pascal statement, which you will see in the next section.

Conclusion

This section presented the **IF–THEN–ELSE** and showed how it could be sequenced to perform a variety of decision-making tasks. In the next section, you will see another Pascal statement that allows for more power in creating branch blocks. Test your understanding of this section by trying the following section review.

5–2 Section Review

1   State the meaning of a branch block.
2   What is the difference between a closed and open branch?
3   Explain the difference between the **IF–THEN** and the **IF–THEN–ELSE** statements.
4   What is a sequenced branch? Give an example.
5   Explain the use of semicolons in **IF–THEN–ELSE** statements.

5–3   The CASE Statement

Introduction

This section is an extension of the last section; you will learn more about the decision-making power of a branch block. Recall that program 5–8 used multiple **IF–THEN–ELSE** statements to make a selection from a list of options (the program user could select one of three formulas). There are many other technology problems where it is necessary to make a selection from a list of choices; this is such a common type of programming requirement that Pascal has a special statement made just for it.

Basic Idea

The basic idea behind this new Pascal statement is to use the value of a variable to cause the program to do something. For example, if the value of the variable is 1, the program will print to the screen **THIS IS ONE**; if the value of the variable is 2, the program will print to the screen **THIS IS TWO**, and so on for five values. This could be done with **IF-THEN-ELSE** statements as shown in program 5–9.

Program 5–9

```
PROGRAM Enter__A__Number;
  VAR
      User__Input
  :INTEGER;

  BEGIN
      WRITE('Enter a whole number from 1 to 5 => ');
      READLN(User__Input);

        IF User__Input = 1 THEN
            WRITELN('THIS IS ONE.')
        ELSE
        IF User__Input = 2 THEN
            WRITELN('THIS IS TWO.')
        ELSE
        IF User__Input = 3 THEN
            WRITELN('THIS IS THREE.')
        ELSE
        IF User__Input = 4 THEN
            WRITELN('THIS IS FOUR.')
        ELSE
        IF User__Input = 5 THEN
            WRITELN('THIS IS FIVE.')
        ELSE
    END.
```

Program 5–9 would work, but it is awkward to program because of all the different possible choices. Luckily, Pascal has an easier way of doing the same thing.

Building a CASE

In Pascal there is a statement called the **CASE-OF** which is

CASE (Expression) **OF** (CaseClause)

where `Expression` = any Pascal expression that is of an ordinal type such as INTEGER, BOOLEAN, or CHAR. will not work with REAL numbers. This is sometimes called a **case selec-tor**.

`CaseClause` = a list of statements each preceded by one or more constants (called **case constants**) or the reserved word **ELSE**.

The structure of the **CASE-OF** is shown in figure 5–3.

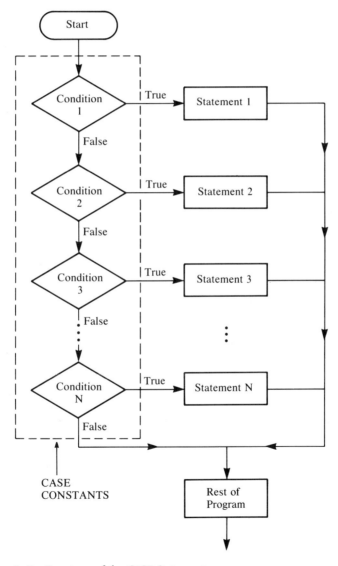

Figure 5–3   Structure of the CASE Statement

Program 5–10 shows how program 5–9 could be implemented using the **CASE-OF** structure rather than the **IF-THEN-ELSE**.

Program 5–10

```
PROGRAM Enter_A_Number;
  VAR
    User_Input
  :INTEGER;

    BEGIN
      WRITE('Enter a whole number from 1 to 5 => ');
      READLN(User_Input);

      CASE User_Input OF
        1 : WRITELN('THIS IS ONE.')
        2 : WRITELN('THIS IS TWO.')
        3 : WRITELN('THIS IS THREE.')
        4 : WRITELN('THIS IS FOUR.')
        5 : WRITELN('THIS IS FIVE.')
      END;   {of case.}

    END.
```

Notice that the CASE statement ends with its own **END;** statement. Also notice how much less programming is required using the **CASE-OF** statement rather than all those **IF-THEN-ELSE** statements. You can also use compound statements with the **CASE-OF** statement.

Compounding the CASE

Consider program 5–11.

Program 5–11

```
PROGRAM Ohms_Law_Solver;
  VAR
    User_Selection
  :INTEGER;

    Voltage,
    Current,
    Resistance
  :REAL;
```

Program 5–11   *continued*

```
      BEGIN
         WRITELN('Select which formula by number:');
         WRITELN('1] V = IR   2] I = V/R   3] R = V/I');
         WRITE('Your selection => ');
         READLN(User_Selection);

      CASE User_Selection OF
         1 : BEGIN
                WRITELN('That is the voltage formula.');
                WRITE('Give me the value of the current => ');
                READLN(Current);
                WRITE('Give me the value of the resistance => ');
                READLN(Resistance);
                Voltage := Current * Resistance;
                WRITELN('The voltage is ',Voltage,' Volts.');
             END;
         2 : BEGIN
                WRITELN('That is the current formula.');
                WRITE('Give me the value of the voltage => ');
                READLN(Voltage);
                WRITE('Give me the value of the resistance => ');
                READLN(Resistance);
                Current := Voltage/Resistance;
                WRITELN('The current is ',Current,' amps.');
             END;
         3 : BEGIN
                WRITELN('That is the resistance formula.');
                WRITE('Give me the value of the voltage => ');
                READLN(Voltage);
                WRITE('Give me the value of the current => ');
                READLN(Current);
                Resistance := Voltage/Current;
                WRITELN('The resistance is ',Resistance,' ohms.');
             END;
      END;   {of CASE.}
   END.
```

Note in program 5–11 that compound statements can be a part of the CASE–OF statement. A more elegant way of presenting program 5–11 is shown in program 5–12.

Program 5–12   (pp. 261–263)

```
   PROGRAM Ohms_Law_Solver;


   PROCEDURE Voltage_Formula;
 {-------------------------------------------------------------}
```

Program 5–12   *continued*

```
    VAR
        Voltage,
        Current,
        Resistance
    :REAL;

        BEGIN
            WRITELN('That is the voltage formula.');
            WRITE('Give me the value of the current => ');
            READLN(Current);
            WRITE('Give me the value of the resistance => ');
            READLN(Resistance);
            Voltage := Current * Resistance;
            WRITELN('The voltage is ',Voltage,' Volts.');
{----------------------------------------------------------------}
        END;   {of procedure Voltage_Formula}

  PROCEDURE Current_Formula;
{----------------------------------------------------------------}
    VAR
        Voltage,
        Current,
        Resistance
    :REAL;

        BEGIN
            WRITELN('That is the current formula.');
            WRITE('Give me the value of the voltage => ');
            READLN(Voltage);
            WRITE('Give me the value of the resistance => ');
            READLN(Resistance);
            Current := Voltage/Resistance;
            WRITELN('The current is ',Current,' amps.');
{----------------------------------------------------------------}
        END;   {of procedure Current_Formula}

  PROCEDURE Resistance_Formula;
{----------------------------------------------------------------}
    VAR
        Voltage,
        Current,
        Resistance
    :REAL;

        BEGIN
            WRITELN('That is the resistance formula.');
            WRITE('Give me the value of the voltage => ');
            READLN(Voltage);
            WRITE('Give me the value of the current => ');
            READLN(Current);
```

Program 5–12 *continued*

```
            Resistance := Voltage/Current;
            WRITELN('The resistance is ',Resistance,' ohms.');
{-----------------------------------------------------------------}
        END;  {of procedure Resistance_Formula}

  PROCEDURE Make_Selection;
{-----------------------------------------------------------------}
    VAR
        User_Selection;
    :INTEGER;

    BEGIN
        WRITELN('Select which formula by number:');
        WRITELN('1] V = IR   2] I = V/R   3] R = V/I');
        WRITE('Your selection => ');
        READLN(User_Selection);
          CASE User_Selection OF
              1 : Voltage_Formula;
              2 : Current_Formula;
              3 : Resistance_Formula;
          END;  {of CASE.}
{-----------------------------------------------------------------}
      END; {of procedure Make_Selection}

{Main programming sequence.}
{=================================================================}
  BEGIN
    Make_Selection;
  END.
{=================================================================}
```

Look at the **CASE-OF** statement in program 5–12. It is now using procedures instead of compound statements. Programming in this manner greatly simplifies the **CASE-OF** statement.

Other Options

The **CASE-OF** statement offers some helpful options, illustrated in program 5–13.

Program 5–13

```
PROGRAM Case_Options;
 VAR
   User_Input
 :INTEGER;
```

Program 5–13 *continued*

```
BEGIN
  WRITE('Give me a whole number from 1 to 100 => ');
  READLN(User_Input);

  CASE User_Input OF
    2,4,6,8 : WRITELN('That''s an even digit.');
    1,3,5,9 : WRITELN('That''s an odd digit.');
     10..50 : WRITELN('The number is from 10 to 50.');
  ELSE
    WRITELN('The number is larger than 50.');
  END;   {of case statement}
END.
```

Note in program 5–13 that you can use more than one case selector for the same statement as long as they are separated by commas. You can also use a range of values (10..50) as case selectors; the **CASE-OF** statement can also include an **ELSE**. These options give the **CASE-OF** statement a great deal of flexibility and provide a powerful option for developing programs with multiple decisions.

Conclusion

This section presented the **CASE-OF** statement and various methods of implementing it. You saw the advantages of using this rather than the **IF-THEN-ELSE** when there must be several selections available to the program user. Test your understanding of this section by trying the following section review.

5–3 Section Review

1 State the main advantage of the **CASE-OF** statement.
2 Can the **IF-THEN-ELSE** statement do the same thing as the **CASE-OF** statement? Explain.
3 Define the term case selector. What must be observed when using it?
4 Define the term case clause. What can be contained within the case clause?

5–4   Strings and Related Things

Introduction

In chapter two, you learned that a string was a group of characters, usually consisting of those from the keyboard. In this section you will be introduced to some important points concerning the use of strings in Turbo Pascal, to be used in the next section of this chapter.

A Character String

A string of characters is a series of one or more characters taken from the ASCII code, enclosed by apostrophes. For example, the following are character strings.

`'Uncle Pascal' 'Two = 2' '$%&()?' ''`

A character string of length zero is known as a **null string**. Thus `''` is a null string. The most experience you have had using strings up to this point has been with the **WRITE** and **WRITELN** procedures. Program 5–14 illustrates the use of character strings.

Program 5–14

```
PROGRAM String_Demo;
  VAR
    Value
  : INTEGER;

  BEGIN
    WRITELN('Uncle Pascal');
    WRITELN('');  {<== A null string.}
    Value := 2 + 2;
    WRITELN(Value);  {<== Not a string, a variable.}
    WRITELN('Value');  {<== A string, not a variable.}
    WRITELN('Value := 3 + 3');  {<== A string, no computation!}
    WRITELN( 3 + 5 );   {<== Not a string, but a computation!}
    WRITELN('Adding 2 + 3 gives ', 2 + 3);
  END.
```

When program 5–14 is executed the output will be

```
Uncle Pascal

4
Value
Value := 3 + 3
8
Adding 2 + 3 gives 5
```

Observe that the null string prints nothing to the screen; the blank line in the output is the null string. Now note that the numerical value of the variable Value was printed; the identifier Value was not treated as a string but as a variable, because it was not enclosed in apostrophes. When it was enclosed in apostrophes, however, (the next line), it was no longer treated as a variable, but as a string, and the string **Value** was printed on the screen. Note that **Value := 3 + 3** was treated as a string and not a computation because it was enclosed within apostrophes. However, in the next line, **3 + 5** was not enclosed in apostrophes and therefore was a computation (the answer 8 was printed to the screen). Note the last **WRITELN** statement and its combination of string and non-string information.

You can see the versatility Turbo Pascal offers in dealing with strings. By understanding these rules, you can create some very powerful and useful Pascal programs.

Defining a String

Standard Pascal does not support a separate string data type. Therefore, you must construct a special routine to handle strings. However, most practical Pascal systems such as Turbo Pascal do contain built-in procedures that allow you to work easily with strings. You declare a Turbo Pascal string as follows.

VAR
 String_Variable : **STRING**[N];

where String_Variable = any legal Pascal variable
                 STRING = a reserved TURBO word to indicate that the variable will be of type string
                       N = a whole number indicating the maximum number of characters that will be in the string; N can have a value from 1 to 255; if N is omitted the value 255 is assumed.

Program 5–15 demonstrates the use of a string variable in Turbo Pascal.

Program 5–15

```
PROGRAM String_Variable;
  VAR
    User_Input
 :STRING[3];

   BEGIN
     WRITE('Please enter your name ==> ');
     READLN(User_Input);
     WRITELN('Since I can only store three ')
     WRITELN('your name is ',User_Input);
   END.
```

Note that the type identifier **STRING**[3] has allowed room for only three characters. Thus, when the program is executed and you enter a string with more than three characters, only the first three characters will be used.

```
Please enter your name ==> Pascal Student
Since I can only store three
your name is Pas
```

To eliminate this problem, most programmers use **STRING**[79]. Even though the IBM screen is 80 characters wide, keeping it to 79 allows the cursor to stay on the same line and avoids an unexpected carriage return. Allocating more space to the string is shown in program 5–16.

Program 5–16

```
PROGRAM String_Variable;
  VAR
    User_Input
  :STRING[79];
  BEGIN
    WRITE('Please enter your name ==> ');
    READLN(User_Input);
    WRITELN('I can now store a whole line of characters.');
    WRITELN('I now know your whole name ',User_Input);
  END.
```

Now, if program 5–16 were executed, most names would be stored in their entirety.

Please enter your name ==> **A Big Pascal Student String**
I can now store whole a whole line of characters.
I now know your whole name **A Big Pascal Student String**

It isn't necessary to use **STRING**[255] for storing single-line strings as this tends to waste memory (you will see why). For most string variables, the type identifier **STRING**[79] is quite satisfactory.

String Constants

Turbo Pascal allows an easy way of assigning string constants, using the following method.

CONST
  String_Constant : **STRING**[N] = 'Actual_String';

where **String_Constant** = the string constant identifier to be used in the program

STRING = the reserved word that indicates that the constant is to be a string

N = the length of the string (may be up to 255)

'Actual String' = the string that will be represented by the string constant identifier

Program 5–17 (p. 268) illustrates this use of string constants.

Program 5–17

```
PROGRAM String__Constants;
  CONST
    Heading : STRING[12] = 'Turbo Pascal';
    Lesson  : STRING[15] = 'Lesson Number 3';
    Space   : STRING[1]  = ' ';
    Return  : STRING[79] = 'Press RETURN/ENTER to continue...';

  BEGIN
    WRITELN('This is a lesson about ',Heading);
    WRITELN(Space);
    WRITELN(Lesson);
    WRITELN;
    WRITELN;
    WRITELN(Return);
    READLN;
    END.
```

When program 5–17 is executed, the output will be

```
This is a lesson about Turbo Pascal

Lesson Number 3


Press RETURN/ENTER to continue. . .
```

Note that the **WRITELN** statement containing the identifier Space causes a carriage return in exactly the same manner as does a **WRITELN;** statement by itself. Also note from the constant declarations that the size of **STRING** can be longer than the actual string; it should not be smaller than the actual string because the whole string constant would not be displayed.

Stringing Strings Together

In Turbo Pascal you can join strings together; this is called **concatenation**. The Turbo function is

CONCAT(S$_1$, S$_2$, . . . S$_N$);

        where **CONCAT** = reserved Turbo word to indicate that string concatenation
                is to take place
      S$_1$, S$_2$, . . . S$_N$ = individual strings to be concatenated

The process of concatenation is illustrated in program 5–18.

Program 5–18

```
PROGRAM Stringing_Strings;
  VAR
    User_Input,
    New_String
  :STRING[79];

  BEGIN
    WRITE('To demonstrate concatenation, give me a word ==> ');
    READLN(User_Input);
    WRITELN;
    New_String := CONCAT(User_Input,'—',User_Input);
    WRITELN('How is this: ',New_String);
  END.
```

Observe that when program 5–18 is executed, the output is

```
To demonstrate concatenation, give me a word ==> Tom
How is this Tom—Tom
```

Note that the **CONCAT** function can have string identifiers (variables or constants) along with an actual string (the **'—'**) as part of the concatenation process.

Conclusion

This section introduced some of the fundamental concepts used in working with strings. In the next section, you will discover many useful built-in Turbo functions and procedures for working with strings. For now, test your understanding of this section by trying the following section review.

5–4 Section Review

1 Define the term string.
2 What is meant by a character string? Give an example.
3 Explain what is meant by a null string.
4 How do you declare a string variable in Turbo Pascal?
5 What is string concatenation? How is this done in Turbo Pascal?

5–5   String Magic

Introduction

In this section you will see the applications of many useful built-in Turbo Pascal string procedures and functions. Many artificial intelligence programs use the techniques of string manipulation presented here. Your technical programs will be more user-friendly if you can communicate to the program user with words as well as numbers.

COPY Function

The first string function to be investigated in this section is the **COPY** function. Essentially, this function returns a selected part of a string. It is

COPY(String__Name, From, To**);**

where **COPY** = the Turbo Pascal reserved word to call the copy function
String__Name = the string from which the selected part is to be taken
From = an integer that tells Pascal how many character places to count from the beginning to where the selected part of the string is to start
To = an integer that tells how many characters of the string are to be selected

An application of the **COPY** function is illustrated in program 5–19.

Program 5–19

```
PROGRAM Copy__String;
   CONST
     Metric__Units : STRING[79] = 'micro milli kilo mega ';
   VAR
     New__String
   :STRING[79];
   BEGIN
     New__String := COPY(Metric__Units, 1, 5);
     WRITELN(New__String);
     New__String := COPY(Metric__Units, 7, 5);
     WRITELN(New__String);
     New__String := COPY(Metric__Units, 11, 9);
     WRITELN(New__String);
     New__String := COPY('Metric__Units', 1, 8);
     WRITELN(New__String);
   END.
```

When program 5–19 is executed, the output will be

```
micro
milli
i kilo me
micro mi
```

Observe that spaces are counted as well as characters. This function is useful for pulling part of a string of characters for further processing.

DELETE Procedure

The **DELETE** procedure is used to delete a part of a string. Note that this is a procedure and not a function. The main difference from a programming standpoint is that a function must be equated to a variable; a procedure, however, is simply called. It can return a value to the calling procedure through a variable parameter. Thus, the form of the **DELETE** procedure is

DELETE (String__Name, From, To);

where **DELETE** = the Turbo Pascal reserved word that calls the procedure
String__Name = the string from which characters are to be deleted
From = an integer whose value determines the number of characters from the beginning of the string where the deletion starts
To = an integer whose value determines the number of characters to actually be deleted

Program 5–20 illustrates an application of the **DELETE** procedure.

Program 5–20

```
PROGRAM Delete_String;   {Deletes a substring from a string.}
   CONST
     Original_String1 : STRING[79] = 'Pascal for Technology';
     Original_String2 : STRING[79] = 'Applied Pascal';
   VAR
     New_String
   : STRING[79];
   BEGIN
     DELETE(Original_String1, 4, 12);
     WRITELN(Original_String1);
     DELETE(Original_String2, 5, 7);
     WRITELN(Original_String2);
   END.
```

When executed, program 5–20 yields

```
Pasnology
Applical
```

and again note that spaces are counted as well as characters.

INSERT Procedure

The **INSERT** procedure is the opposite of the **DELETE** procedure. This procedure inserts a given string into another string starting at a specified location. This built-in Turbo procedure has the following form.

```
INSERT(Inserted_String, Target_String, From);
```

where **INSERT** = the Turbo Pascal reserved word that calls the procedure for inserting one string into another

`Inserted_String` = the string to be inserted

`Target_String` = the string into which the insertion is to take place

`From` = an integer that indicates the number of the character
(or space) in Target_String from which the insertion
is to take place

An application of the **INSERT** procedure is illustrated in program 5–21.

Program 5–21

```
PROGRAM Insert_String;
 CONST
   First_String  : STRING[79] = 'Applied for Technology';
   Second_String : STRING[79] = 'Uncle Pascal ';
 BEGIN
   INSERT(Second_String, First_String, 9);
   WRITELN(Second_String);
   WRITELN(First_String);
END.
```

When program 5–21 is executed, the output will be

```
Uncle Pascal
Applied Uncle Pascal for Technology
```

Note that `Second_String` had a space as its last character; this was
necessary to produce a space between the words `Pascal` and `for`.

LENGTH Function

The **LENGTH** function returns an integer whose value is the number of characters in
the given string. The form of this built-in Turbo function is

```
LENGTH(String_Name);
```

where **LENGTH** = the built-in Turbo Pascal reserved word to get the number of
characters of a given string

`String_Name` = the string whose characters will be counted

Program 5–22 demonstrates an application of the **LENGTH** function.

Program 5–22

```
PROGRAM Length_of_String;
 VAR
   User_Input
 :STRING[79];
   String_Length
 :INTEGER;
 BEGIN
 WRITELN('Give me a string of not more than 5 characters => ');
   READLN(User_Input);
   String_Length := LENGTH(User_Input);
   IF String_Length > 5 THEN WRITELN('That''s too long.')
   ELSE
   WRITELN('Thank you.');
   WRITELN('The string length is ',String_Length,' characters.');
 END.
```

An example of executing program 5–22 is

```
Give me a string of not more than 5 characters => 12345
Thank you.
The string length is 5 characters.
```

Another example of program execution is

```
Give me a string of not more that 5 characters => How is this?
That's too long.
The string length is 12 characters.
```

Observe from the indicated output that spaces and "?" are counted as part of the character string.

POS Function

The **POS** function is another built-in Turbo Pascal function. It returns the value of the location where one string starts within another string. This function has the following form.

```
POS(String_Sample, Target_String);
```

        where **POS** = the built-in Turbo Pascal reserved word for finding the position of the occurrence of one string inside another

`String_Sample` = the sample string which will be searched for

`Target_String` = the string in which the search is to be done

Program 5–23 (p. 274) is an example of an application of the **POS** function.

Program 5–23

```
PROGRAM String_Position;
  CONST
    Select_List : STRING[79] = 'Tom Dick Harry';
  VAR
    User_Name
 :STRING[79];
    Position
 :INTEGER;
  BEGIN
    WRITE('Enter your name to gain access => ');
    READLN(User_Name);
 IF POS(User_Name, Select_List) <> 0 THEN WRITELN('Welcome to Pascal!')
    ELSE
    WRITELN('Sorry, access is denied.');
    Position := POS(User_Name, Select_List);
    WRITELN('Your name starts at position ',Position);
  END.
```

When program 5–23 is executed, the resulting output is

```
Enter your name to gain access => Tom
Welcome to Pascal!
Your name starts at position 1
```

However, any series of strings will return a value if found within the target string. Consider the following program execution.

```
Enter your name to gain access => ick
Welcome to Pascal!
Your name starts at position 6
```

STR Procedure

The **STR** procedure is a built-in Turbo Pascal procedure that will convert an integer into a string. It has the form

STR(Number, String_Name);

where **STR** = the Turbo Pascal reserved word for converting a whole number into a string

Number = the whole number to be converted into a string

String_Name = the string that will contain the character string from the whole number

An application of the **STR** procedure is demonstrated in program 5–24.

Program 5–24

```
PROGRAM String_Procedure;
  CONST
     Combination_Codes : STRING[79] = '1234 0987 4756';
  VAR
    User_String_Code
  :STRING[79];
    Correct_Code,
    User_Input
  :INTEGER;
  BEGIN
    WRITE('Enter your combination => ');
    READLN(User_Input);
    Correct_Code := User_Input + 2 ;
    STR(Correct_Code, User_String_Code);
    IF POS(User_String_Code, Combination_Codes) <> 0 THEN
    WRITELN('Your combination is correct!')
    ELSE
    WRITELN('Sorry, that is not a correct combination.');
  END.
```

A sample execution of program 5–24 is

```
Enter your combination => 1232
Your combination is correct.
```

Note that 2 is added to the value entered by the program user. The resulting answer is then converted to a string variable so that the **POS** function can test if the resulting string is in the target string called Combination_Codes. Note that there are many different numbers which can be entered by the program user that will produce a "Your combination is correct" result.

```
Enter your combination => 2
Your combination is correct.
```

This happened because the string 4 is in the target string.

VAL Procedure

The last of the Turbo built-in procedures is the **VAL** procedure. This does just the opposite of what the **STR** function did. This procedure converts a string into a whole number. This can only be done if the string to be converted contains numerical characters with no decimal points, commas, or spaces. This procedure has the following form.

VAL(String__Number, Number, Code);

where **VAL** = the built-in Turbo Pascal reserved word for the **VAL** procedure

String__Number = a string that contains number characters which is to be converted to a whole number

Number = a REAL or INTEGER type variable that will contain the converted whole number

Code = an INTEGER that will have a value of zero if the conversion is successful

An application is demonstrated in program 5–25.

Program 5–25

```
PROGRAM Value_of_String;
  VAR
    Code,
    Number_Value,
    Final_Value
  :INTEGER;
    User_Input
  :STRING[79];
  BEGIN
      WRITE('Input characters or numbers => ');
      READLN(User_Input);
      VAL(User_Input, Number_Value, Code);
      IF Code <> 0 THEN
      WRITELN('They are not numbers I can convert.')
    ELSE
      BEGIN
        Final_Value := Number_Value * 2;
        WRITELN('Two times the value of your input is ',Final_Value);
      END;
  END.
```

Note in program 5–25 that the Code variable is used as a test to tell the program user if the input string was successfully converted to a whole number. An example of execution of the above program is

Input characters or numbers => **A character!**
They are not numbers I can convert.

Inputting a string of number characters produces the following result.

Input characters or numbers => **12**
Two times the value of your input is 24.

Conclusion

This section presented the built-in Turbo Pascal string functions and procedures. You saw that there are many applications for strings in Turbo Pascal. The next section will present the important concept of the Pascal `UNIT`, which you have already seen used. Test your understanding of this section by trying the following section review.

5–5 Section Review

1 Which built-in Turbo Pascal function can be used to extract the middle name of a person and store it in a separate variable?
2 Is a function or procedure used to determine the length of a string? What is its name?
3 State the purpose of the `POS` function.
4 Explain the difference between the `VAL` procedure and the `STR` procedure.

5–6   Program Debugging and Implementation— Turbo Pascal Units

Introduction

The concept of a program library has already been mentioned in this text. Here, procedures, functions, or other frequently-used Pascal items such as constants and variables can be stored. Any future programs that you write can make reference to this library and use any procedure, function, constant, or variable inside the library. This gives a great saving in programming time; as such, the concept of a library is used extensively in many Pascal programs.

Basic Idea

The basic idea of a Pascal library is shown in figure 5–4 (p. 278); you can see that if you know what is inside the library, you can use it in any program you wish to write.

A Pascal library, then, is nothing more than procedures, functions and declarations that may be used by other programs; the main difference between a library and a regular Pascal program is that the library type can have parts of it used by other Pascal programs.

Suppose you must write several different Pascal programs that use the same formulas over and over again. They could be programs for drafting technology that work with the mensuration formulas for a circle (its area and circumference). Rather than writing these formulas over and over, you could create one program containing these formulas, save it on the disk as a library program and then simply call formulas from other programs you develop. This process is illustrated in figure 5–5 (p. 279).

You may not want to do this for just two formulas, but the examples have been kept simple so that new coding concepts can be emphasized. Once you know how, you can develop your own library containing various formulas you would use in your

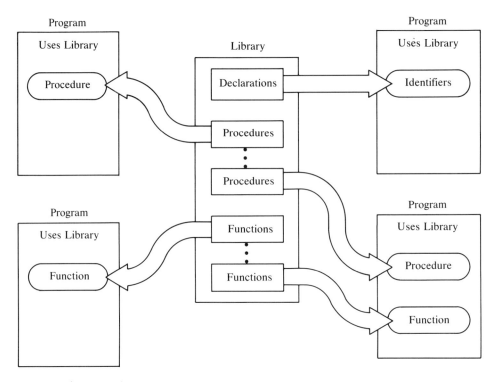

Figure 5–4 Basic Idea of a Pascal Library

own technology area. Turbo Pascal has a very easy way of creating a library; it is called a **UNIT**, and its form differs slightly from a standard Pascal program.

The Unit

Figure 5–6 shows the basic structure of a Turbo Pascal unit. Note the differences between the unit and a standard Pascal program.

Functions in a Unit

A sample unit will be developed that will contain two functions. One function will compute the area of a circle; the other will compute its circumference. You will then see how to use this unit from another Pascal program. Program 5–26 (p. 280) shows the unit as it is entered into the Turbo editor.

There are two important points about using program 5–26 as a **UNIT**. First, the name you use to save the unit must be the same as the identifier you gave the unit (**CircleF**, in this example). This means that the unit identifier must meet the requirements of DOS for naming files.

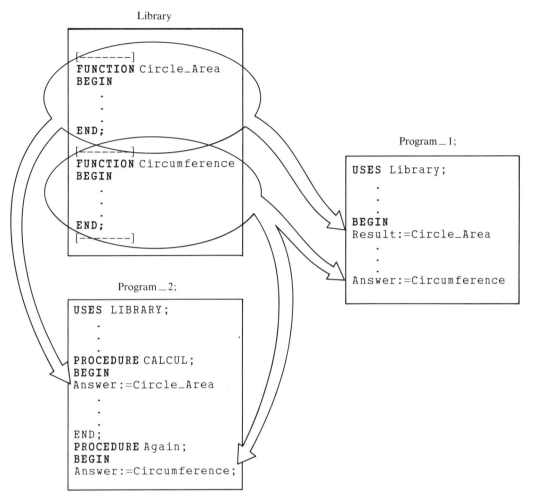

Figure 5-5   Using Formulas for a Circle

```
UNIT (IDENTIFIER);
INTERFACE
USES (LIST OF OTHER UNITS) ← Optional
        (Procedures, functions, and
        other declarations)

IMPLEMENTATION
        (Body of Procedures
        and Functions)

END.
```

Figure 5-6   Basic Structure of a Turbo Pascal Unit

Program 5–26

```
UNIT CircleF;

INTERFACE

    FUNCTION Circle_Area(Radius : REAL):REAL;
    FUNCTION Circumference(Radius : REAL):REAL;

IMPLEMENTATION

    FUNCTION Circle_Area;
        BEGIN
            Circle_Area := PI * Radius * Radius;
        END;

    FUNCTION Circumference;
        BEGIN
            Circumference := 2 * PI * Radius;
        END;

END.
```

Second, the compiler option must be changed to save the compiled program to the disk. When you write the Pascal program that will use this unit, the unit must be on the disk in a compiled form.

You can change the compiler option by using ALT-C, D. The destination will be changed from memory to disk. You can then compile the unit.

When the compiled unit is saved to the disk it will have the extension .TPU meaning Turbo Pascal Unit. The compiled file name of program 5–26 will be CIRCLEF.TPU. You can check the directory to see if this is so after the program has been successfully compiled (ALT-F, D). Note that the compiled form of the unit must be on the same disk as the program that will use it.

Next, you should create a Pascal program that will use the compiled unit (CIRCLEF.TPU). A sample program using this unit is shown in program 5–27.

Program 5–27

```
PROGRAM About_Circles;

  USES CircleF;    {The UNIT file name.}

  VAR
    Radius,
    Answer
  :REAL;
```

Program 5–27   *continued*

```
BEGIN
  WRITELN
  WRITE('Give me the circle radius => ');
  READLN(Radius);

  Answer := Circle_Area(Radius); {This is in the UNIT}
  WRITELN('The circle area is ',Answer,' square units.');

  Answer := Circumference(Radius); {Also in the UNIT}
  WRITELN('The circle circumference is ',Answer,' units.');
END.
```

Be sure to change the compile option to save in memory rather than to disk unless you want the compiled version of the Pascal program saved as well. In either case, when you run the above program it will now use the `CIRCLEF.TPU UNIT` you created to compute the area and circumference of the circle.

Other Unit Capabilities

You can also use a Turbo Pascal unit to store constants that you want to use in programs you develop. For example, once you have all the metric conversions stored in a unit (called `METRIC.TPU`, for example), you can use them in any future Turbo Pascal program.

The metric conversion unit is illustrated in program 5–28.

Program 5–28

```
UNIT Metric;

  INTERFACE

    CONST
      InchToCentimeter = 2.54;
      FeetToMeter = 0.3048;
      GallonToLiter = 3.785;

  IMPLEMENTATION

    BEGIN

    END.
```

Note that the unit in program 5–28 does not contain any functions or procedures—it only contains constants. This unit is also compiled to the disk as `METRIC.TPU`. Program 5–29 shows how the unit could be used.

Program 5–29

```
PROGRAM Convert__To__Metric;

  USES
   Metric;

  VAR
   Inches,
   Centimeters
  :REAL;

   BEGIN
    WRITELN;
    WRITELN('Give me a measurement in inches,');
    WRITE('and I''ll convert it to centimeters => ');
    READLN(Inches);

    {Do conversion:}
     Centimeters := InchToCentimeter * Inches;

  WRITELN('There are ',Centimeters,' centimeters in ',Inches,' inches.');

  END.
```

Procedures in the Unit

Program 5–30 is an example of a unit that contains a procedure. The procedure takes an input containing a number which may or may not have a metric prefix and converts it to the actual value. As an example, 3.2m would be converted to 0.0032 (3.2 milli = 3.2×10^{-3} = 0.0032). This unit can then be used by any program in which the program user is expected to input this kind of notation. Observe the use of the built-in Turbo Pascal string functions and procedures presented in the last section.

Program 5–30

```
UNIT Convert;

INTERFACE

 TYPE
  String79 = STRING[79];

  PROCEDURE Change (InputValue : String79; VAR OutValue : REAL;
                    VAR Success : BOOLEAN);

IMPLEMENTATION
```

Program 5–30 *continued*

```
PROCEDURE Change;

  VAR

  TempValue,
  Multiplier
:REAL;

  Converted,
  Multiplier
:INTEGER;

  Metric
:STRING[1];

  BEGIN
      {Get the last character from the input string.}
      String_Length := LENGTH(InputValue);
      Metric := COPY(InputValue, String_Length, String_Length);
      IF (Metric = 'u') OR
         (Metric = 'm') OR
         (Metric = 'k') OR
         (Metric = 'M')
      THEN
      {Remove last character from input string.}
       DELETE(InputValue, String_Length, String_Length);
      {Convert number string to a real number.}
       VAL(InputValue, TempValue, Converted);
     IF Converted = 0 THEN {Conversion was successful.}
      BEGIN
        Success := TRUE;
        {Convert metric character to multiplier.}
        IF Metric = 'u' THEN
           Multiplier := 0.000001
        ELSE
           IF Metric = 'm' THEN
           Multiplier := 0.001
        ELSE
           IF Metric = 'k' THEN
           Multiplier := 1000.0
        ELSE
           IF Metric = 'M' THEN
           Multiplier := 1000000.0
        ELSE
            Multiplier := 1;
          {Create number}
          OutValue := TempValue * Multiplier;
       END
    ELSE
     BEGIN {Conversion was not successful.}
      Success := FALSE;
      WRITELN('Your input cannot be processed.');
     END;
  END;  {Of procedure Change}
END.  {of implementation}
```

Program 5–31 uses the unit from program 5–30.

Program 5–31

```
PROGRAM Convert_Them;

  USES
   Convert;

  VAR
   Input_String
  :STRING[79];

   Out_Value
  :REAL;

   Successful
  :BOOLEAN;

  BEGIN
    WRITELN;
    WRITE('Give me a value with a metric prefix (u, m, k or M): ');
    READLN(Input_String);

    {Call the procedure in the unit.}
    Change(Input_String, Out_Value, Successful);
    IF Successful = TRUE THEN
      WRITELN('The converted value is ',Out_Value);
  END.
```

Conclusion

In this section you saw how to create your own Pascal library for your field of technical specialty using the Turbo Pascal feature of creating units. A powerful, useful, and time-saving group of special functions, procedures and/or constants can be conveniently stored and then used by any future programs you may develop. Check your understanding of this section by trying the following section review.

5–6 Section Review

1 Explain the concept of a library as used in Pascal.
2 State how a library can be developed using Turbo Pascal.
3 What is the main difference between a Turbo Pascal unit and a Pascal program?
4 Explain what is needed in a program that uses a unit.

Summary

1 Relational operators are symbols that indicate a relation between two quantities.
2 Relational operators can be used as the expression part of an **IF-THEN** statement.

3   The result of a relational operator is either TRUE or FALSE.

4   The **IN** [**. . .**] will have either a TRUE or FALSE result and can therefore be used with an **IF–THEN** statement.

5   Boolean algebra uses variables with only two conditions: TRUE or FALSE.

6   The **AND** operator in Boolean algebra means that its result will be TRUE only if the variables are both TRUE; otherwise, the result is FALSE.

7   The **OR** operator in Boolean algebra means that its result will be FALSE only if the variables are both FALSE; otherwise, the result is TRUE.

8   The **NOT** operator changes the value of a statement or variable (changes TRUE to FALSE or FALSE to TRUE).

9   The **XOR** (exclusive OR) operator in Boolean algebra means that its result will be TRUE only if one variable or the other is TRUE. If both variables are TRUE, or both variables FALSE, its outcome will be FALSE.

10   Relational and Boolean operators may be combined in a Pascal statement.

11   A Pascal expression may contain more than one Boolean operator.

12   The **IF–THEN–ELSE** statement in Pascal is an example of a closed branch.

13   **IF–THEN–ELSE** statements may be sequenced so that the program is capable of performing multiple branches.

14   The **CASE** statement in Pascal is an easy way of creating a program with multiple choices.

15   The **CASE** statement must be of an ordinal type such as an INTEGER, BOOLEAN or CHAR. It cannot be a REAL or a string.

16   A **CASE** statement can be a compound Pascal statement with a **BEGIN** and **END** or it can be a procedure call.

17   The **CASE** expression may consist of multiple terms such as 1,2,3 or 1..5.

18   A character string consists of one or more characters taken from the ASCII code and enclosed by apostrophes.

19   A string variable is declared in Turbo Pascal with **STRING**[N], where N is an INTEGER from 1 to 255.

20   Turbo Pascal has several built-in functions and procedures that allow different types of string manipulation.

21   Turbo Pascal uses the unit as the method of storing important programming information in a library.

Interactive Exercises

Directions

These exercises require that you have access to a computer and software that supports Pascal, specifically, the Turbo Pascal Development System, version 4.0, from Borland International. They are provided here to give you valuable experience and immediate feedback on what the concepts and commands introduced in the chapter will do—they are also fun.

Exercises

1   See if you can predict what will happen when program 5–32 (p. 286) is executed. Then try the program to test your prediction.

Program 5–32

```
PROGRAM One;
 CONST
  True_One = TRUE;
  False_One = FALSE;

  VAR
   Value_1,
   Value_2
  :BOOLEAN;

  BEGIN

    IF True_One THEN
     WRITELN('First statement.');
    IF False_One THEN
     WRITELN('Second statement.');
    IF True_One AND True_One
     THEN WRITELN('Third Statement.');
    IF True_One AND False_One
     THEN WRITELN('Fourth Statement.');
    IF True_One OR False_One
     THEN WRITELN('Fifth Statement.');
    IF NOT False_One
     THEN WRITELN('Sixth Statement.');
    IF NOT True_One
      THEN WRITELN('Seventh Statement.');
    IF NOT(False_One) OR NOT(True_One)
      THEN WRITELN('Eighth Statement.');

  END.
```

2 For program 5–33, what number will cause only the first sentence to be displayed? Only the second sentence? Only the third? All three sentences to be displayed?

Program 5–33

```
PROGRAM Two;
 VAR
  Your_Number
 :INTEGER;

  BEGIN
    WRITE('Input a whole number => ');
    READLN(Your_Number);
```

Program 5–33   *continued*

```
     IF (Your_Number <= 7) AND (Your_Number > 0)
      THEN WRITELN('First sentence.');

     IF (Your_Number = 3) OR (Your_Number = 8)
      THEN WRITELN('Second sentence.');

     IF (Your_Number >= 6) OR (Your_Number = 3)
      THEN WRITELN('Third sentence.');
 END.
```

3   Will program 5–34 compile? If not, why not? What are the minimum changes you need to make for it to compile?

Program 5–34

```
PROGRAM Three;
   VAR
    Input
   :STRING[79];

    BEGIN
      WRITE('Give me a letter from a to d => ');
      READLN(Input);

      IF Input IN ['a'..'d'] THEN
        WRITELN('Thanks for following directions!')
      ELSE
        WRITELN('Please follow directions.');

    END.
```

4   Will program 5–35 compile? If not, why not? What are the minimum changes you need to make so it will compile?

Program 5–35

```
PROGRAM Four;
   VAR
    Input
   :REAL;
```

Program 5–35   *continued*

```
BEGIN
  WRITE('Give me a number from 1 to 5 => ');
  READLN(Input);

  IF Input IN [1..5] THEN
    WRITELN('Thanks for following directions!')
  ELSE
    WRITELN('Please follow directions.');

END.
```

5 Will program 5–36 compile? If not, why not? What are the minimum changes you need to make for it to compile?

Program 5–36

```
PROGRAM Five;
  VAR
  User_Input
  :STRING[79];

  BEGIN
    WRITE('Give me a letter from a to d => ');
    READLN(User_Input);

      CASE User_Input OF
        'a' : WRITELN('First letter of the alphabet.');
        'b' : WRITELN('That was the second letter.');
        'c' : WRITELN('You selected the third letter.');
        'd' : WRITELN('The fourth letter was your selection.')
      ELSE
            WRITELN('Please follow directions.');

      END;

  END.
```

6 Will program 5–37 compile? If not, why not? Under what conditions will the program not function as expected?

Program 5-37

```
PROGRAM Six;
   VAR
    Your_Name
   :STRING[3];

   BEGIN
     WRITE('Give me your first name => ');
     READLN(Your_Name);
     WRITELN('Thank you ',Your_Name);

   END.
```

7 Will program 5–38 compile? If not, why not? Predict what the program will do, then test your prediction. Explain how the program does what it does.

Program 5-38

```
PROGRAM Seven;
   VAR
    First_Name,
    Last_Name
   :STRING[79];

   BEGIN
     WRITE('Give me your first name => ');
     READLN(First_Name);
     WRITE('Give me your last name => ');
     READLN(Last_Name);

     IF First_Name < Last_Name THEN
       WRITELN('Alphabetically, your first name comes ',
               'before your second name.')
     ELSE
       WRITELN('Alphabetically, your second name comes ',
               'before your first name.');

   END.
```

Pascal Commands

Table 5–4   Relational Operators

Symbol	Meaning	TRUE Examples	FALSE Examples
>	Greater than.	5 > 3	3 > 5
		C > A	f > j
>=	Greater than or equal to.	6 >= 6	18 >= 25
		E >= B	D >= X
<	Less than.	3 < 5	5 < 3
		A < C	Y < M
<=	Less than or equal to.	10 <= 10	110 <= 55
		C <= C	g <= b
=	Equal to.	35 = 35	35 = 50
		t = t	T = t
IN	Is a member of	3 IN [1..4]	5 IN [20..100]
		e IN [a..z]	F IN [A..E]

IF (Expression) **THEN** (Statement)

where anytime (Expression) is TRUE, (Statement) will be executed.

IF (Expression) **THEN** (Statement$_1$) **ELSE** (Statement$_2$).

where anytime (Expression) is TRUE, (Statement$_1$) will be executed; otherwise, (Statement$_2$) is executed.

CASE (Expression) **OF** (CaseClause)

where **Expression =** any Pascal expression that is of an ordinal type such as INTEGER, BOOLEAN, or CHAR. Will not work with REAL numbers. This is sometimes called a case selector.

Table 5–5   Boolean Operators

Boolean operator	Example	Meaning
NOT	Y = NOT B	Y is the opposite of B, thus if B is TRUE, Y is FALSE.
AND	Y = A AND B	Y is TRUE only if both A and B are TRUE; otherwise Y is FALSE.
OR	Y = A OR B	Y is TRUE if A is TRUE or B is TRUE or if A AND B are TRUE. Otherwise, Y is FALSE.
XOR	Y = A XOR B	Y is TRUE if A is TRUE or B is TRUE. Otherwise, Y is FALSE.

`CaseClause` = a list of statements, each preceded by one or more constants (called case constants) or the reserved word **ELSE**

String Variables

VAR
 `String_Variable : `**STRING**`[N];`

where `String_Variable` = any legal Pascal variable
STRING = a reserved Turbo word to indicate that the variable will be of type string
N = a whole number indicating the maximum number of characters that will be in the string; N can have a value from 1 to 255

String Constants

CONST
 `String_Constant : `**STRING**`[N] = 'Actual_String';`

where `String_Constant` = the string constant identifier to be used in the program
STRING = the reserved word to indicate that the constant is to be a string
N = the length of the string (up to 255)
`'Actual_String'` = the string that will be represented by the string

Concatenation

CONCAT $(S_1, S_2, \ldots S_N);$

where **CONCAT** = reserved Turbo word to indicate that string concatenation is to take place
$S_1, S_2, \ldots S_N$ = individual strings to be concatenated

COPY Function

COPY`(String_Name, From, To);`

where **COPY** = the Turbo Pascal reserved word to call the copy function
`String_Name` = the string from which the selected part is to be taken
`From` = an INTEGER that tells Pascal how many character places to count from the beginning to where the selected part of the string is to start
`To` = an INTEGER that tells how many characters of the string are to be selected

DELETE Procedure

DELETE `(String_Name, From, To);`

where **DELETE** = the Turbo Pascal reserved word that calls the procedure

`String_Name` = the string from which characters are to be deleted

From = an INTEGER whose value determines the number of characters in the string from where the deletion starts

To = an INTEGER whose value determines the number of characters to actually be deleted

INSERT Procedure

`INSERT(Inserted_String, Target_String, From);`

where **INSERT** = the Turbo Pascal reserved word that calls the procedure for inserting one string into another

`Inserted_String` = the string to be inserted

`Target_String` = the string into which the insertion is to take place

From = an INTEGER that indicates the number of the character in Target_String from which the insertion is to take place

LENGTH Function

`LENGTH(String_Name);`

where **LENGTH** = the built-in Turbo Pascal reserved word to get the number of characters of a given string

`String_Name` = the string whose characters will be counted

POS Function

`POS(String_Sample, Target_String);`

where **POS** = the built-in Turbo Pascal reserved word for finding the position of the occurrence of one string inside another

`String_Sample` = the sample string which will be searched for

`Target_String` = the string in which the search is to be done

STR Procedure

`STR(Number, String_Name);`

where **STR** = the Turbo Pascal reserved word for converting a whole number into a string

`Number` = the whole number to be converted into a string

`String_Name` = the string that will contain the character string from Number

VAL Procedure

`VAL(String_Number, Number, Code);`

where **VAL** = the built-in Turbo Pascal reserved word for the value procedure

`String_Number` = a string that contains number characters which is to be converted to a whole number

`Number` = a REAL or INTEGER variable that will contain the converted whole number

`Code` = an INTEGER that will have a value of zero if the conversion was successful

Turbo Pascal Unit Structure

UNIT (Identifier);

INTERFACE

USES (List of other units)
(Procedures, Functions and other declarations.)

IMPLEMENTATION
(Body of Procedures and Functions)

END.

Self-Test

Directions

Figure 5–7 (p. 294) is a flowchart for troubleshooting a hypothetical robot. Flowcharts of this type are very common in industry. They are used to guide the technician through a series of steps in which the next step depends upon a previous measurement or other observation. This type of problem lends itself very easily to programming—specifically, programming that requires the use of branch statements.

Program 5–39 (p. 294) is an implementation of the flowchart for troubleshooting the robot. All of the questions in the self-test refer to this program and flowchart.

Questions

1 Which procedure performs the light check? What kind of branching is used to do this?
2 Which of the program procedures uses a **CASE** statement?
3 How many different procedures are used in the program?
4 What does the **USES CRT** mean in the program?
5 How many open branches are used in the program? What procedures contain them?
6 How many procedures use **IF–THEN–ELSE** statements? Which ones are they?
7 What procedures pass values back to the calling procedure? What kind of a variable is this called?
8 Explain why a **CASE** statement was not used in the procedure **TP＿1＿ Measurement**.
9 Referring to the procedure **TP＿1＿Measurement**, there are two **END** statements that are not followed by the semicolon. Explain why.
10 Which procedures use logic statements? What logic statements are used?

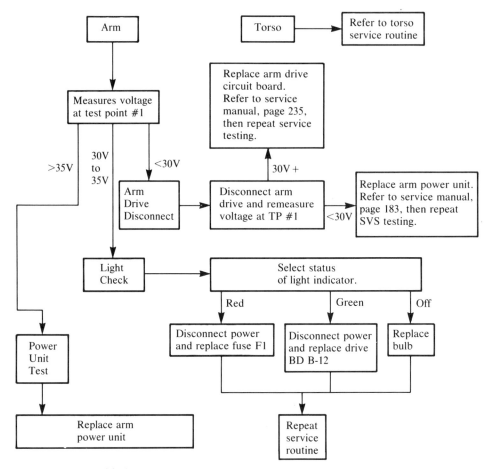

Figure 5–7   Troubleshooting Chart for Robot

Program 5–39

```
PROGRAM Robot_TroubleShooter;

{****************************************************************
            Developed By:  A. Robotics Student
                 Date:  December 1992
 ****************************************************************
   Program Description:  This program was developed to demonstrate
 an application of different types of branch blocks found in
 Pascal.
       The program simulates computer-aided troubleshooting
 (CAT) by leading the program user through the trouble-
 shooting exercise of a hypothetical robot.  The program user
 plays the game by entering choices and "measurements" as if
 the robot were actually being tested.
 ****************************************************************}
```

Program 5-39   *continued*

```
{Library Units}
 USES
   CRT;


 TYPE
   String79 = STRING[79];
 {*****************************************************************
                        Procedures Used:
 ----------------------------------------------------------------}
 {PROCEDURE Main_Programming_Sequence;

        DESCRIPTION:
            This procedure calls all of the primary procedures
            used in this program.  It is the first procedure given
            in order to make the structure of the program easy
            to read and follow.

        PARAMETERS:
            None.

        SAMPLE CALL:
            BEGIN
              Main_Programming_Sequence;
            END.
 ---------------------------------------------------------------}
   PROCEDURE Program_Title;
   FORWARD;
   {
     DESCRIPTION:
        This procedure displays the program title and waits for
        the program user to press the RETURN/ENTER key.

     PARAMETERS:
        None.

     SAMPLE CALL:
         Program_Title;
   ---------------------------------------------------------------}
   PROCEDURE Explain_Program;
   FORWARD;
   {
     DESCRIPTION:
        This procedure explains the purpose of the program to
        the program user.

     PARAMETERS
        None.

     SAMPLE CALL:
        Explain_Program;
   ---------------------------------------------------------------}
   PROCEDURE Start_Testing_Sequence;
   FORWARD;
   {
```

Program 5–39   *continued*

```
      DESCRIPTION:
          This procedure controls the testing sequence for the
          robot troubleshooting program.

      PARAMETERS:
          None.

      SAMPLE CALL:
          Start_Testing_Sequence;
 ------------------------------------------------------------}
    PROCEDURE Torso_Service;
    FORWARD;
    {
      DESCRIPTION:
          This procedure tells the program user to refer to the
      Torso Service Manual since no programming is available
      for troubleshooting this section of the robot.

      PARAMETERS:
          None.

      SAMPLE CALL:
          Torso_Service;
  -----------------------------------------------------------}
    PROCEDURE Arm_Service;
    FORWARD;
    {
      DESCRIPTION
          This procedure has the user measure the voltages at
      TP-1.

      PARAMETERS:
          None.

      SAMPLE CALL:
          Arm_Service;
   ----------------------------------------------------------}
    PROCEDURE TP_1_Measurement(Value : REAL; VAR Next : INTEGER);
    FORWARD;
    {
      DESCRIPTION
          This procedure routs the program user to the "Power
      Unit Test" or "Light Check" or "Arm Disconnect Drive"
      according to the results of the voltage reading at
      TP-1.

      PARAMETERS:
          Value (input) = The voltage reading at TP-1.
          Next (output) = The next measurement to be made.

      SAMPLE CALL:
          TP_1_Measurement(Voltage, Next_Process);
  -----------------------------------------------------------}
    PROCEDURE Power_Unit;
    FORWARD;
    {
```

Program 5-39 *continued*

```
        DESCRIPTION
            This procedure instructs the program user to replace
        the "Arm Power Unit".

        PARAMETERS:
            None.

        SAMPLE CALL:
            Power_Unit;
    -------------------------------------------------------------}
      PROCEDURE Light_Check;
      FORWARD;
      {
        DESCRIPTION
            This procedure guides the program user in checking the
        status of the indicator light and gives directions for
        proceeding based upon the observation.

        PARAMETERS:
            None.

        SAMPLE CALL:
            Light_Check;
    -------------------------------------------------------------}
      PROCEDURE Arm_Drive_Disconnect;
      FORWARD;
      {
        DESCRIPTION
            This procedure instructs the program user in the
        next step of the servicing procedure based upon the
        voltage reading at this test point.

        PARAMETERS:
            None.

        SAMPLE CALL:
            Arm_Drive_Disconnect;
    -------------------------------------------------------------}
{**************************************************************}
 PROCEDURE Main_Programming_Sequence;
{============================================================}

 BEGIN

    Program_Title;    {Displays the title of Robot Troubleshooter.}
    Explain_Program; {Explains the program to the user.}
    Start_Testing_Sequence; {Controls the testing subroutines.}

    {============================================================}
 END; {of procedure Main_Programming_Sequence}

 PROCEDURE Program_Title;
{------------------------------------------------------------}

  BEGIN

    ClrScr; {Clears the screen.}
    Window(15,5,79,24);
```

Program 5–39 *continued*

```
    WRITELN('*********** ROBOT TROUBLESHOOTER ************');
    WRITELN('*                                           *');
    WRITELN('*                for the                    *');
    WRITELN('*                                           *');
    WRITELN('*             XR-4 HYPOTHETICAL             *');
    WRITELN('*                                           *');
    WRITELN('*               Developed by:               *');
    WRITELN('*                                           *');
    WRITELN('*                Your Name                  *');
    WRITELN('*                                           *');
    WRITELN('*********************************************');

    Window(1,1,80,24);  {Restore to full screen.}
    GOTOXY(15,23);
    WRITE('Press RETURN/ENTER to continue...');
    READLN;
{------------------------------------------------------------}
  END;  {of procedure Program_Title}

  PROCEDURE Explain_Program;
{------------------------------------------------------------}
    BEGIN

      ClrScr;  {Clear the screen.}
      Window(15,5,79,24);

      WRITELN('This program will assist you in troubleshooting');
      WRITELN('the XR-4 Hypothetical Robot.');
      WRITELN;
      WRITELN('Just follow the directions on the screen and');
      WRITELN('make the required measurements.  Then enter the');
      WRITELN('measurement values.  The program will direct you');
      WRITELN('as to what to do next.');

      Window(1,1,80,23);
      GOTOXY(15,23);
      WRITELN('Press RETURN/ENTER to continue...');
      READLN;
{------------------------------------------------------------}
    END;  {of procedure Explain_Program}

  PROCEDURE Start_Testing_Sequence;
{------------------------------------------------------------}
    VAR
    User_Input
    :CHAR;

    BEGIN

    ClrScr;
    WRITELN;
    WRITELN('Enter the letter of the section you are troubleshooting:');
    WRITELN;
    WRITE('T)orso   A)rm => ');
    READLN(User_Input);
```

Program 5–39 *continued*

```
        IF (User_Input = 'T') OR (User_Input = 't') THEN
          Torso_Service
        ELSE
          Arm_Service;
      {----------------------------------------------------------}
        END;  {of procedure Start_Testing_Sequence}

      PROCEDURE Torso_Service;
      {----------------------------------------------------------}
        BEGIN

          ClrScr;
          GOTOXY(5,10);
          WRITELN('Refer to Torso Service Manual.');
      {----------------------------------------------------------}
        END;  {of procedure Torso_Service.}

  PROCEDURE Arm_Service;
{----------------------------------------------------------}
      VAR
        Voltage_TP1
       :REAL;
        Result
       :INTEGER;
      BEGIN

        ClrScr;
        WRITELN;
        WRITE('Measure voltage at Test Point #1 => ');
        READLN(Voltage_TP1);

        TP_1_Measurement(Voltage_TP1, Result);

        CASE Result OF
          1 : Arm_Drive_Disconnect;
          2 : Light_Check;
          3 : Power_Unit;
        END; {of Case}
    {----------------------------------------------------------}
      END;

    PROCEDURE TP_1_Mcasurement{(Value : REAL; VAR Next : INTEGER)}
    {----------------------------------------------------------}
        BEGIN

          IF Value < 30 THEN
            BEGIN
              WRITELN;
              WRITELN('Problem is in the arm drive unit.');
              WRITELN('Disconnect the arm drive and remeasure');
              WRITELN('the voltage at TP #1.');
              Next := 1;  {Value to return to calling procedure.}
            END

          ELSE
```

Program 5–39   *continued*

```
        IF (Value >= 30) AND (Value <= 35) THEN
            BEGIN
             WRITELN;
             WRITELN('Light check should be made.');
             WRITELN('Determine the status of the light indicator.');
             Next := 2;  {Value to return to calling procedure.}
            END

        ELSE

        IF Value > 35 THEN
            BEGIN
             WRITELN;
             WRITELN('Power unit needs testing.');
             WRITELN('Replace the Arm Power Unit.');
             Next := 3;  {Value to return to calling procedure.}
            END;
{------------------------------------------------------------------}
    END;  {of procedure TP_1_Measurement.}

    PROCEDURE Power_Unit;
{------------------------------------------------------------------}
    BEGIN

      WRITELN;
      WRITELN('Power unit test has failed.');
      WRITELN('Replace => Arm Power Unit.');
{------------------------------------------------------------------}
    END;  {of procedure Power_Unit}

    PROCEDURE Light_Check;
{------------------------------------------------------------------}
    VAR

      Status
    :INTEGER;

    BEGIN

      WRITELN;
      WRITELN('Indicate status of light by selecting one of');
      WRITELN('the following numbers:  ');
      WRITELN;
      WRITELN('1] RED   2] GREEN   3] Off');
      WRITELN;
      WRITE('Light status number => ');
      READLN(Status);

      CASE Status OF
        1 : WRITELN('Disconnect power and replace fuse F1');
        2 : WRITELN('Disconnect power and replace drive BD B-12');
        3 : WRITELN('Replace bulb.');
      END;  {of case.}
```

Program 5–39 *continued*

```
      WRITELN;
      WRITELN('Repeat service routine.');
  {----------------------------------------------------------------}
    END;   {of procedure Light_Check}

  PROCEDURE Arm_Drive_Disconnect;
  {----------------------------------------------------------------}
    VAR
      Voltage
    :REAL;

    BEGIN
      WRITELN;
      WRITELN('Disconnect arm drive and remeasure voltage ');
      WRITELN('at TP #1.');
      WRITELN;
      WRITE('Enter value of voltage => ');
      READLN(Voltage);

      IF Voltage >= 30 THEN
        BEGIN
          WRITELN;
          WRITELN('Replace arm drive circuit board.');
          WRITELN('Refer to Service Manual page 235,');
          WRITELN('then repeat service testing.');
        END
      ELSE
        BEGIN
          WRITELN;
          WRITELN('Replace arm power unit.');
          WRITELN('Refer to Service Manual page 183.');
          WRITELN('Then repeat SVS testing.');
        END;
  {----------------------------------------------------------------}
    END;   {of procedure Arm_Drive_Disconnect}

BEGIN
  Main_Programming_Sequence;
END.
```

Problems

General Concepts

Section 5–1

1 Describe a relational operator. Give an example.
2 Explain in words what is meant by
    A) 15 < = This   B) 27 >= That
3 Which of the following is TRUE?
    A) 4 IN[1..12]   B) d IN[A..X]
4 Explain what is meant by the Pascal statement
    IF Yes_ THEN Do_This;
5 Describe the unique features of Boolean algebra.

6 Explain the meaning of the following Boolean operators.
A) **AND**   B) **OR**   C) **NOT**   D) **XOR**

7 Can relational and Boolean operators be combined? Give an example.

Section 5–2

8 Explain the meaning of the Pascal command
```
IF This THEN That ELSE Something_Different;
```

9 For the above command, which identifiers are of type BOOLEAN? Which one(s) is/are a procedure?

10 State what is meant by nesting **IF–THEN–ELSE** statements. Give an example.

11 What is the rule for using semicolons in **IF–THEN–ELSE** statements?

Section 5–3

12 Explain, in words, the meaning of a Pascal **CASE** statement.

13 In a **CASE** statement what is the expression? The caseclause? What is the important rule regarding the case selector?

14 When is the **CASE** statement used in Pascal?

15 Give an example of each of the following in a **CASE** caseclause statement.
A) Procedure B) Function C) Compound Statement

16 Explain what is meant by
```
A) CASE This OF
     'a', 'b', 'z' : Do_That;
B) CASE Another OF
     A..M : Some_More;
```

Section 5–4

17 State what is meant by a character string. Give an example.

18 What is a null string?

19 Describe the difference between the following Pascal statements.
```
WRITELN('Value := 2 + 5');
WRITELN('Value := ', 2 + 5);
```

20 Explain how a string variable is declared in Turbo Pascal. How does standard Pascal handle strings?

21 Explain how a string constant is declared in Turbo Pascal. Give an example.

22 Describe what is meant by concatenation as applied to strings. How is this done in Turbo Pascal?

Section 5–5

23 Describe the built-in Turbo function that will return a selected part of a string.

24 What Turbo Pascal function or procedure is available to remove a selected part of a string? Give an example.

25 State the Turbo Pascal function or procedure that will put a string into a selected position of another string. Give an example.

26 Explain the difference between the built-in Turbo Pascal **LENGTH** function and the **POS** function. Give an example of each.

27 Describe the difference between the built-in Turbo Pascal procedures **STR** and **VAL**. Give an example of each.

Section 5–6

28 Explain the concept of a program library as it applies to Pascal.

29 Does Turbo Pascal have a user-created library available? Explain.
30 Describe the major difference between a Turbo Pascal unit and a Turbo Pascal program.
31 What may be contained within a Turbo Pascal unit?
32 Give an example of how a unit is used by a program.
33 Explain the difference between the **INTERFACE** and **IMPLEMENTATION** parts of a unit.

Program Analysis

34 Is there a problem in program 5–40? If so, state what it is.

Program 5–40

```
PROGRAM Problem__34;
    BEGIN
     IF 3 => 5 THEN
    WRITELN('You got it!');
    END.
```

35 Will program 5–41 compile? If not, why not?

Program 5–41

```
PROGRAM Problem__35;
    BEGIN
        IF 8 > 7 AND 3 < 5
        THEN
        WRITELN('This is problem 35.');
    END.
```

36 Analyze program 5–42 and predict what you think it will do. If there is a problem, what are the necessary corrections?

Program 5–42

```
PROGRAM Problem__36;
    BEGIN
      IF NOT(3 > 1) AND (5 >= 2) OR NOT(2 = 2)
    THEN WRITELN('This is problem 36.')
    ELSE WRITELN('This is still the same problem.');
    END.
```

37   Will program 5–43 compile? If not, what are the minimum changes necessary?

Program 5–43

```
PROGRAM Problem__37;
    BEGIN
      IF (10 > 5) THEN WRITE('A true statement.');
      ELSE WRITELN('A false statement.');
      END.
```

38   Determine what the logic statement in program 5–44 will cause the program to do.

Program 5–44

```
PROGRAM Problem__38;
    BEGIN
      IF (12 < 24) XOR (5 = 6)
      THEN WRITELN('Top banana!')
      ELSE
      WRITELN('Didn''t happen.');
      END.
```

39   If there is an error in program 5–45, determine the minimum changes required to correct it. If there is no error, what do you predict the program will do?

Program 5–45

```
PROGRAM Problem__39;
   CONST
     One = 1;
     Two = 2;
     BEGIN
       CASE One OF
         1..5 : WRITELN('This was displayed.')
       ELSE
         WRITELN('Something else was displayed.');
       END.
```

40   If there is an error in program 5–46, determine the minimum changes required to correct it. If there is no error, what do you predict the program will do?

Program 5–46

```
PROGRAM Problem_40;
   CONST
     One = 1.0;
     Two = 2.0;
     BEGIN
        CASE One OF
          1..5 : WRITELN('This was displayed.')
        ELSE
          WRITELN('Something else was displayed.');
        END.
```

41 Determine if there is a compile-time error in program 5–47. If there is, determine the minimum changes required to make the program compile.

Program 5–47

```
PROGRAM Problem_41;
    VAR
      Character : STRING[0];
    BEGIN
          WRITELN('The character is ',Character);
    END.
```

42 Determine if there is a problem in program 5–48. If there is, state the minimum changes required to make the program function.

Program 5–48

```
PROGRAM Problem_42;
    CONST
      This_Word : STRING[224];
    BEGIN
       WRITELN('This is the word:  ',This_Word);
    END.
```

Program Design

When developing these programs use the structure developed in the Pascal program used for the self-test. To summarize, this means

1. All procedures and functions are declared **FORWARD** and documented.
2. The first declared procedure presents the program sequence.
3. The traditional main program sequence (between the final **BEGIN** and **END.**) calls only one procedure, the first procedure explained in 2 above.

Electronics Technology

43   Develop a Pascal program that will compute the power dissipation of a resistor where the user input is the value of the resistor and the current in the resistor. The program is to warn the user when the power dissipation is above 1 watt. The power dissipation of a resistor is

$$P = I^2 R$$

where P = power dissipation in watts
      I = current in amps
      R = resistance in ohms

44   Create a Pascal program that will convert an input number into the resistor color code. The relationship is

```
0 = Black 1 = Brown 2 = Red 3 = Orange
4 = Yellow 5 = Green 6 = Blue 7 = Violet
8 = Gray 9 = White
```

45   Write a Pascal program that determines between what standard capacitor values the value of a given capacitor is. The program user inputs a capacitor value and the program will indicate between what two standard values it belongs. For the sake of simplicity, assume the standard capacitor values to be used by the program are

0.001 μF	0.1 μF	10 μF	1000 μF
0.0015 μF	0.15 μF	15 μF	1500 μF
0.0022 μF	0.22 μF	22 μF	2200 μF
0.0033 μF	0.33 μF	33 μF	3300 μF

46   Develop a Turbo Pascal unit that contains functions for all three forms of Ohm's law. Then develop a Pascal program that uses the unit to solve for any of the three forms of Ohm's law selected by the program user. The three forms of Ohm's law are

$$V = IR \quad I = V/R \quad R = V/I$$

where V = voltage in volts
      I = current in amps
      R = resistance in ohms

Business Applications

47   Create a Pascal program where the price of the following items is already entered.

Soap = $12.50   Asphalt = $27.59   Glue = $2.33
Gum = $0.57

The program user will select an item by number and quantity. The program will then compute the total cost.

Computer Science
48 Develop a Pascal program that demonstrates to the program user all forms of branching used by Turbo Pascal.

Drafting Technology
49 Create a Pascal program in which the program user inputs the name of a figure and the program presents the formula used for calculating its area. Choose a rectangle, triangle, circle, and parallelogram as the figures.

Agriculture Technology
50 Develop a Pascal program that will simulate computer-aided troubleshooting of a hypothetical diesel engine. The troubleshooting chart is shown in figure 5–8.

Health Technology
51 Create a Pascal program that will allow the program user to convert from or to metric. The user can then choose length, volume, or weight for the conversion.

Manufacturing Technology
52 Develop a Pascal program that will compute the weight of a machined part. The program user can select a cylinder, rectangle, or cone form, and can specify whether it is made of

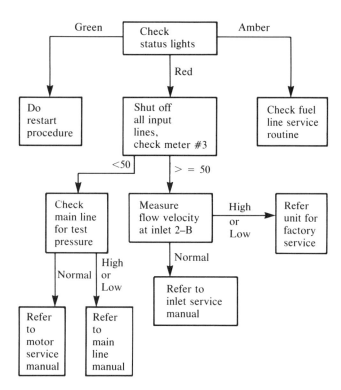

Figure 5–8   Troubleshooting Chart for Diesel Engine

copper, aluminum, or steel. The program must automatically compute the volume and then use the density of the selected material to compute the total weight.

Business Applications

53 Modify the program you developed for problem 47 so that the user can input the value of each item with a leading dollar sign ($).

Computer Science

54 Create a flowchart of the program you developed for problem 48.

Drafting Technology

55 Develop a Turbo Pascal unit that contains the formulas for the area of the figures used in the program you developed for problem 49, then modify your program by adding **USES** for the created unit.

Agriculture Technology

56 Develop a Turbo Pascal unit that contains the conversion constants for square feet to acres, acres to square feet, feet to miles, and miles to feet. Then create a Pascal program that uses the unit.

Health Technology

57 Create a Turbo Pascal unit that contains the conversion of Fahrenheit to Centigrade and Centigrade to Fahrenheit. Then make a Pascal program that uses the UNIT.

Manufacturing Technology

58 Write a Turbo Pascal unit that will contain the formulas and densities of the materials for the objects in problem 52, then modify the program so that it uses the unit.

6 Types and Loops

Objectives

This chapter will give you the chance to learn:

1 The meaning and use of Pascal types.
2 The basic concepts and reasons for loop blocks.
3 The development of a Pascal program using the **FOR** statement.
4 Methods for nesting **FOR** loops.
5 The development of a Pascal program using the **REPEAT** statement.
6 The development of a Pascal program using the **WHILE** statement.
7 Various methods of nesting using combinations of the above looping statements.
8 Program debugging techniques.

Key Terms

Real
Integer
Boolean
Byte
Scalar
Ordinal

Predecessor
Successor
Loop
Branch
Nested Loops

Outline

This chapter will show you how to create your own types of data. You will also see how to organize this data and to use some interesting built-in Turbo Pascal features. After this important material on data types, you will be introduced to the power of the Pascal programming loop.

Here you will discover how to have the computer do the same task over and over again. Think of it—you will be able to create a computer program that will solve the same problem as many times as you want with different values each time. Using this technique, you can create tables of useful information with a wide range of technical applications.

You will also see how to use a Pascal type. Understanding this powerful concept will open a whole new area of programming in Pascal to you. You now have enough background in Pascal programming so that this exciting idea will have meaning to you.

You will see that Pascal offers several ways of causing the computer to repeat the same process. When you complete this chapter, you will be able to use all three of the fundamental programming techniques available—action, branching, and looping. Let's get started!

6–1   Pascal Types

You have seen that Pascal has variables and constants that must have a type definition such as REAL, INTEGER, BOOLEAN, BYTE, CHAR, and STRING[N].

You may be surprised to find that Pascal lets you define your own type of data. This means your variables and constants can have a whole new kind of meaning—something never thought of by the developers of Pascal. Doing this will also make your programs easier to read, modify, and understand.

Basic Idea

To get an idea of how defining your own data type and using it in a program will make it easier to understand, let us first look at program 6–1, a Pascal program using the **CASE** statement.

Program 6–1

```
PROGRAM An_Example;

 PROCEDURE Check_Lights(Condition : INTEGER);

  BEGIN
   CASE Condition OF
    1 : WRITELN('Check system pressure and continue test.');
    2 : WRITELN('System OK.');
    3 : WRITELN('Check system fuse and continue test.');
   END;  {of Case}

 END;  {of procedure}

 PROCEDURE Test_It;
  VAR
   Test_Input
  :INTEGER;

  BEGIN
   WRITELN('1] Red  2] Green  3] Off');
   WRITE('Select light condition by number => ');
   READLN(Test_Input);

    Check_Lights(Test_Input);

  END;

BEGIN
  Test_It;

END.
```

The procedures in program 6–1 could be used in a computer-aided trouble-shooting program similar to the robot troubleshooting program presented in the last chapter. Observe that the procedure Check_Lights has an INTEGER-type value passed to it by the procedure Test_It. This INTEGER value may be a 1, 2, or 3 and will cause an appropriate output to appear on the monitor. Look at the **CASE** statement of the first procedure.

```
CASE Condition OF
 1 : WRITELN('Check system pressure and continue test.');
 2 : WRITELN('System OK.');
 3 : WRITELN('Check system fuse and continue test.');
END; {of Case}
```

It may not be clear to everyone reading this **CASE** statement exactly what 1, 2, or 3 represent. It would be more descriptive if the **CASE** statement could be written

```
CASE Condition OF
  Red : WRITELN('Check system pressure and continue test.');
Green : WRITELN('System OK.');
  Off : WRITELN('Check system fuse and continue test.');
  END; {of Case}
```

Writing the **CASE** statement in this way makes it very clear exactly what conditions will cause a particular statement to appear on the monitor. Recall, however, that the case constants must have an ordinal arrangement. You can change this by defining a new type in your Pascal program, as shown in program 6–2.

Program 6–2

```
PROGRAM Typed;

 TYPE
  Light_Status = (Red, Green, Off);

 PROCEDURE Check_Lights(Condition : Light_Status);
   BEGIN
    CASE Condition OF
      Red    : WRITELN('Check system pressure and continue test.');
      Green  : WRITELN('System OK.');
      Off    : WRITELN('Check system fuse and continue test.');
    END;  {of Case}
  END;  {of procedure}

 PROCEDURE Test_It;
   VAR

    Test_Input
   :INTEGER;

    Reading
   :Light_Status;

   BEGIN
      WRITELN('1] Red  2] Green  3] Off');
      WRITE('Select light condition by number => ');
      READLN(Test_Input);
       CASE Test_Input OF
         1 : Reading := Red;
         2 : Reading := Green;
         3 : Reading := Off;
       END;

       Check_Lights(Reading);

    END;

 BEGIN
   Test_It;
 END.
```

Program 6–2 will compile and run successfully. From the program user's standpoint, there will be no difference in the operation of the two programs. However, from the programmer's standpoint, or for anyone else who may have to understand your program, it is much clearer. Look at the changes that were introduced.

```
TYPE
 Light_Status = (Red, Green, Off);
```

This was the type definition. Here, Pascal is told that a new type called **Light_Status** is being defined which can have only three values (Red, Green, or Off). You can now define data in terms of this new type.

```
VAR

 Test_Input

:INTEGER;

 Reading
:Light_Status;
```

Note the variable **Reading**. It is defined as the type **Light_Status**. This means that **Reading** can now have any one of these three values: Red, Green, or Off.

```
CASE Test_Input OF
 1 : Reading := Red;
 2 : Reading := Green;
 3 : Reading := Off;
END;
```

This **CASE** statement assigned a corresponding value (Red, Green, or Off) to the variable **Reading**; this value was then passed on to the called procedure

```
Check_Lights(Reading);
```

and from there, was used by the **CASE** statement of the called procedure.

```
CASE Condition OF
  Red : WRITELN('Check system pressure and continue test.');
Green : WRITELN('System OK.');
  Off : WRITELN('Check system fuse and continue test.');
END; {of Case}
```

The kind of data types you define in Pascal are all **scalar**. There are some important things you should know about scalars.

Scalar Data Types

A scalar data type has a distinct set of possible values. These values are considered to be ordered in a specific way; that is, they can be put into a one-to-one correspondence with an INTEGER sequence. The scalar types for Turbo Pascal are

INTEGER—Maximum value 32767, minimum value −32768

BOOLEAN—Two values: TRUE or FALSE.

   CHAR—All letters of the alphabet, symbols, and number symbols 0 to 9.

   BYTE—Values from 0 to 255.

All of these data types have a one-to-one correspondence with an integer sequence. For example, in the CHAR type, there are no values between A and B. The value A comes before B and the value C comes after B. The following are not scalar types.

   REAL—Theoretically there are an infinite number of values between any two numbers (i.e. between 1 and 2: 1.1, 1.2, 1.21, 1.212, etc.).

STRING—Theoretically there are infinite possibilities in the combinations available.

Look at the type **Light__Status** that was defined to have three values: Red, Green, and Off. Since this must be scalar (remember, all types defined by you will be scalar), this means that Red comes before Green and Green comes before Off. It also means that there are no other values between Red, Green, and Off. The values Red, Green, and Off have a one-to-one correspondence with 0, 1, and 2. This is demonstrated by some unique Pascal built-in functions.

The ORDinal Function

The ORDinal function in Pascal returns the ordinal number of an ordinal-type expression. The ordinal function in Pascal is

ORD(N)

where N = an ordinal-type expression

Program 6–3 gives an example of the use of this function.

Program 6–3

```
PROGRAM Ordinal__Demo__1;
  BEGIN
    WRITELN('Ordinal value of 0 is ',ORD(0));
    WRITELN('Ordinal value of 1 is ',ORD(1));
    WRITELN('Ordinal value of a is ',ORD('a'));
    WRITELN('Ordinal value of b is ',ORD('b'));
    WRITELN('Ordinal value of TRUE is ',ORD(TRUE));
    WRITELN('Ordinal value of FALSE is ',ORD(FALSE));
  END.
```

When program 6–3 is executed, the monitor will display

```
Ordinal value of 0 is 0
Ordinal value of 1 is 1
Ordinal value of a is 97
```

```
Ordinal value of b is 98
Ordinal value of TRUE is 1
Ordinal value of FALSE is 0
```

The ordinal values of 0 and 1 are 0 and 1 respectively, because both of these numbers are integers. The ordinal values of the characters are 97 for 'a' and 98 for 'b', their ordinality in the ASCII code used by your computer. The ordinal values of TRUE and FALSE are 1 and 0 respectively.

What happens when you try to take the ordinal value of data that is not ordinal? Consider program 6–4—it will not compile.

Program 6–4

```
PROGRAM Ordinal_Demo_2;
  BEGIN
   WRITELN(ORD(1.0));
   WRITELN(ORD('applied'));
  END.
```

Program 6–4 will not compile because the number 1.0 is of type REAL (because of the decimal point), and has no ordinality. The same is true for the string **'applied'**; recall that strings have no ordinality.

Now consider the ordinality of a type defined by you.

Program 6–5

```
PROGRAM Ordinal_Demo_3;
 TYPE
   Light_Status = (Red, Green, Off);

  BEGIN
   WRITELN(ORD(Red));
   WRITELN(ORD(Green));
   WRITELN(ORD(Off));
  END.
```

When program 6–5 is executed the monitor will display

```
0
1
2
```

The first value of the new type has an ordinality of 0, the second an ordinality of 1, and the third an ordinality of 2. Thus, this program confirms that a user-defined type will be ordinal.

The Predecessor Function

Another Pascal function is the PREDecessor function, defined as

PRED(N)

where N = any ordinal-type expression

Program 6–6 gives an example.

Program 6–6

```
PROGRAM Predecessor_Demo_1;
  BEGIN
   WRITELN(PRED(0));
   WRITELN(PRED(1));
   WRITELN(PRED('a'));
   WRITELN(PRED('b'));
   WRITELN(PRED(TRUE));
   WRITELN(PRED(FALSE));
  END.
```

When program 6–6 is executed, the monitor displays

```
-1
0
'
a
FALSE
TRUE
```

The number that precedes 0 is −1, and the number that precedes 1 is 0. The character that precedes 'a' is the apostrophe ('). This is the ordinality of these two characters in the ASCII code. As you would expect, the character 'a' precedes the character 'b', and FALSE precedes TRUE, while TRUE precedes FALSE.

What happens if you try to take the predecessors of numbers that are not ordinal? Look at program 6–7.

Program 6–7

```
PROGRAM Predecessor_Demo_2;
 TYPE
  Light_Status = (Red, Green, Off);

  BEGIN
   WRITELN(PRED(1.0));
   WRITELN(PRED(Off));
  END.
```

As you might expect, program 6–7 will not compile because the predecessor function will work only with ordinal values.

What do you think program 6–8 will do?

Program 6–8

```
PROGRAM Predecessor_Demo_3;
 TYPE
  Light_Status = (Red, Green, Off);

  BEGIN
   WRITELN(PRED(Green));
   WRITELN(PRED(Off));
  END.
```

Program 6–8 will not compile either because Turbo Pascal only allows certain predefined types to be displayed on the screen. You are correct in thinking that the predecessor of Green is Red; however, these are not string variables, and they are not numbers. They are the values for a brand-new, never-heard-of-before type! You can, however, display their ordinal values.

Program 6–9

```
PROGRAM Predecessor_Demo_2;
 TYPE
  Light_Status = (Red, Green, Off);

  BEGIN
   WRITELN(ORD(PRED(Green)));
   WRITELN(ORD(PRED(Off)));
  END.
```

When program 6–9 is executed, the monitor will now display

0
1

The SUCCessor Function

The SUCCessor function in Pascal is just the opposite of the PREDecessor function. The successor function returns the ordinal value of the following piece of data. The successor function is

SUCC(N)

where N = an ordinal-type expression

As an example, consider program 6–10.

Program 6–10

```
PROGRAM Successor_Demo_1;
  BEGIN
   WRITELN(SUCC(0));
   WRITELN(SUCC(1));
   WRITELN(SUCC('a'));
   WRITELN(SUCC('b'));
   WRITELN(SUCC(TRUE));
   WRITELN(SUCC(FALSE));
  END.
```

When executed, program 6–10 will display on the monitor

```
1
2
b
c
TRUE
TRUE
```

The only surprise here may be the two TRUEs in a row. You would think that the successor of TRUE would be FALSE. The successor of TRUE was displayed as TRUE, because any ordinal value assigned to a Boolean expression that is different from 0 is considered to be TRUE. The only ordinal value considered to be FALSE by Pascal for a Boolean type is 0. Any other value will be considered TRUE. This is demonstrated by program 6–11.

Program 6–11

```
PROGRAM Successor_Demo_2;
  BEGIN
   WRITELN(ORD(SUCC(TRUE)));
   WRITELN(ORD(SUCC(FALSE)));
  END.
```

When program 6–11 is executed, the monitor will display

```
2
1
```

Both of these ordinal values are considered to be TRUE for a Boolean type. What do you think program 6–12 will display on the monitor?

Program 6–12

```
PROGRAM Successor_Demo_3;
 TYPE
   Light_Status = (Red, Green, Off);
  BEGIN
  WRITELN(ORD(SUCC(Red)));
  WRITELN(ORD(SUCC(Green)));
  END.
```

The monitor display would be

1
2

You should never attempt to get the predecessor or successor of a value that is not defined in your new data type, because the results are not predictable.

An Application Program

Consider program 6–13. It converts the first two color bands of the resistor color code to their numerical value. It defines a new data type which is the resistor color code itself. The ordinal value of each color represents the value of the digit represented by that color.

Program 6–13

```
PROGRAM Resistor_Color_Code;
 TYPE
   Color_Code = (Black, Brown, Red, Orange, Yellow, Green, Blue,
                 Violet, Gray, White);

  FUNCTION Resistor_Digits (FirstBand, SecondBand : Color_Code):INTEGER;
   VAR
    First_Digit,
    Second_Digit
   :INTEGER;

  BEGIN
   First_Digit := ORD(FirstBand);
   Second_Digit := ORD(SecondBand);
    Resistor_Digits := (ORD(First_Digit) * 10 + ORD(Second_Digit));
  END;

  PROCEDURE Use_It;
   VAR
    Color
:Color_Code;
    Digit_Value
:INTEGER;
```

Program 6–13 *continued*

```
BEGIN
   Digit_Value := Resistor_Digits(Red, Green);
   WRITELN('The digits are ',Digit_Value);
END;

BEGIN
  Use_It;
END.
```

When program 6–13 is executed, it displays on the monitor the value 25. This is the numerical value of the first two color bands of a resistor if their colors are Red and Green respectively.

Conclusion

This section presented the concept of data types. More importantly, here you saw how you could define your own data type. You got a deeper understanding of the meaning of ordinality by seeing examples of the **ORD**, **PRED**, and **SUCC** Pascal functions.

Using your own defined data types will make your Pascal programs more readable and easier to modify. Check your understanding of this section by trying the following section review.

6–1 Section Review

1 Explain what is meant by a data type. Give an example.
2 Explain what is meant by a scalar. Give an example.
3 What does the term ordinal mean as applied to Pascal?
4 Give an example of a non-ordinal variable.
5 Can a user-defined type be displayed directly on the monitor? Explain.

6–2   Introductory Concepts

Discussion

Loops are the third and last kind of programming block. Recall the Boehm and Jacopini completeness theorem that stated that there are only three kinds of programming blocks needed for any computer program, no matter how complex: action blocks, branch blocks, and loop blocks. This section will introduce you to the reasons for using loop blocks in programming, what loops look like, and how they differ from branch blocks.

General Idea

Suppose you wanted to know the distance a falling object would fall toward the earth after a certain number of seconds (neglecting air resistance). The equation for the distance traveled by an object falling toward the earth, neglecting air resistance, is

$$S = 0.5 \text{ } at^2$$

where S = distance traveled by the object in feet

a = acceleration of the object ($32/\text{ft/sec}^2$)

t = time, in seconds, the object is falling

For example, suppose you wanted the distance that a falling object would cover for the first three seconds of its fall and you specifically wanted to know its distance at 1 second, 2 seconds, and 3 seconds. Program 6–14 shows one way these values could be computed.

Program 6–14

```
PROGRAM Solve_Distance_1;
  CONST
   a = 32;   {Acceleration due to gravity.}

  FUNCTION Distance(Time : REAL):REAL;
   BEGIN
      Distance := 0.5 * a * Time * Time;
   END;

  PROCEDURE Calculate_It;
   VAR
    Answer
   :REAL;

     BEGIN
       Answer := Distance(1.0);
       WRITELN('Distance at 1.0 seconds is ',Answer:3:3,' feet.');
       Answer := Distance(2.0);
       WRITELN('Distance at 2.0 seconds is ',Answer:3:3,' feet.');
       Answer := Distance(3.0);
       WRITELN('Distance at 3.0 seconds is ',Answer:3:3,' feet.');
     END;

BEGIN
 Calculate_It;
END.
```

When program 6–14 is executed, it will display

```
Distance at 1 seconds is 16.00 feet
Distance at 2 seconds is 64 feet.
Distance at 3 seconds is 144.00 feet.
```

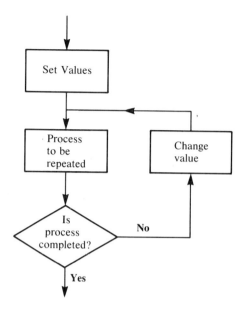

Figure 6–1   Idea of Looping

You will get the answers you were seeking for the three different values of time. What is happening in program 6–14 is what you would do to solve the same problem if you were using a pocket calculator, pencil, and paper. You would do the same thing that is being done in the function of the program—and you would do it three times for each of the three values of time, recording your answer each time. It could be said that you were looping; that is, once you solved the problem for Time = 1, you would loop back and solve the same problem over again for Time = 2. You would repeat this process until you calculated for the last value of Time required, which in this case was Time = 3. In other words, you would go back and repeat the same process three times; this idea is shown in figure 6–1.

What You Need to Know

To do a looping problem such as the distance problem, there are four things you need to know.

1. The value for starting the loop.
2. Action to be taken within the loop.
3. The value for ending the loop.
4. Increments used for each pass through the loop.

Figure 6–2 illustrates this process.

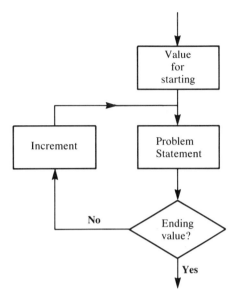

Figure 6–2   Steps in a Program Loop

Going back to the distance program, here is how each of the four items required for a loop is answered.

1. The value for starting the loop.
   The starting value of Time = 1.
2. Action to be taken within the loop.
   Solve for the distance an object falls each second.
3. The value for ending the loop.
   The ending value of Time = 3.
4. Increments used for each pass through the loop.
   Time is incremented by 1 second, each pass.
   (Time = 1, 2, and 3 seconds.)

In this example, Time could just as easily have been increased by 2 each time and solved for Time = 1, Time = 3, and Time = 5 (with the ending value changed to 5). Remember, then, that this type of loop block requires four pieces of information: where to start, what to do, when to end, and how much to change each time.

Loop and Branch Compared

Recall that the difference between a loop block and a branch block is that in a loop block the program has the potential to go back and repeat the same process. In a branch block, the program always advances forward and makes a decision between alternatives. The flowcharts of the two are compared in figure 6–3 on page 324.

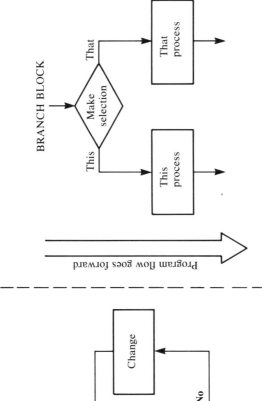

Figure 6–3 Comparing the Loop and Branch Block

Why Loops Are Needed

Program 6–14 is not very efficient. It can be made more efficient by converting it to a loop block, but first let's see why the program is not efficient. Suppose you wanted a program to give you the distance that an object would fall every millisecond (one thousandth of a second) for the first three seconds. Doing this using the same method as in program 6–14 would require thousands of program lines as shown in Program 6–15.

Program 6–15

```
PROGRAM Solve_Distance_1;
  CONST
   a = 32;   {Acceleration due to gravity.}

  FUNCTION Distance(Time : REAL):REAL;
  BEGIN
     Distance := 0.5 * a * Time * Time;
  END;

  PROCEDURE Calculate_It;
   VAR
    Answer
   :REAL;

     BEGIN
       Answer := Distance(0.001);
       WRITELN('Distance at 1.0 millisecond is ',Answer:3:3,' feet.');
       Answer := Distance(0.002);
       WRITELN('Distance at 2.0 milliseconds is ',Answer:3:3,' feet.');
       Answer := Distance(0.003);
       WRITELN('Distance at 3.0 milliseconds is ',Answer:3:3,' feet.');
                            .
                            .
                            .
                   (And so on for thousands of lines)
                            .
                            .
                            .
     END;

BEGIN
 Calculate_It;
END.
```

Obviously, this method of programming is not that efficient; it would require hours, if not days, of programming time and would try the patience of any programmer. There has to be a better way; that is what the rest of the chapter is about.

Conclusion

In the next section, you will see the structure of a loop block. In this section you saw what is needed for a loop block and why the method used here is not practical. Check your understanding of this section by trying the following section review.

6–2 Section Review

1 Name the three different kinds of programming blocks.
2 State why program 6–15 is not practical.
3 Define a programming loop.
4 Explain the difference between a loop block and a branch block.

6–3   Repeating Things

Introduction

In the last section, you were shown why loop blocks were necessary and what they had to contain. You also saw the difference between a loop block and a branch block. In this section, you will learn how to construct a loop block; you'll also be introduced to an important and powerful loop statement.

Armed with the knowledge presented in this section plus what you have learned in previous chapters, you will be able to use the full power of the computer in terms of the three types of programming blocks used to solve any problem: action, branch, and loop blocks.

Fundamental Loop Structure

Here, you will see a practical solution to the distance problem encountered in the last section in which you needed to compute the distance an object would fall every second for three seconds. Recall that you needed to know four things in order to work this problem. Here is an outline of what you want to do.

1. Start with Time = 1 Second.
2. Repeat the following:
   (Formula for the distance)
   (Display the answer)
3. Increase by 1 more second.
4. Stop repeating after Time = 3 Seconds.

Looking at this scheme, you can see that all of the necessary loop ingredients are there; you know what value to start with, you know the action to be taken (formula, display answer), you know how much you will increase the value of Time by each pass through the loop (increase by 1 second), and you know when the loop will end (after Time = 3 seconds).

There is a statement in Pascal that allows you to loop easily. It is called the **REPEAT–UNTIL**. Its structure is

REPEAT(Statements)**UNTIL**(Expression)

where Statements **=** any legal Pascal statement

Expression **=** when this is TRUE, the repeating of the Statements will stop and the program will move forward

Program 6–16 is an example of using the **REPEAT–UNTIL** statement. The program is not very useful, because the repeating will never stop, since the value of the variable Time never changes.

Program 6–16

```
PROGRAM Loop_Forever;
 CONST
   a = 32;   {Acceleration due to gravity.}

 FUNCTION Distance(Time : REAL) : REAL;
  BEGIN
    Distance := 0.5 * a * Time * Time;
  END;

 PROCEDURE Loop_It;
  VAR
   Answer
:REAL;
   Time
:INTEGER;

   BEGIN

    Time := 1;   {This is the value for starting the loop.}

    REPEAT   {This is the start of the loop.}

      Answer := Distance(Time);   {This is what will be repeated.}
      WRITELN('Distance at ',Time,' seconds is ',Answer:2:2,' feet.');

    UNTIL Time = 3;   {This is when the loop will finish.}

   END;

BEGIN
  Loop_It;
END.
```

When program 6–16 is executed, the output will forever be

```
Distance at 1 seconds is 16.00 feet.
Distance at 1 seconds is 16.00 feet.
```

```
Distance at 1 seconds is 16.00 feet.
Distance at 1 seconds is 16.00 feet.
                    .
                    .
                    .
```

The problem with program 6–16 is that one of the four steps for creating a loop was omitted—incrementing! Program 6–16 did not change the value of the variable Time.

An Important Concept

There is a neat trick in programming that lets you change the value of a variable in one easy program line.

```
Time := Time + 1;
```

This program line is not an algebraic expression (that is why the `:=` sign is used rather than the `=`.) It is an instruction to the computer. The instruction means

> To the value stored in the memory
> location designated by the variable Time,
> add 1 and store the result back in the same
> memory location.

This important concept is illustrated in figure 6–4.
To increase the variable Time by two, you only need

```
Time := Time + 2;
```

or to increase it by 0.5,

```
Time := Time + 0.5;
```

or to decrease it by 3 each time through the loop,

```
Time := Time - 3;
```

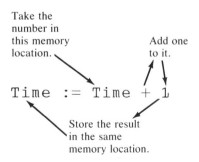

Figure 6–4   Meaning of Time := Time + 1

Put this instruction inside the loop in your program, following all the other statements you want the loop to do. To get an idea of what happens inside such a loop, consider program 6–17.

Program 6–17

```
PROGRAM This_Loop_Works;
  VAR
   Counter
 :INTEGER;

   BEGIN

     Counter := 1;   {Initialize counter}

     REPEAT          {Start loop here.}

       WRITELN('Value of counter = ',Counter);
       Counter := Counter + 1;   {Increment counter.}

     UNTIL Counter = 5;   {End loop here.}

   END.
```

When program 6–17 is executed, the output will be

```
Value of counter = 1
Value of counter = 2
Value of counter = 3
Value of counter = 4
```

Note where the loop stops—right when the value of the variable counter = 5. Also note that the statements inside the loop will always be executed at least once. This can create a problem as shown in program 6–18.

Program 6–18

```
PROGRAM This_Loop_Has_A_Problem;
  VAR
   Counter
 :INTEGER;

   BEGIN

     Counter := 1;   {Initialize counter}
```

Program 6–18   *continued*

```
    REPEAT          {Start loop here.}

      WRITELN('Value of counter = ',Counter);
      Counter := Counter + 1;  {Increment counter.}

    UNTIL Counter = 1;  {End loop here.}

  END.
```

You may be surprised to find that the above program will never stop! When it is executed, the output will show the value forever incrementing.

```
Value of counter = 1
Value of counter = 2
Value of counter = 3
Value of counter = 4
              .
              .
              .
```

One way to prevent this is to change the **UNTIL** to

UNTIL Counter > 1;

Now the program will pass through the loop only once, because the statement will stay TRUE anytime the variable `Counter` is larger than 1.

You can also cause the **REPEAT–UNTIL** loop to count things backwards, as shown in program 6–19.

Program 6–19

```
PROGRAM This_Loop_Goes_Backwards;
  VAR
   Counter
 :INTEGER;

   BEGIN

     Counter := 5;  {Initialize counter}

     REPEAT          {Start loop here.}

       WRITELN('Value of counter = ',Counter);
       Counter := Counter — 1;  {Decrement counter.}

     UNTIL Counter <= 1;  {End loop here.}

   END.
```

When program 6–19 is executed, it will output

```
Value of counter = 5
Value of counter = 4
Value of counter = 3
Value of counter = 2
```

If the **UNTIL** condition were changed to

UNTIL Counter < 1;

the output would have included one more line.

```
Value of counter = 1
```

Falling Body Revisited

The falling-body program can now be put into a **REPEAT-UNTIL** loop and solved for the distance every second, starting at 1 second and continuing to 5 seconds.

Program 6–20

```
PROGRAM Falling_Body;

  CONST
    a = 32;   {Acceleration due to gravity.}

  FUNCTION Distance(Time : REAL) : REAL;
  BEGIN
    Distance := 0.5 * a * Time * Time;
  END;

  PROCEDURE Do_It;
  VAR
   Answer
 :REAL;
   Time
 :INTEGER;

   BEGIN

    Time := 1;     {Initialize timer.}

    REPEAT
      Answer := Distance(Time);
      WRITELN('Distance at ',Time,' seconds = ',Answer:3:3,' feet.');
      Time := Time + 1  {Increment Time}
    UNTIL Time > 5;

  END;

BEGIN
  Do_It;
END.
```

When program 6–20 is executed, the output will be

```
Distance at 1 seconds = 16 feet.
Distance at 2 seconds = 64 feet.
Distance at 3 seconds = 144 feet.
Distance at 4 seconds = 256 feet.
Distance at 5 seconds = 400 feet.
```

and the loop will properly terminate at 5 seconds.

Conclusion

This section introduced the **REPEAT–UNTIL** loop used in Pascal. In the next section you will be introduced to another method of doing loops, increasing your options for implementing this powerful concept. Check your understanding of this section by trying the following section review.

6–3 Section Review

1 Describe a fundamental loop structure.
2 Explain the meaning of `Counter := Counter + 1;`.
3 Why is it poor practice to use an equality to terminate a **REPEAT–UNTIL** loop (ie. `UNTIL Variable = 2;`)?
4 What is the minimum number of times a **REPEAT–UNTIL** loop will be executed?
5 State the preferred method of terminating a **REPEAT–UNTIL** loop.

6–4   WHILE Things are Happening

Introduction

In the last section, you learned the structure of the **REPEAT–UNTIL** loop. In this section you will see another loop structure available to you in Pascal. Having these options in loop structure increases the versatility of your programs and gives you options not available in all programming languages.

Basic Idea

The second way of creating a loop block in Pascal is the **WHILE–DO** loop. The structure of this loop is

WHILE (expression) **DO** (statement)

where (expression) = any legal expression that will produce a Boolean value
(statement) = a single or compound statement with **BEGIN–END**

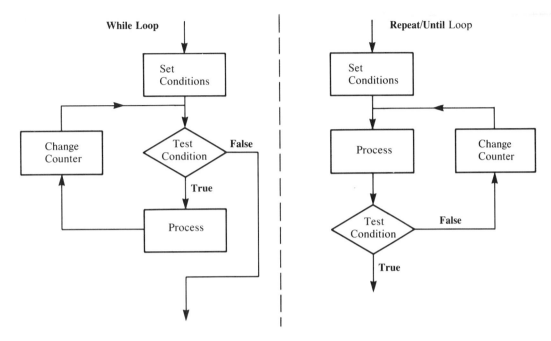

Figure 6–5   Comparison of WHILE with REPEAT-UNTIL

Figure 6–5 compares the **WHILE-DO** and **REPEAT-UNTIL** loops.

Note that the **REPEAT-UNTIL** loop evaluates its ending condition after it has executed its enclosed statements. However, the **WHILE-DO** loop evaluates its ending condition before executing the enclosed statements. Also note that the statements within the **REPEAT-UNTIL** do not require a **BEGIN-END** to indicate the extent of the loop, but the **WHILE-DO** loop does require a **BEGIN-END** to indicate the extent of the loop. An application of this important loop structure is illustrated in program 6–21.

Program 6–21

```
PROGRAM While_Loop_Demo;

 VAR
  User_Input
:REAL;

  BEGIN
```

Program 6–21   *continued*

```
    WRITE('Give me a number from 1 to 10 = ');
    READLN(User_Input);

    WHILE (User_Input < 1) OR (User_Input > 10) DO
      BEGIN
        WRITELN('Please make your number between 1 and 10.');
        WRITELN('Try again.');
        WRITE('Enter a number from 1 to 10 = ');
        READLN(User_Input);
      END;   {of While/Do loop}
  END.
```

The intent of the above program is to get the program user to enter a number between 1 and 10. If a value larger or smaller is entered the error message within the **WHILE–DO** loop will be displayed. Such a loop is referred to as an **error-trapping routine**. Its purpose is to force the program user to enter a limited range of values. This technique has many important applications in technology programs, as you will see.

When program 6–21 is executed, if the program user inputs a value larger than 10, the display will be

```
Give me a number from 1 to 10 = 12
Please make your number between 1 and 10.
Try again.
Enter a number from 1 to 10 = 15
Please make your number between 1 and 10.
Try again.
Enter a number from 1 to 10 = 0
                  •
                  •
                  •
```

(This process will continue WHILE the
program user inputs a number less than 1 or
greater than 10.)

You can see why the **WHILE–DO** loop used this way is referred to as a **sentinel loop**. In program 6–21, you have no idea how many times the program user will enter incorrect data—if at all. It wouldn't make sense to use a **REPEAT–UNTIL** loop for this application, since the **error trap** would be displayed at least once, even if the user entered the correct range of values.

As a Counting Loop

The **WHILE-DO** loop can be used as a counting loop as shown in program 6–22.

Program 6–22

```
PROGRAM Counting_Loop;
 VAR
  Counter
  :INTEGER;

  BEGIN
    Counter := 0;          {Initial conditions.}

      WHILE Counter <> 5 DO       {Final conditions}
        BEGIN
          WRITELN('Counter is = ',Counter);      {Body of loop}
          Counter := Counter + 1;             {Incremental value}
        END;

  END.
```

When program 6–22 is executed the output will be

```
Counter is = 0
Counter is = 1
Counter is = 2
Counter is = 3
Counter is = 4
```

Note that the loop contains the four necessary ingredients for a counting loop: initial value, process to be repeated, final value, and increment. Be careful with this loop, because you could run into one that would repeat itself more times than expected.

Program 6–23

```
PROGRAM Problem_Loop;
 VAR
  Counter
  :INTEGER;

  BEGIN
    Counter := 0;          {Initial conditions.}
```

Program 6–23 *continued*

```
    WHILE Counter <> 5 DO        {Final conditions}
      BEGIN
        WRITELN('Counter is = ',Counter);      {Body of loop}
        Counter := Counter + 2;          {Incremental value}
      END;

  END.
```

The loop in program 6–23 will repeat forever because the value of the variable Counter will never be 5. Since it is being incremented by 2 each time its value will be 0, 2, 4, 6, 8, . . . and so on. Good programming practice would have you change the **WHILE–DO** conditions to

WHILE Counter **<=** 5 **DO**

Conclusion

The **WHILE–DO** loop is preferred over that of the **REPEAT–UNTIL** because it checks the loop condition *before* execution. This section showed an application as an error-trapping routine. In the next section, you will see a third type of Pascal loop. For now, test your understanding of this section with the following section review.

6–4 Section Review

1 Describe the structure of the **WHILE–DO** loop.
2 What is the main difference between the **WHILE–DO** loop and the **REPEAT–UNTIL** loop?
3 Explain what is meant by an error-trapping routine. Why is a **WHILE–DO** loop used?
4 State the precaution that should be taken when constructing a **WHILE–DO** loop.

6–5   FOR One Thing TO Another

Introduction

You have been shown two types of loop structures in Pascal. The **WHILE–DO** means "Do something while a condition is TRUE". The **REPEAT–UNTIL** means "Repeat something until a condition is TRUE". In this section, you will learn about a third type of loop structure which essentially means "Do something a specific number of times".

The FOR Statement

In Pascal, the third type of loop structure is called the **FOR–TO**. Its structure is

FOR(variable) **:=** (expression$_1$) **TO**(expression$_2$)**DO**(statement)

where (`variable`) = any ordinal variable identifier
(`expression`$_1$) = the beginning value of the (variable)
(`expression`$_2$) = the ending value of the (variable)
(`statement`) = any legal Pascal statement which must be enclosed by
BEGIN-END if more than one statement is used.

Program 6–24 is an example of a **FOR-TO** loop in Pascal.

Program 6–24

```
PROGRAM For_Loop;
  VAR
    Counter;
 :INTEGER;

     BEGIN

        FOR Counter := 1 TO 5 DO
          BEGIN
            WRITELN('This is a compound statement.');
            WRITELN('The value of the counter is: ');
            WRITELN(Counter);
          END;   {of For-To}

     END.
```

When program 6–24 is executed, the output will be

```
This is a compound statement.
The value of the counter is
1
This is a compound statement.
The value of the counter is
2
This is a compound statement.
The value of the counter is
3
This is a compound statement.
The value of the counter is
4
This is a compound statement.
The value of the counter is
5
```

Note that program 6–24 shows that all four conditions for a counting loop are given: the beginning value is 1, the ending value is 5, the body of the loop is defined by its own **BEGIN-END** statement, and by implication the incremental value is 1. Thus for a **FOR-TO** loop in Pascal, the only increment you can have is 1!

Decrementing with the FOR-TO

The **FOR-TO** loop can also be used for decrementing. The structure is

FOR(variable) :=(expression₁)**DOWNTO**(expression₂)**DO**(statement)

This loop is illustrated in program 6–25.

Program 6–25

```
PROGRAM For__Loop;
  VAR
    Counter;

 :INTEGER;

     BEGIN

        FOR Counter := 5 DOWNTO 1 DO
          BEGIN
            WRITELN('This is a compound statement.');
            WRITELN('The value of the counter is: ');
            WRITELN(Counter);
          END;   {of For-To}

        END.
```

When program 6–25 is executed, the output will be

```
This is a compound statement.
The value of the counter is
5
This is a compound statement.
The value of the counter is
4
This is a compound statement.
The value of the counter is
3
This is a compound statement.
The value of the counter is
2
This is a compound statement.
The value of the counter is
1
```

If the ordinal value of the first expression of an incrementing **FOR-TO** is greater than that of the second expression, the statement is never executed, as in program 6–26.

Program 6–26

```
PROGRAM For_Loop;
  VAR
    Counter;
 :INTEGER;

      BEGIN

        FOR Counter := 5 TO 1 DO
          BEGIN
            WRITELN('This is a compound statement.');
            WRITELN('The value of the counter is: ');
            WRITELN(Counter);
          END;  {of For-To}

      END.
```

When program 6–26 is executed, nothing will be displayed on the screen, because the counter was told to start at 5 and count up to 1. This means that a **FOR–TO** loop in Pascal will never go into endless repeats as **REPEAT–UNTIL** and **WHILE–DO** loops can.

Using any Ordinal Value

It's important to point out that the **FOR–TO** loop in Pascal can use any ordinal value. In the first section of this chapter, you were introduced to ordinal types, meaning you can use a CHAR for the type of (variable) to be used as the counter, as illustrated by program 6–27.

Program 6–27

```
PROGRAM Character_Loop;
  VAR
   Counter
  :CHAR;

    BEGIN

      FOR Counter := 'A' TO 'E' DO
        WRITELN('The counter is ',Counter);

    END.
```

When program 6–27 is executed, the output will be

```
The counter is A
The counter is B
The counter is C
The counter is D
The counter is E
```

Observe the type of variable used in program 6–27. It is a CHARacter, not a number, because the (variable) used in a Pascal **FOR-TO** loop can be any ordinal type—recall from the first section of this chapter that letters of the English alphabet are included. This means that a **FOR-TO** loop could also be performed with a BOOLEAN type as shown in program 6–28.

Program 6–28

```
PROGRAM Boolean_Loop;

VAR
 Counter
:BOOLEAN;

  BEGIN

      FOR Counter := FALSE TO TRUE DO
          WRITELN('The counter is ',Counter);

  END.
```

Execution of program 6–28 would yield

```
The counter is FALSE
The counter is TRUE
```

This means that you can use your own data types as the variable in a **FOR-TO** loop. Recall the data type you used in the first section of this chapter for the resistor color code. You could use this to create a **FOR-TO** loop as illustrated in program 6–29.

Program 6–29

```
PROGRAM Your_Own_Type_Loop;
  TYPE
    Color = (Black, Brown, Red, Orange, Yellow, Green,
         Blue, Violet, Gray, White);
```

Program 6–29 *continued*

```
VAR
 Counter
:Color;

  BEGIN

      FOR Counter := Black TO Orange DO
      WRITELN('The counter is ',ORD(Counter));

  END.
```

Execution of program 6–29 will produce

```
The counter is 0
The counter is 1
The counter is 2
The counter is 3
```

Program 6–29 could be made to sequence through any part of the values defined for the type Color, and could also decrement. Recall that this data type can not be displayed directly to the screen and therefore needs the Pascal **ORD** function to have its ordinality displayed.

Conclusion

This section presented the third kind of Pascal loop structure, the **FOR–TO**. This loop is available more for convenience than for necessity. It is a convenient way of creating a loop that will increment or decrement any ordinal value by 1. Test your understanding of this section by trying the following section review.

6–5 Section Review

1 Name the three different kinds of loops available in Pascal.
2 State the unique properties of the Pascal **FOR–TO** loop.
3 Explain how to make the **FOR–TO** loop decrement.
4 What will happen if the beginning value of the **FOR–TO** loop is larger than the ending value when the loop is to increment?
5 State the reason for **FOR–TO** loops in Pascal.

6–6   Nested Loops

Discussion

This section shows the concept of **nested loops**. You will see how to place one loop inside another loop, and find that doing this can produce some surprising results.

General Idea

A nested loop is simply one loop inside another. Program 6–30 shows a nested loop using **FOR–TO** loops.

Program 6–30

```
PROGRAM Nested_Loop_1;
 VAR
  Outer_Value,
  Inner_Value
:INTEGER;

   BEGIN

    FOR Outer_Value := 1 TO 2 DO
     BEGIN

       WRITELN('Outer value = ',Outer_Value);

       FOR Inner_Value := 1 TO 5 DO
        BEGIN
           WRITELN(' Inner value = ',Inner_Value);
        END;   {Inner loop.}

     END;   {Outer loop.}

   END.
```

When program 6–30 is executed, it will output

```
Outer value = 1
  Inner value = 1
  Inner value = 2
  Inner value = 3
  Inner value = 4
  Inner value = 5
Outer value = 2
  Inner value = 1
  Inner value = 2
  Inner value = 3
  Inner value = 4
  Inner value = 5
```

Nested Loop Structure

Note that in the output of program 6–30, the **outer loop** counts once, then the **inner loop** goes through its complete count, and only then does the outer loop count once again. This nesting of one loop inside another should be formatted in such a way that it is clear which loop is the outside loop and which is the inside loop. Observe the indentation used by program 6–30 to indicate this.

```
FOR Outer_Value := 1 TO 2 DO
BEGIN
   WRITELN('Outer value = ',Outer_Value);

   FOR Inner_Value := 1 TO 5 DO
    BEGIN
       WRITELN('Inner value = ',Inner_Value);
    END; {Inner loop.}
END; {Outer loop.}
```

When structuring nested loops it is recommended that the **BEGIN** and **END** of each loop be lined up vertically. Note from the structure of the segment of program 6–30 shown that the **BEGIN** of the outer loop lines up vertically with the **END** of the outer loop. The same is true of the **BEGIN** and **END** parts of the inner loop. It is also a good idea to indicate the purpose of each **END;** statement (such as {Inner loop} or {Outer loop}.

Nesting with REPEAT-UNTIL

You can also nest loops with the Pascal **REPEAT-UNTIL** loop, as illustrated in program 6–31.

Program 6–31

```
PROGRAM Nested_Loop_2;    {This loop has a problem}
 VAR
  Outer_Value,
  Inner_Value
:INTEGER;

   BEGIN

     {Initialize variables.}
      Outer_Value := 0;
      Inner_Value := 0;

   REPEAT

      WRITELN('Outer value = ',Outer_Value);

      REPEAT

         WRITELN(' Inner value = ',Inner_Value);
         Inner_Value := Inner_Value + 1;
```

Program 6–31   *continued*

```
      UNTIL Inner_Value >= 5;    {Inner loop}

      Outer_Value := Outer_Value + 1;

   UNTIL Outer_Value > 2;  {Outer loop.}
END.
```

There are several important points to make about using nested **REPEAT-UNTIL** loops. First, you should initialize the variables. Unlike a **FOR-TO** loop, there is no telling what the variables `Inner_Value` and `Outer_Value` will be when the loop starts. Therefore, you set each to a value that will make the loop work the way you want it. In this case both were set to zero.

There is another precaution to be taken when nesting a **REPEAT-UNTIL**. Look at the output of program 6–31.

```
Outer value = 0
 Inner value = 0
 Inner value = 1
 Inner value = 2
 Inner value = 3
 Inner value = 4
Outer value = 1
 Inner value = 5
Outer value = 2
 Inner value = 6
```

This may not be the output you would expect, but it occurs because the inner loop variable had an ending value of 4 the first time through the loop. The second time through the inner loop, the inner loop variable was now 5 (remember a **REPEAT-UNTIL** will always go through the loop at least once). After the outer loop was invoked again the inner loop variable increased its value to 6. You need a way of **resetting** the inner loop variable back to its original value when the inner loop is completed. This is accomplished by adding the program line shown in program 6–32.

Program 6–32

```
PROGRAM Nested_Loop_3;    {The problem is corrected}
 VAR
  Outer_Value,
  Inner_Value
 :INTEGER;
```

Program 6–32 *continued*

```
BEGIN

  {Initialize variables.}
   Outer_Value := 0;
   Inner_Value := 0;

  REPEAT

     WRITELN('Outer value = ',Outer_Value);

     REPEAT

        WRITELN(' Inner value = ',Inner_Value);
        Inner_Value := Inner_Value + 1;

     UNTIL Inner_Value >= 5;  {Inner loop}

     Inner_Value := 0; {Reset inner value.}
     Outer_Value := Outer_Value + 1;

  UNTIL Outer_Value > 2;   {Outer loop.}

END.
```

Observe the additional program line that resets the variable `Inner_Value` to zero. Now when program 6–32 is executed, the output is

```
Outer value = 0
 Inner value = 0
 Inner value = 1
 Inner value = 2
 Inner value = 3
 Inner value = 4
Outer value = 1
 Inner value = 0
 Inner value = 1
 Inner value = 2
 Inner value = 3
 Inner value = 4
Outer value = 2
 Inner value = 0
 Inner value = 1
 Inner value = 2
 Inner value = 3
 Inner value = 4
```

Nesting WHILE-DO Loops

Observe program 6–33.

Program 6–33

```
PROGRAM Nested_Loop_5;
 VAR
  Outer_Value,
  Inner_Value
:INTEGER;

   BEGIN

     {Initialize variables.}
      Outer_Value := 0;
      Inner_Value := 0;

     WHILE Outer_Value < 3 DO

       BEGIN

         WRITELN('Outer value = ',Outer_Value);

         WHILE Inner_Value < 5 DO
          BEGIN
           WRITELN(' Inner value = ',Inner_Value);
           Inner_Value := Inner_Value + 1;
          END;   {Inner loop.}

         Inner_Value := 0;  {Reset inner value.}
         Outer_Value := Outer_Value + 1;

       END;   {Outer loop}

    END.
```

Note that while the **WHILE-DO** loop requires that both variables be initialized and that the inner loop variable be reset to zero. These steps are necessary for the same reasons they were needed in the **REPEAT-UNTIL** loop.

Mixing Nested Loops

You can mix different types of Pascal loops, as shown in program 6–34.

Program 6–34

```
PROGRAM Nested_Loop_6;
 VAR
  Outer_Value,
  Inner_Value
:INTEGER;

   BEGIN

     {Initialize variables.}
      Outer_Value := 0;
      Inner_Value := 0;

    WHILE Outer_Value < 3 DO

      BEGIN

        WRITELN('Outer value = ',Outer_Value);

        REPEAT
          WRITELN('Inner value = ',Inner_Value);
          Inner_Value := Inner_Value + 1;
        UNTIL Inner_Value > 5;   {Inner loop.}

       Inner_Value := 0;   {Reset inner value.}
       Outer_Value := Outer_Value + 1;

      END;   {Outer loop}

   END.
```

Conclusion

This section presented the concept of nested loops. You saw that program structure enables visualization of what these loops were doing. You also saw that the "cleanest" nested loop is a **FOR-TO** loop; when **REPEAT-UNTIL** or **WHILE-DO** loops are nested, you must keep track of the values of the loop variables in order to achieve what you want the loop to do. Check your understanding of this section by trying the following section review.

6–6 Section Review

1 What is meant by a nested loop?
2 Explain what type of program structure should be used when developing a nested loop.
3 What precautions must you take when using a **REPEAT-UNTIL** in a nested loop?
4 State the advantage of using the **FOR-TO** in a nested loop.

6–7   A Loop Application

Introduction

This section presents a practical application of Pascal loops in which the values of an electrical circuit are calculated for a range of values. This type of calculation gives the program user an insight into technology problems seldom available without the use of the computer.

Application Program

The application program calculates the impedance of a series LRC circuit. One of the interesting phenomena of this circuit is that its total opposition to current flow (called the impedance) changes with the applied frequency. The point of minimum circuit impedance is called the resonant frequency. A schematic of such a circuit with a graph of its impedance and frequency change is shown in figure 6–6.

Program 6–35 is the application program for the impedance problem graphed in figure 6–6.

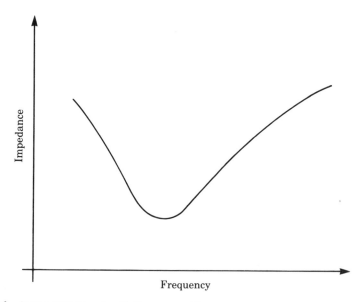

Figure 6–6   Series LRC Circuit with Frequency Plot

Program 6-35

```
PROGRAM Series_Resonant_Circuit;
{*************************************************************
             Developed by: An Electronic Student
                   Date:   December 1995
*************************************************************
   This program will calculate the impedance of a series
   resonant circuit consisting of a resistor, capacitor and
   inductor.  The program will then ask for the lowest
   frequency and the highest frequency as well as the change
   in frequency for each calculation.  The value of the
   circuit impedance is then calculated for each frequency.
*************************************************************}
  {Library Units}
  USES
    CRT;

  {Global Variables}
   TYPE
    String79 = STRING[79];

   VAR
     Repeat_Program
  :BOOLEAN;
{*************************************************************
               Procedures Used:
----------------------------------------------------------}
{PROCEDURE Main_Programming_Sequence;

   DESCRIPTION
       This procedure calls all of the primary procedures
       used in this program.  It is the first procedure given
       in order to make the structure of the program easy
       to read and follow.

     PARAMETERS:
       None.

     SAMPLE CALL:
       BEGIN
         Main_Programming_Sequence;
       END.
-----------------------------------------------------------}
  PROCEDURE Explain_Program;
  FORWARD;
  {
     DESCRIPTION:
         This procedure explains the purpose and operation
         of the program to the program user.

     PARAMETERS:
       None.

    SAMPLE_CALL:
       Explain_Program;
-------------------------------------------------------}
```

Program 6–35 *continued*

```
    PROCEDURE Get_Values;
    FORWARD;
    {
       DESCRIPTION
            This procedure gets the values of the circuit
            resistance, capacitance, inductance, highest
            frequency, lowest frequency, and incremental
            change in frequency for program calculations.

       PARAMETERS:
         None.

       SAMPLE CALL:
            Get_Values;
  ------------------------------------------------------------}
   PROCEDURE Display_Values(Resistance, Inductance, Capacitance,
                            Starting_Frequency,
                            Ending_Frequency,
                            Change_in_Frequency : REAL);

   FORWARD;
   {
      DESCRIPTION:
           This procedure calls the program functions that
           do the calculations and displays the answers on
           the monitor.

      PARAMETERS:
            Resistance (input) = Circuit resistance in ohms.
            Inductance (input) = Circuit inductance in henrys.
           Capacitance (input) = Circuit capacitance in farads.
    Starting_Frequency (input) = Beginning frequency for calculations.
      Ending_Frequency (input) = Ending frequency for calculations.
Change_in_Frequency (input) = Change in frequency for new calculation.

      SAMPLE CALL:
           Display_Values(10,1E-3,1E-6,1E3,10E3,1E3);

      RESULT:
           The circuit impedance for a frequency range from
           1000 Hz to 10000 Hz in steps of 1000 Hz each.
  ------------------------------------------------------------}
   FUNCTION Impedance(Inductor, Capacitor, Resistor,
                      Frequency : REAL) : REAL;
   FORWARD;
   {
      DESCRIPTION
           This function calculates the total impedance of
           the circuit.

      ROUTINES USED:
         None.

      PARAMETERS:
            Inductor (input) = Value of the circuit inductor in henrys.
           Capacitor (input) = Value of the circuit capacitor in farads.
            Resistor (input) = Value of the circuit resistor in ohms.
           Frequency (input) = Value of the applied frequency in Hertz.
```

Program 6–35 *continued*

```
        SAMPLE CALL:
            Total_Z := Impedance(1E-3,1E-6,10,1E3);

        RESULT:
            Total_Z = 1.53198018E+2
 -------------------------------------------------------------------}
   PROCEDURE Program_Repeat;
   FORWARD;
   {
        DESCRIPTION
            This procedure asks the program user if the program
            is to be repeated.

        PARAMETERS:
            None.

        SAMPLE CALL:
            Program_Repeat;
 -------------------------------------------------------------------}

{******************************************************************}
 PROCEDURE Main_Programming_Sequence;
{==================================================================}

    BEGIN

        Explain_Program;    {Explain program to user.}

    REPEAT
        Get_Values;         {Get circuit values from user.}
        Program_Repeat;     {Ask if program is to be repeated.}
    UNTIL Repeat_Program = FALSE;

{==================================================================}
    END;  {of procedure Main_Programming_Sequence}

  PROCEDURE Explain_Program;
{------------------------------------------------------------------}

    BEGIN

        ClrScr;    {Clear the screen.}
        WRITELN;
        WRITELN('This program will compute the impedance of a series');
        WRITELN('LRC circuit for a range of frequencies.');
        WRITELN;
        WRITELN('You must enter the value of the resistor in ohms');
        WRITELN('value of the inductor in henrys, and the capacitor');
        WRITELN('in farads.');
        WRITELN;
        WRITELN('The program will then ask for the minimum frequency');
        WRITELN('to begin calculation, the maximum frequency to end');
        WRITELN('calculation, and the incremental frequency change');
        WRITELN('for each calculation.');
        WRITELN;
        WRITE('Press RETURN/ENTER to continue.');
        READLN;
{------------------------------------------------------------------}
```

Program 6–35 *continued*

```
  END;  {of procedure Explain_Program}

    PROCEDURE Get_Values;
{-------------------------------------------------------------}
    VAR
     Resistor,         {Circuit resistance}
     Capacitor,        {Circuit capacitance}
     Inductor,         {Circuit inductance}
     Start_Frequency, {Starting frequency for calculations}
     End_Frequency,   {Ending frequency for calculations}
     Change_Frequency {Incremental change in frequency}
    :REAL;

      BEGIN

        ClrScr;   {Clear the screen.}
        WRITELN;
        WRITELN('Enter the value of the:');
        WRITELN;
        WRITE('Resistor in ohms = ');
        READLN(Resistor);
        WRITE('Capacitor in farads = ');
        READLN(Capacitor);
        WRITE('Inductor in henrys = ');
        READLN(Inductor);
        WRITE('Starting frequency in hertz = ');
        READLN(Start_Frequency);
        WRITE('Ending frequency in hertz = ');
        READLN(End_Frequency);
        WRITE('Change in frequency in hertz = ');
        READLN(Change_Frequency);

        Display_Values(Resistor, Inductor, Capacitor,
                    Start_Frequency, End_Frequency,
                    Change_Frequency);
  {-------------------------------------------------------------}
      END;  {of procedure Get_Values}

  PROCEDURE Display_Values{(Resistance, Capacitance, Inductance,
                    Start_Frequency, End_Frequency,
                    Change_Frequency : REAL)};
{-------------------------------------------------------------}
    VAR
      Total_Impedance,
      Frequency
    :REAL;

      BEGIN

      ClrScr;  {Clear the screen}
      {Initialize variables}
      Frequency := Starting_Frequency; {Starting value of calculations.}

      REPEAT     {Begin the loop.}
```

Program 6–35   *continued*

```
        Total_Impedance := Impedance(Inductance,Capacitance,Resistance,
                           Frequency);
        WRITELN('Frequency = ',Frequency:3:3,' Hz Impedance = ',
                Total_Impedance:3:3,' ohms.');
        Frequency := Frequency + Change_in_Frequency; {Increment}
      UNTIL Frequency > Ending_Frequency;
  {------------------------------------------------------------}
    END; {of procedure Display_Values}

  PROCEDURE Program_Repeat;
  {------------------------------------------------------------}
  VAR
    Desire
  :CHAR;

    BEGIN

        GOTOXY(5,25);  {Display message at bottom of screen.}
        WRITE('Do you want to repeat the program (Y/N)? => ');
        READLN(Desire);
         IF (Desire = 'N') OR (Desire = 'n') THEN
          Repeat_Program := FALSE
         ELSE
          Repeat_Program := TRUE;
  {------------------------------------------------------------}
    END;  {of procedure Repeat_Program}

FUNCTION Impedance{(Inductor,Capacitor,Resistor,Frequency
                   :REAL):REAL};
  {------------------------------------------------------------}
  VAR
     XL,         {Inductive reactance.}
     XC          {Capacitive reactance.}
  :REAL;

    BEGIN

      XL := 2 * PI * Frequency * Inductor;
      XC := 1/(2 * PI * Frequency * Capacitor);

      Impedance := SQRT(SQR(XL - XC) + SQR(Resistor));
  {------------------------------------------------------------}
    END;  {of function Impedance}

  BEGIN
    Main_Programming_Sequence;
  END.
```

A sample execution of program 6–35 will produce

```
Enter the value of the:

Resistor in ohms = 10
Capacitor in farads = 1E—6
```

```
Inductor in henrys = 1E-3
Starting frequency in hertz = 1E3
Ending frequency in hertz = 1E4
Change in frequency in hertz = 1E3
[Screen is cleared.]
Frequency = 1000.000 Hz  Impedance = 153.198 ohms.
Frequency = 2000.000 Hz  Impedance = 67.753 ohms.
Frequency = 3000.000 Hz  Impedance = 35.634 ohms.
Frequency = 4000.000 Hz  Impedance = 17.743 ohms.
Frequency = 5000.000 Hz  Impedance = 10.009 ohms.
Frequency = 6000.000 Hz  Impedance = 14.995 ohms.
Frequency = 7000.000 Hz  Impedance = 23.482 ohms.
Frequency = 8000.000 Hz  Impedance = 31.975 ohms.
Frequency = 9000.000 Hz  Impedance = 40.131 ohms.
Frequency = 10000.000 Hz Impedance = 47.970 ohms.
```

Program Analysis

There are several important points to note about program 6–35. First, note that the program user has the option of repeating the program. This is accomplished through the procedure **Program__Repeat** and is called from the main programming sequence.

```
PROCEDURE Main__Programming__Sequence;
{===============================================================}

   BEGIN

      Explain__Program; {Explain program to user.}

   REPEAT
      Get__Values; {Get circuit values from user.}
      Program__Repeat; {Ask if program is to be repeated.}
   UNTIL Repeat__Program = FALSE;
{===============================================================}
   END; {of procedure Main__Programming__Sequence}
```

Note that the **REPEAT–UNTIL** loop starts after the program has been explained to the program user; it is usually safe to assume that the user understands the operation of the program after the first time through, and need not suffer through the same explanation again.

The other important aspect of the program is that the program user selects the starting, ending, and incremental values for the program. This was achieved by using variables for the initial and final value of the loop as well as the incremental value.

```
{Initialize variables}
Frequency := Starting__Frequency; {Starting value of calculations.}

REPEAT     {Begin the loop.}
```

```
Total_Impedance := Impedance(Inductance,Capacitance,Resistance,
                             Frequency);
WRITELN('Frequency = ',Frequency:3:3,' Hz Impedance = ',
        Total_Impedance:3:3,' ohms.');
Frequency := Frequency + Change_in_Frequency; {Increment}
UNTIL Frequency > Ending_Frequency;
```

Observe that the initial value of the loop **Impedance** contained a value selected by the program user. This was also true of the final value of the loop **Ending_Frequency**. The incremental value of the loop **Change_in_Frequency** contained a user-selected value as well. Thus, in this program, not only were the circuit values selected by the program user, but just as important, the range of circuit variables for investigating the circuit were also selected by the user.

Conclusion

The ability to observe technology problems over a range of values is a powerful computer simulation tool. The applications program demonstrated in this section showed many of the important features of designing such a program in Pascal. You have come a long way in your ability to create programs that analyze problems in technology. Check your understanding of this section by trying the following section review.

6–7 Section Review

1 Explain in your own words what the applications program presented in this section will do.
2 How many procedures are contained in this program? Name them.
3 Describe the types of loops used in this program.
4 Explain why the whole program is not repeated when the program user repeats the program.
5 Are there any nested loops in this program? If so, what procedure or function contains them?

6–8   Program Debugging and Implementation— Preparing User Input

Introduction

In this section, you will learn a way to prevent a problem common to Pascal programs, a problem that occurs when the program user is to input a number and something other than a number is input. This can happen when the user accidentally presses the wrong keys, doesn't understand what is wanted, or intentionally enters other characters (such as commas with the number). This section will first develop a series of programs to show you the potential problem and then one method of solving it.

An Input Problem

Consider program 6–36.

Program 6–36

```
PROGRAM Possible_Input_Problem;
 VAR
   User_Input
 :INTEGER;

   BEGIN

     WRITE('Please input a whole number = ');
     READ(User_Input);

     WRITELN('Thank you.');

   END.
```

If the program user does exactly what the program asks, there will be no problem.

```
Please input a whole number = 5
Thank you.
```

But if the program user inputs something other than a whole number, such as numbers separated by commas or a string, the program will terminate with a screen message.

```
Please input a whole number = Whole Number
Runtime error 106 at 0000:005E.
```

This type of message is simply stating that the program expected a numerical value and not a string. The error message and the termination of the program usually have the effect of confusing the user and there is no indication of what to do from that point on. Program 6–37 shows a way of taking care of this problem.

Program 6–37

```
PROGRAM Reduced_Input_Problem;
 VAR
   User_Input
 :STRING[79];

   BEGIN
```

Program 6–37 *continued*

```
        WRITE('Please input a whole number = ');
        READ(User_Input);
        WRITELN('Thank you.');

    END.
```

Inputting with Strings

Note from program 6–37 that the input variable has been changed to a type string. Now, no matter what the user input, the program will accept it.

```
Please input a whole number = 12,367.23
Thank you.
```

and another example:

```
Please input a whole number = Whole Number
Thank you.
```

In either case, the input will not cause a run-time error. However, if you now want to do something with these numbers, there is another problem. Look at program 6–38.

Program 6–38

```
PROGRAM New_Problem;
 VAR
   Number_1,
   Number_2
 :STRING[79];

   Answer
 :REAL;

   BEGIN

      WRITELN('Give me two numbers and I''ll add them:');
      WRITELN;
      WRITE('First number = ');
      READLN(Number_1);
      WRITE('Second number = ');
      READLN(Number_2);

        Answer := Number_1 + Number_2;

      WRITELN('The sum of the two numbers is ',Answer);

    END.
```

When you try to compile the above program you will get a **type-mismatch error** because you are mixing STRINGs with REALs in the program line

```
Answer := Number_1 + Number_2;
```

and thus the program will not compile. You need a method of converting the input string into a number; this is accomplished with the Turbo Pascal **VAL** procedure as shown in program 6–39.

Program 6–39

```
PROGRAM Solution_to_Problem;
 VAR
   String_1,
   String_2
 :STRING[79];

   Number_1,
   Number_2,
   Answer
 :REAL;

   Conversion_Success,
   First_Try
 :INTEGER;

   BEGIN

    REPEAT

      WRITELN('Give me two numbers and I''ll add them:');
      WRITELN;
      WRITE('First number = ');
      READLN(String_1);
      WRITE('Second number = ');
      READLN(String_2);

   {Convert the strings to numbers}

          VAL(String_1, Number_1, Conversion_Success);
             First_Try := Conversion_Success;
          VAL(String_2, Number_2, Conversion_Success);

       IF (First_Try <>0 ) OR (Conversion_Success <> 0) THEN
          BEGIN
            WRITELN;
            WRITELN('Please enter numerical values only,');
            WRITELN('such as 23 or 0.23 or 3E4.');
            WRITELN('Do not use commas with numbers.');
            WRITELN;
          END;

     UNTIL (First_Try = 0) AND (Conversion_Success = 0);
```

Program 6–39   *continued*

```
      Answer := Number_1 + Number_2;

  WRITELN('The sum of the two numbers is ',Answer:3:2);

END.
```

Note that program 6–39 uses the built-in Turbo procedure **VAL**, which was presented in the last chapter. Recall it was defined as

`VAL(String_Number, Number, Code);`

where `String_Number` = a string that contains number characters to be converted to a number

`Number` = a REAL or INTEGER-type variable that will contain the converted number

`Code` = an INTEGER number that will return a value of zero if the conversion is successful

Note that for both inputs, the value of the conversion success is stored in

`First_Try` and `Conversion_Success`.

`{Convert the strings to numbers}`

```
VAL(String_1, Number_1, Conversion_Success);
   First_Try := Conversion_Success;
VAL(String_2, Number_2, Conversion_Success);
```

There is also a **REPEAT-UNTIL** loop that includes the user input part of the program and will not allow a computation to take place or an answer to be displayed until the conversion of both inputs is successful (that both inputs can be converted to numbers).

Conclusion

This section presented the important concept of protecting the user input from an accidental run-time error. Professionally-developed programs have this important feature and you should consider using this technique in programs you develop. Check your understanding of this section by trying the following section review.

6–8 Section Review

1   State one of the common problems that can be encountered when the program user is required to input a number into a running Pascal program.
2   Explain what happens when a string is input for a number in a Pascal program.
3   What kind of Pascal variable should be used for getting the value of a user input number in order to prevent the type of input error in question 2 above?

4   What built-in Turbo procedure is used to convert a string into a number? Is this converted number REAL or INTEGER?

Summary

1   Pascal allows you to define your own data type.
2   A user-defined data type is always a scalar quantity.
3   A scalar quantity can be put into one-to-one correspondence with an INTEGER sequence. This is referred to as an ordinal quantity.
4   Type REAL is not a scalar quantity because you can have an infinite number of values between any two given values of this type.
5   The ORDinal function returns the ordinal value of any ordinal-type Pascal expression.
6   CHARacters are ordinals because they have a definite sequence and a one-to-one correspondence with an INTEGER sequence.
7   The PREDecessor function will return the ordinal value that precedes the current one.
8   The SUCCessor function will return the ordinal value that succeeds the current one.
9   User-defined types can be used in a **CASE** statement because they are all ordinal.
10   For programming a loop you need to know the starting value, the ending value, what is to be repeated, and the incremental value.
11   The difference between a program loop and a program branch is that a loop has the potential of going back and repeating a part of the program whereas a branch will always move forward in the program.
12   One method of creating a loop in Pascal is to use the **REPEAT-UNTIL** statement.
13   `Variable := Variable + 1` means to increase by one the value of the number stored in the memory location indicated by `Variable`.
14   The **REPEAT-UNTIL** loop will always perform the body of the loop at least once.
15   It's important to observe that the counting variable does achieve the **UNTIL** condition of a **REPEAT-UNTIL** loop or the loop will continue indefinitely until interrupted by the program user.
16   Another method of creating a loop in Pascal is with the **WHILE-DO** loop.
17   The **WHILE-DO** loop does not have to perform the body of the loop if the **WHILE** condition is FALSE. This is the main difference between this kind of loop and the **REPEAT- UNTIL**.
18   The same precaution is necessary with the counting variable in a **WHILE-DO** loop as with a **REPEAT-UNTIL** loop.
19   The **FOR-TO** loop is the third kind of loop that is available in Pascal.
20   The **FOR-TO** loop is used in Pascal as a convenience. It will increment or decrement the counting variable in units of one only.
21   Loops may be contained inside each other; this is called nesting.

Interactive Exercises

Directions

These exercises require that you have access to a computer and software that supports Pascal, specifically the Turbo Pascal Development System, version 4.0, from Borland International. They are provided here to give you valuable experience and immediate feedback on what the concepts and commands introduced in the chapter will do. They are also fun.

Exercises

1 Program 6–40 uses a user-defined type. See what happens when you try to compile it.

Program 6–40

```
PROGRAM TYPE_1;
 TYPE
  Letters = (A,B,C);

 VAR
  Input
 :Letters;

   BEGIN
      WRITE('Give me a capital letter from A to C = ');
      READ(Input);
   END.
```

2 Program 6–41 is similar to program 6–40. The difference is that the user-defined type is supposed to be output to the monitor. See what happens now when you try to compile it.

Program 6–41

```
PROGRAM Type_2;
 TYPE
  Letters = (A,B,C);

 VAR
  Output
 :Letters;

   BEGIN
      Output := B;
      WRITELN('The value of Output is = ',Output);
      READ(Input);
   END.
```

3 Remember that a user-defined type will always be a scalar quantity. Predict what program 6–42 (p. 362) will do, then try it.

Program 6-42

```
PROGRAM Type_Loop;
 TYPE
  Letters = (A,B,C);

 VAR
  Output
 :Letters;

   BEGIN
        FOR Output := A TO C DO
          BEGIN
           WRITELN('Will this loop work?');
          END;
       END.
```

4   Note the difference between program 6–43 and program 6–42. The loop is now counting backwards. What is your prediction for this one? Give it a try.

Program 6-43

```
PROGRAM Type_Loop_Again;
 TYPE
  Arrangement = (First, Second, Third, Fourth, Fifth, Last);

 VAR
  New_Variable
 :Arrangement;

   BEGIN
        FOR New_Variable := Fourth DOWNTO First DO
          BEGIN
           WRITELN(ORD(New_Variable));
          END;
     END.
```

5   Program 6–44 shows a method of getting the user-defined type values displayed on the monitor. Think it will work? Give it a try and see.

Program 6-44

```
PROGRAM Type_Loop_Again_2;
 TYPE
  Arrangement = (First, Second, Third, Fourth, Fifth, Last);
```

Program 6–44 *continued*

```
VAR
 New_Variable
:Arrangement;

  BEGIN
        FOR New_Variable := First TO Third DO
          BEGIN
            CASE New_Variable OF
             First : WRITELN('First');
             Second : WRITELN('Second');
             Third : WRITELN('Third');
             Fourth : WRITELN('Fourth');
            END;  {Case}
          END;   {loop}
  END.
```

6 What do you think will happen with program 6–45? You may be surprised at the result!

Program 6–45

```
PROGRAM More_Types;
 TYPE
  Arrangement = (First, Second, Third, Fourth, Fifth, Last);

 VAR
  New_Variable
 :Arrangement;

  BEGIN
        FOR New_Variable := First TO Fourth DO
          BEGIN
            CASE New_Variable OF
             First : WRITELN(PRED(First));
             Second : WRITELN(PRED(Second));
             Third : WRITELN(PRED(Third));
             Fourth : WRITELN(PRED(Fourth));
            END;  {Case}
          END;   {loop}
  END.
```

7 Now that you know what program 6–45 does, you may have a good idea of what program 6–46 (p. 364) will do. Try it and see if you're correct.

Program 6–46

```
PROGRAM More_Types_2;
 TYPE
  Arrangement = (First, Second, Third, Fourth, Fifth, Last);

 VAR
  New_Variable
 :Arrangement;

   BEGIN
        FOR New_Variable := First TO Fourth DO
          BEGIN
            CASE New_Variable OF
             First : WRITELN(PRED(ORD(First)));
             Second : WRITELN(PRED(ORD(Second)));
             Third : WRITELN(PRED(ORD(Third)));
             Fourth : WRITELN(PRED(ORD(Fourth)));
            END;  {Case}
          END;  {loop}
   END.
```

8   Look at the difference between program 6–47 and program 6–46. You should have no problem predicting this one, but how about the first statement after the **CASE** statement? What will that do?

Program 6–47

```
PROGRAM More_Types_3;
 TYPE
  Arrangement = (First, Second, Third, Fourth, Fifth, Last);

 VAR
  New_Variable
 :Arrangement;

   BEGIN
        FOR New_Variable := First TO Fourth DO
          BEGIN
            CASE New_Variable OF
             First : WRITELN(SUCC(ORD(First)));
             Second : WRITELN(SUCC(ORD(Second)));
             Third : WRITELN(SUCC(ORD(Third)));
             Fourth : WRITELN(SUCC(ORD(Fourth)));
            END;  {Case}
          END;  {loop}
   END.
```

9   Note that in program 6–48 the arrangement of **ORD** and **SUCC** has been turned around in some statements but not in others. This usually makes a good exam question. Think you know what will happen?

Program 6–48

```
PROGRAM More_Types_4;
 TYPE
  Arrangement = (First, Second, Third, Fourth, Fifth, Last);

 VAR
  New_Variable
 :Arrangement;

   BEGIN
        FOR New_Variable := First TO Fourth DO
          BEGIN
            CASE New_Variable OF
             First : WRITELN(SUCC(ORD(First)));
             Second : WRITELN(ORD(SUCC(Second)));
              Third : WRITELN(ORD(SUCC(Third)));
             Fourth : WRITELN(SUCC(ORD(Fourth)));
            END;  {Case}
          END;  {loop}
   END.
```

10   Program 6–49 shows a powerful feature of Turbo Pascal. When you do this program, you will get a nice little "happy face" on the screen; the number following the **#** sign represents the character ASCII code value. Try this program using different numbers (between 1 and 255) following the **#** sign.

Program 6–49

```
PROGRAM What_Does_it_Do;

 BEGIN
  WRITELN(#1);
 END.
```

11   Program 6–50 (p. 366) may just knock your socks off, especially if you have no idea what it will do. It's a great way of getting anything and everything you want onto the monitor.

Program 6–50

```
PROGRAM What_Does_it_Do_2;
 VAR
 Counter
:INTEGER;

  BEGIN
    Display := 0;

      FOR Counter := 1 TO 255 DO
        BEGIN
          WRITE(CHAR(Counter));
        END;
    END.
```

12 Study program 6–51 carefully before you try to compile it. Understanding what is happening here can help prevent future frustrations.

Program 6–51

```
PROGRAM Counter_1;
VAR
 Counter
:REAL;

  BEGIN
   FOR Counter := 1 TO 255 DO
       BEGIN
           WRITE('Count value is ',Counter);
       END;

  END.
```

13 Note in program 6–52 that the counting variable type has been changed to BYTE. What do you think will happen now? Another good exam question!

Program 6–52

```
PROGRAM Counter_2;
VAR
 Counter
:BYTE;
```

Program 6–52 *continued*

```
BEGIN
    FOR Counter := 1 TO 255 DO
      BEGIN
        WRITE('Count value is ',Counter);
      END;
END.
```

14 Program 6–53 looks almost the same as program 6–52. But will it really work? Try it to make sure you know.

Program 6–53

```
PROGRAM Counter_3;
VAR
 Counter
:BYTE;

  BEGIN
      FOR Counter := 250 TO 300 DO
        BEGIN
          WRITE('Count value is ',Counter);
        END;
    END.
```

15 **KEYPRESSED** is a Turbo Pascal reserved word. It is of type BOOLEAN and remains FALSE until any key on the keyboard is pressed; then it becomes TRUE. It's a handy programming feature. Give program 6–54 a try and see what it does.

Program 6–54

```
PROGRAM New_Command;
  USES CRT;
    BEGIN
      REPEAT
        WRITE('*');
      UNTIL KEYPRESSED;
    END.
```

16 A neat application of a **FOR-TO** loop is to cause a predictable program delay. Turbo Pascal provides a built-in procedure called **DELAY**, defined as
DELAY(time :WORD);

where time = the appropriate delay time in milliseconds (must be of type WORD)

The Turbo Pascal type WORD is like a positive INTEGER. Its value ranges from 0 to 65535.

Note what program 6–55 does. Try changing the value of the delay to get a feel for the number you program in and the corresponding delay it actually creates.

Program 6–55

```
PROGRAM New_Procedure;
  USES CRT;

  PROCEDURE DELAY(Time : WORD);
    VAR
      Counter
    :INTEGER;
      BEGIN
        FOR Counter := 1 TO Time DO
      END;

  BEGIN
    REPEAT
      DELAY(5000);
      WRITE('*');
    UNTIL KEYPRESSED;
  END.
```

Pascal Commands

User-Defined Type

```
TYPE
  Variable = (Value₁, Value₂ . . . Valueₙ);
```

where Variable = any legal Pascal identifier
$Value_N$ = any legal Pascal identifier that will now represent the ordinal values of the variable

```
REPEAT(Statements)UNTIL(Expression)
```

where (Statements) = any legal Pascal statement
(Expression) = when this is TRUE, the repeating of the statements will stop and the program will move forward

```
WHILE(expression)DO(statement)
```

where (expression) = any legal expression that will produce a Boolean value
(statement) = may be a single statement or a compound statement with **BEGIN-END**.

```
FOR(variable):= (expression₁)TO(expression₂)DO(statement)
```

where (variable) = any ordinal variable identifier
$(expression_1)$ = the beginning value of the variable
$(expression_2)$ = the ending value of the variable

(statement) = any legal Pascal statement which must be enclosed by **BEGIN-END** if more than one statement is used

---

Self-Test

Directions

The program for this self-test was written by a beginning Pascal student. It may contain some compile-time or run-time errors. It is a study of different methods of looping using different counter types and all three Pascal methods of looping. Answer the questions for this self-test by referring to program 6–56.

Program 6–56

```
PROGRAM Types_and_Loops;
  USES
    CRT;

  TYPE
    New_Kind = (One, Two, Three, Four, Five, Six, Seven, Eight, Nine,
               Ten);
    My_Type = (Applied, Pascal, For_, Technology);

  VAR
    Counter_1,
    Counter_2,
    Counter_3,
    User_Selection
    :INTEGER;

    Stepper_1,
    Stepper_2,
    Stepper_3
    :REAL;

    Other_1
    :New_Kind;

    Second_1
    :My_Type;

    User_Choice
    :CHAR;

{-------------------------------------------------------}
    PROCEDURE Explain_Program;
    FORWARD;

    PROCEDURE User_Loop_Selection;
    FORWARD;

    PROCEDURE Loop_1;
    FORWARD;

    PROCEDURE Loop_2;
    FORWARD;
```

Program 6–56   *continued*

```
      PROCEDURE Loop_3;
      FORWARD;

      PROCEDURE Loop_4;
      FORWARD;

      PROCEDURE Loop_5;
      FORWARD;

      PROCEDURE Loop_6;
      FORWARD;

      PROCEDURE Loop_7;
      FORWARD;

      PROCEDURE Loop_8;
      FORWARD;

      PROCEDURE Loop_9;
      FORWARD;

      PROCEDURE Loop_10;
      FORWARD;

      PROCEDURE Loop_11;
      FORWARD;

      PROCEDURE Loop_12;
      FORWARD;

      PROCEDURE Program_Repeat;
      FORWARD;
   {--------------------------------------------------------}

   PROCEDURE Main_Sequence;
      BEGIN
         Explain_Program;
        REPEAT
          User_Loop_Selection;
          Program_Repeat;
        UNTIL (User_Choice = 'n') OR (User_Choice = 'N');
      END;   {of procedure Main_Sequence}

   PROCEDURE Explain_Program;
      BEGIN
         ClrScr;   {Clear the screen}
         WRITELN;
         WRITELN('This is a demonstration program about different');
         WRITELN('types of program loops in Pascal.');
         WRITELN;
         WRITELN('It is intended to be used as a learning aid');
         WRITELN('so that the user can see the actual effect');
         WRITELN('of the different types of loops used in Pascal.');
         WRITELN;
         WRITELN('Since this program is part of a Self-Test, the');
         WRITELN('loops are identified only by number and not by');
         WRITELN('the type of loop actually used.');
         WRITELN;
```

Program 6–56   *continued*

```
          WRITE('Press the RETURN/ENTER key to continue...');
          READLN;
      END;

PROCEDURE User_Loop_Selection;
    BEGIN
      ClrScr;  {Clear the screen.}
    REPEAT
      WRITELN;
      WRITELN('Select loop by number:');
      WRITELN;
      WRITELN('1] 2] 3] 4] 5] 6] 7] 8] 9] 10] 11] 12]');
      WRITELN;
      WRITE('Your selection => ');
      READLN(User_Selection);
    UNTIL User_Selection IN [1..12];

    CASE User_Selection OF
         1 : Loop_1;
         2 : Loop_2;
         3 : Loop_3;
         4 : Loop_4;
         5 : Loop_5;
         6 : Loop_6;
         7 : Loop_7;
         8 : Loop_8;
         9 : Loop_9;
        10 : Loop_10;
        11 : Loop_11;
        12 : Loop_12;
    END;   {of Case}
END;  {of procedure User_Loop_Selection.}

PROCEDURE Program_Repeat;
  BEGIN
    GOTOXY(3,25);
    WRITE('Do you want to repeat the program (Y/N) = ');
    READLN(User_Choice);
  END;  {of procedure Repeat_Program}

PROCEDURE Loop_1;
  PROCEDURE Compute_It(First, Second : REAL);
    BEGIN
      WRITELN('The value of the counting variable is = ',First);
      WRITELN('Value of the other counting variable is = ',Second);
    END;  {of procedure Compute_It}

  BEGIN
    Stepper_1 := 0;
    Stepper_2 := 5;
    Compute_It(Stepper_1, Stepper_2);
    Stepper_1 := 1.5;
    Stepper_2 := 3.5;
    Compute_It(Stepper_1, Stepper_2);
    Stepper_1 := 3.25;
    Stepper_2 := 2.75;
    Compute_It(Stepper_1, Stepper_2);
  END;  {of procedure Loop_1}
```

Program 6–56   *continued*

```
   PROCEDURE Loop_2;
    BEGIN
      Stepper_1 := 4;
        WHILE Stepper_1 < 5 DO
         BEGIN
          WRITELN('Stepper value = ',Stepper_1);
          Stepper_1 := Stepper_1 + 0.1;
         END;   {of While}
    END;  {of procedure Loop_2}

   PROCEDURE Loop_3;
    BEGIN
      Stepper_1 := 4;
        WHILE Stepper_1 < 4 DO
         BEGIN
          WRITELN('Stepper value = ',Stepper_1);
          Stepper_1 := Stepper_1 + 0.5;
         END;   {of While};
    END;  {of procedure Loop_3}

   PROCEDURE Loop_4;
    BEGIN
      Stepper_1 := 4;
       REPEAT
        WRITELN('Stepper value = ',Stepper_1);
        Stepper_1 := Stepper_1 - 0.5;
       UNTIL Stepper_1 < 1;
    END;  {of procedure Loop_4}

   PROCEDURE Loop_5;
    BEGIN
     Other_1 := One;
       WHILE Other_1 < Ten DO
        BEGIN
         WRITELN('Counter value = ',ORD(Other_1));
         Other_1 := SUCC(Other_1);
        END;
    END;  {of procedure Loop_5}

   PROCEDURE Loop_6;
    BEGIN
      Second_1 := Applied;
        WHILE Second_1 < Technology DO
         BEGIN
          WRITELN('Counter value = ',ORD(Second_1));
          Second_1 := SUCC(Second_1);
         END;
    END;  {of procedure Loop_6}

   PROCEDURE Loop_7;
    BEGIN
      Counter_1 := 0;
       REPEAT
        Counter_1 := Counter_1 + 2;
        WRITELN('Counter value = ',Counter_1);
       UNTIL Counter_1 = 5;
    END;  {of procedure Loop_7}
```

Program 6-56   *continued*

```
PROCEDURE Loop_8;
 BEGIN
   FOR Second_1 := Technology DOWNTO Applied DO
     BEGIN
       WRITELN('Ordinal value of counter = ',ORD(Second_1));
     END;
 END;  {of procedure Loop_8}

PROCEDURE Loop_9;
 BEGIN
   FOR Counter_2 := 5 TO 1 DO
     BEGIN
       WRITELN('Counter value = ',Counter_2);
     END;
 END;  {of procedure Loop_9}


PROCEDURE Loop_10;
 BEGIN
   FOR Counter_3 := 1 TO 5 DO
     BEGIN
       Counter_3 := 5;
       WRITELN('Value of counter = ',Counter_3);
     END;
 END;  {of procedure Loop_10}

PROCEDURE Loop_11;
 BEGIN
   FOR Counter_1 := 1 TO 2 DO
     BEGIN
      FOR Counter_2 := 1 TO 3 DO
        BEGIN
         FOR Counter_3 := 1 TO 4 DO
           BEGIN
             WRITELN('Counter 1 = ',Counter_1,' Counter 2 = ',
                     Counter_2,' Counter 3 = ',Counter_3);
           END;  {Third loop}
           WRITELN;
        END;     {Second loop}
        WRITELN;
     END;         {First loop}
 END;  {of procedure Loop_11}

PROCEDURE Loop_12;
 BEGIN
   Stepper_1 := 1;
   Other_1 := One;
   WHILE Stepper_1 < 10 DO
     REPEAT
      FOR Counter_1 := 5 DOWNTO 1 DO
        BEGIN
         Stepper_1 := Stepper_1 + 1;
         Other_1 := SUCC(Other_1);
         WRITELN('Real counter = ',Stepper_1,' Integer counter = ',
                 Counter_1);
        END;
     UNTIL Other_1 = Ten;
 END;  {of procedure Loop_12}
```

Program 6–56   *continued*

```
BEGIN
  Main_Sequence;

END.
```

Questions

1 Will the program compile? If not, why not?
2 Are there any loops which will continue until stopped by the program user? If so, which one(s) and why?
3 Which loop has the greatest number of loops when executed (outside of any that must be terminated by the user) and which one has the least?
4 What is the purpose of the **REPEAT-UNTIL** loop in the procedure User_Loop-_Selection?
5 Which of the loops are nested loops? How many levels of nesting do they have?
6 Which loop uses REAL, INTEGER, and user-defined types?
7 Explain how Loop_6 gets incremented. How would you decrement such a loop?
8 How many times will Loop_10 loop? Explain why this happens.
9 Which loop blocks (if any) decrease the value of the variable?
10 Is there any **FOR-TO** loop that uses the variables Stepper_1, Stepper_2, or Stepper_3 as counting variables? If not, why not?

Problems

General Concepts

Section 6–1

1 Name the six different Pascal types presented up to this point.
2 State what is meant by a scalar quantity.
3 Which of the following types are scalar?
A) INTEGER B) REAL C) BOOLEAN D) BYTE E) STRING
4 Explain what is meant by an ordinal-type expression.
5 Describe what the **ORD** function does in Pascal. What would **ORD(a)** be?
6 State the ordinal value of TRUE and FALSE.
7 Explain what is meant by the following Pascal functions:
A) **SUCC** B) **PRED**
8 What will be returned for **SUCC(TRUE)**? Explain why this happens.

Section 6–2

9 What is a program loop? Explain how a loop is different from a program branch.
10 State the four things you need to know in order to create a program loop.
11 Give an example of why programming loops are needed.

Section 6–3

12 Explain the meaning of the following Pascal command.
REPEAT(Statements)**UNTIL**(Expression)

13 Describe good programming practice for setting the value of the (**Expression**) in a **REPEAT-UNTIL** loop.

14 Explain the meaning of the following statement.
Value := Value + 1;

15 State what may be enclosed between the **REPEAT** and **UNTIL** commands in a Pascal program. Is it necessary to have a **BEGIN-END** between these commands if more than one statement is used?

16 What is the minimum number of times a **REPEAT-UNTIL** loop will be executed?

Section 6–4

17 Explain the meaning of the following Pascal command.
WHILE (expression) DO (statement)

18 Compare the **WHILE-DO** loop with the **REPEAT-UNTIL** loop. What are the differences? The similarities?

19 What is an error-trapping routine? Describe its use.

20 Is it necessary to have a **BEGIN-END** following the **WHILE-DO** if more than one statement is used?

21 State the precaution that should be taken when constructing a **WHILE-DO** loop.

Section 6–5

22 Explain the meaning of the following Pascal command.

FOR(variable) := (expression₁)TO(expression₂)DO(statement)

23 What type must the counting variable be in a **FOR-TO** loop?

24 Can you decrement with a **FOR-TO** loop in Pascal? If you can, explain how this is done.

25 By how much can you change the value of the counting variable in a Pascal **FOR-TO** loop?

26 Explain why ordinal values can be used with a **FOR-TO** Pascal loop.

27 State the reason for using a **FOR-TO** loop in Pascal.

Section 6–6

28 Explain what is meant by a nested loop.

29 Describe the type of program structure that should be used with nested loops.

30 What is the advantage of using a **FOR-TO** loop in a Pascal program?

31 For a program that has nested loops, which loop is completed first—the outer loop or the inner loop?

Section 6–7

32 Give an example of how to construct a Pascal program so the program user can repeat the program.

Program Analysis

33 Is there a problem in program 6–57? If so, state what it is.

Program 6–57

```
PROGRAM Problem_33;
  TYPE
    Metric = (Micro, Milli, Kilo, Mega);
```

Program 6–57 *continued*

```
VAR
 Measurements
:Metric;

  BEGIN
    Measurements := Micro;
    WRITELN('One of the measurements is ',Measurements);
  END.
```

34 If you correctly analyzed the bug in program 6–57, there is a similar problem in program 6–58. Determine what it is.

Program 6–58

```
PROGRAM Problem__34;
  TYPE
    Metric = (Micro, Milli, Kilo, Mega);

  VAR
   Measurements
  :Metric;

    BEGIN
      WRITE('Input the metric unit used: ');
      READLN(Measurements);
    END.
```

35 Turbo Pascal has a special feature for the outputting of BOOLEAN variables. Will program 6–59 compile? If not, why not? If it does, why does it?

Program 6–59

```
PROGRAM Problem__35;

 VAR
  Test__Variable
 :BOOLEAN;

  BEGIN
    Test__Variable := TRUE;
    WRITELN('Value of the test variable is ',Test__Variable);
  END.
```

36 Program 6–60 is similar to program 6–59. Carefully analyze program 6–60 and determine if it will compile. If you think it will, why will it? If you think it won't, why not?

Program 6–60

```
PROGRAM Problem__36;

VAR
 Test__Variable
:BOOLEAN;

  BEGIN
    WRITELN('Enter the value of the test variable = ');
    READLN(Test__Variable);
  END.
```

37 Is there a bug in program 6–61? If so, what is it? What would you do to correct it?

Program 6–61

```
PROGRAM Problem__37;
 TYPE
   Shapes = (Square, Rectangle, Circle, Triangle);

 VAR
  Figures
 :Shapes;

   BEGIN
     WRITELN('The ordinal value of Square is = ',ORD(Square));
     WRITELN('The successor of Square is = ',SUCC(Square));
   END.
```

38 Program 6–62 will not compile. Do you know why? What would you do to correct the problem?

Program 6–62

```
PROGRAM Problem__38;
   BEGIN
    REPEAT
      WRITELN('This is a loop?');
    UNTIL KEYPRESSED;
   END.
```

39 Is there a bug in program 6–63? If there is, what do you think it could be?

Program 6–63

```
PROGRAM Problem__39;
 USES CRT;
   BEGIN
     WHILE NOT KEYPRESSED DO
      BEGIN
       WRITELN('This is a loop?');
      END;
    END.
```

40 Will program 6–64 compile? If not, why not?

Program 6–64

```
PROGRAM Problem__40;
USES CRT;
VAR
 Counter
:REAL;
  BEGIN
    FOR Counter := 5 TO 10 DO
      BEGIN
       WRITELN('This is a loop?');
      END;
   END.
```

41 Is there a bug in program 6–65? If so, state what you think it is.

Program 6–65

```
PROGRAM Problem__41;
USES CRT;
VAR
 Counter
:INTEGER;
  BEGIN
    FOR Counter := 5 TO 10 DO;
     BEGIN
      WRITELN('This is a loop?');
     END;
   END.
```

42 What will program 6–66 do if it compiles successfully? Explain why it does what you think it does.

Program 6–66

```
PROGRAM Problem_42;
 USES CRT;
 TYPE
   Counter_Type = (One, Two, Three, Four, Five);
 VAR
  Counter
 :Counter_Type;
    BEGIN
    FOR Counter := Five DOWNTO One DO
     BEGIN
      REPEAT
        WRITELN('This is a loop?');
      UNTIL Counter = Three;
     END;
    END.
```

Program Design

For all of the following programs use a structure that is assigned to you by your instructor.

Electronics Technology

43 Create a Pascal program that will compute the voltage drop across a resistor for a range of current values selected by the program user. The program user is to input the value of the resistor, the beginning and ending current values, and the incremental value of the current. The relationship between the resistor voltage and current is

$$V = IR$$

where V = voltage across the resistor in volts
     I = current in resistor in amps
           R = value of resistor in ohms

44 Develop a Pascal program that will calculate the power dissipated in a resistor for a range of voltage values across the resistor. The program user is to select the value of the resistor and the beginning as well as the ending resistor voltages. The relationship is

$$P = V^2/R$$

where P = power dissipation of the resistor in watts
     V = voltage across the resistor in volts
     R = value of the resistor in ohms

45 Modify the program from problem 44 so that the program user will be warned when the

power dissipation of the resistor exceeds a maximum power dissipation determined by the program user. An example warning could be the monitor display

WARNING! Power exceeds 2 watts!!!

46   Modify the program in problem 44 by adding an inner loop so that the program user may observe the power dissipation for a range of resistors as well as a range of voltage values. The program user would now select the voltage range and resistance range as well as the incremental values for each.

Business Applications

47   Develop a Pascal program that will display the interest compounded annually from one to thirty years. The program user is to input the principal and the rate of interest. The mathematical relationship is

$$Y = A(1 + N)^T$$

where Y = the amount
      A = the principal invested
      N = interest rate
      T = number of years

Computer Science

48   Create a Pascal program that will take any number to any power. The program user is to select both the base and the power.

Drafting Technology

49   You are to develop a Pascal program that will show the relationship of the change in volume of a sphere for a given change in radius. The program user selects the initial, incremental, and ending values of the radius. The mathematical relationship is

$$V = 4/3\pi r^3$$

where V = volume of the sphere in cubic units
      π = the constant pi
      r = radius of the sphere in linear units

Agriculture Technology

50   You have been assigned to construct a program in Pascal that will demonstrate the change in volume of a cylindrical water tank as the amount of water in the tank changes (increases or decreases). The program user is to input the radius and height of the water tank as well as the incremental increase or decrease of the amount of water in the tank.

Health Technology

51   A health science class is to construct a program in Pascal that will compute and display the values of a range of temperatures in degrees Fahrenheit and degrees centigrade. The program user is to select the lowest, highest, and incremental temperature change. The mathematical relationship is

$$F = 9/5C + 32$$

where F = temperature in degrees Fahrenheit
      C = temperature in degrees centigrade

Manufacturing Technology
52 Develop a Pascal program that will show the change in the volume of a metal cone as it is machined. Assume that the machining reduces the height of the metal cone. The program user enters the dimensions of the cone as well as the incremental changes in the cone height. The mathematical relationship is

$$V = h/3(A_1 + A_2 + \sqrt{A_1A_2})$$

where V = volume of the cone in cubic units
      h = height of the cone in linear units
      A_1 = area of the lower base in square units
      A_2 = area of the upper base in square units

Business Applications
53 Modify the program for problem 47 so that the program contains an inner loop that will display the interest for different amounts of money. The program user would now select the range for the principal and the incremental difference as well as entering the rate of interest.

Agriculture Technology
54 Modify the program for problem 48 so that the program user is warned when the water tank is empty or overflowing.

Drafting Technology
55 Create a Pascal program that will display a range of the same lengths in feet, meters, and inches. The program user can select the beginning, ending, and incremental measurements in feet.

Agriculture Technology
56 The relationship between the number of units of fertilizer and the expected crop yield is

$$Y = F/(2^F) + 1$$

where Y = the yield improvement factor
      F = arbitrary units of fertilizer per acre
Develop a Pascal program to find the value of F that produces a maximum value of Y.

Health Technology
57 Create a Pascal program that will display the heartbeat of a patient over a range of beats per minute, beats per hour, and beats per day. The program user can enter the minimum and maximum number of beats per minute and the value of the increment.

Manufacturing Technology
58 Develop a Pascal program that will show the range of weights in pounds and kilograms. The program user can enter the minimum and maximum numbers of the weight along with the increment value either in pounds or kilograms.

Business Applications
59 Write a Pascal program that will compute the sales tax for a range of values and round the result off to the nearest cent. The program user can input the minimum and maximum dollar amounts, and the percentage of the sales tax. The increment is to be determined by the program so that a printout is given for every one-cent change in the total amount.

7 Arranging Information

Objectives

This chapter gives you the opportunity to learn:

1 What an array is.
2 Why arrays are useful.
3 Different ways of representing arrays.
4 The application of arrayed information.
5 Bubble-sorting techniques for numerical data.
6 Bubble-sorting techniques for strings.
7 Applications of sorting techniques.
8 Debugging techniques for run-time errors.
9 The development of a complex program that uses arrays and sorting techniques.

Key Terms

Array
Subscript
Index
Initialize
Congruent Array

Packed Arrays
Bubble Sort
Debug Routine
Auto-Debug

Outline

A very important concept, the **array**, is presented in this chapter. It is an easy way of keeping track of many different things that have similar characteristics; this could include people, resistors, or money. The computer is good at this and Pascal is an especially good language for arrays.

The word array comes from arrangement; an array is a way of arranging things. Once you have your information arranged, you can use it in various useful ways. Here you will learn several ways to arrange data, such as sorting, accessing, or testing it. The techniques presented here will not only give you greater control over the computer but also over the information you put into the computer. You will learn techniques that will help you create powerful programs with broad application to technology problems, ranging from inventory control to counting and sorting items of interest to the program user.

7–1   What is an Array?

Keeping Track of Things

An array is an easy way of keeping track of things. You can think of an array as an arrangement. An array is really a structured data type composed of a specific number of elements of the same type. As you will see, each of these types is accessible by its index.

To get a general idea of the use of arrays, take the example of a small hospital. Suppose you needed a computer program that would keep track of patients in a hospital. The program you were to develop would do the following.

1. Enter the name of the patient and receive the following information:
   Patient location
   Patient illness
   Patient days in hospital
   Patient bill
2. List all the patients
   In the entire hospital
   On one floor
   In the same room
   By number of days in the hospital

By type of medical problems
By amount owed to the hospital
3. Get totals of
Amount owed by all of the patients
Amount owed by patients on the same floor
Amount owed by medical problems
Amount owed by number of days in hospital
4. Get information about
Average medical bill of all patients
Average medical bill of patients on one floor
Average medical bill of patients by medical problems
Average medical bill by number of days in hospital per patient

At first glance, you might think that such a program would have to be very large and complex. You're correct in thinking that this program would be large, at least as measured by the programs you have seen in this text. However, the program would not be complex. Here is the reason why.

Tagging with Subscripts

Look at the hospital layout in figure 7–1 on page 386 and note that

1. There are three floors.
2. Each floor has three rooms.
3. Each room has four beds.
4. Not all of the beds are occupied.
5. Every patient has a name.

Usually, rooms in a hospital are numbered according to the floor. Rooms on the first floor could be numbered 11 for the first, 12 for the second, and so on. For the second floor, the first room could be numbered 21 and the second room 22, while the first room on the third floor could be numbered 31, the second 32, and so on. Each bed of each room could also be numbered. For example, the first bed of the second room of the third floor could be called 321. Figure 7–2 shows this arrangement.

As can be seen in figure 7–2 on page 387, each bed now has its own unique number different from that of all the other beds. Now, each bed can be identified by the use of **subscripts** as shown below.

$$\text{Bed}_{F,R,B}$$

where F = the floor number
R = the room number on the given floor
B = the bed number in the given room

For example,

$$\text{Bed}_{2,1,3}$$

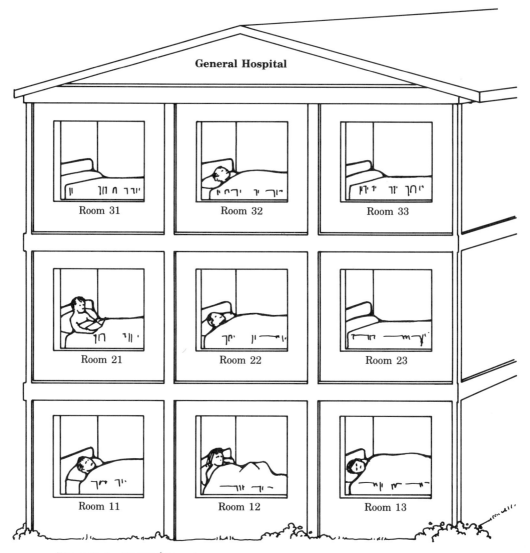

Figure 7–1 Hospital Layout

would mean the second floor, first room, third bed. Thus you can begin to see that an array is a structured data type consisting of a fixed number of components of the same type. Each of these components can be accessed by its subscript (as you will see, this is called an **index**).

You can arrange the subscripts in any fashion you want. As an example, you could let the first subscript represent the room and the second subscript represent the floor. It makes no difference how the subscripts are used, as long as the notation is consistent.

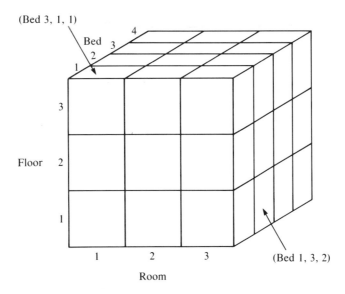

Figure 7–2   Numbering of Rooms and Beds

You can now go one step further. Once each bed is identified with a subscript, the same subscript notation can now be used to identify the patient.

$$Patient_{2,1,3}$$

would mean the patient on the second floor, first room, third bed, and

$$Balance_{2,1,3}$$

could mean the amount of money owed by the same patient.

Both of these can be thought of as ARRAYs because they are structured data types composed of a fixed number of components of the same type (such as Patient or Balance), and each can be directly accessed by its subscript.

Representing Arrays

In Pascal such a representation of objects is called an array. It simply means an arrangement. An array is represented as

`Identifier : ARRAY[index`$_1$`, index`$_2$`, . . . index`$_N$`] OF TYPE`

where `Identifier` = any legal Pascal variable identifier
       `index` = any scalar type or subrange except INTEGER

Note that the index refers to what has been called the subscript. The term index is used because it emphasizes that an ordinal value which identifies a specific component of a data structure is being used.

The number of indices in the array represents the size of the array. As an example, an array with three indices is said to be a three-dimensional array, while an array with two indices is said to be a two-dimensional array. An array with one index is called a one-dimensional array.

Getting back to the hospital program, each patient is an element in the array, and so is the patient's bill, the number of days in the hospital, and the reason for being there. The name of each patient assigned to a specific bed could be represented as a three-dimensional array in Pascal as

VAR
    `Patient__Name = **ARRAY**[1..3, 1..3, 1..4] **OF STRING**[79];`

where `Patient__Name` = the arrayed string variable that contains the patient's name

`1..3` = the first index that will have the subrange values of 1, 2, or 3; these values will represent the floor number for one of the three floors

`1..3` = the second index that will have the subrange values of 1, 2, or 3; these values will represent the room number of one of the three rooms found on each floor

`1..4` = the third index that will have the subrange values of 1, 2, 3, or 4; these values will represent the bed number in one of the four beds found in each room

Hence,

`Patient__Name[2,1,3]`

would mean the memory location reserved for the name of the patient on the second floor, first room, third bed.

This concept could be extended so that another arrayed variable could be defined as

VAR
    `Amount__Owed = **ARRAY**[1..3, 1..3, 1..4] **OF REAL**;`

and the element

`Amount__Owed[2,1,3]`

would mean the amount of money owed by the same patient (the patient on the second floor, first room, third bed).

Another arrayed variable could also be defined that would indicate the number of days the patient has been in the hospital.

VAR
    `Days__Here = **ARRAY**[1..3, 1..3, 1..4] **OF INTEGER**;`

and the element

`Days__Here[2,1,3]`

would mean the number of days in the hospital for the same patient (the one on floor two, room one, bed three). And a third arrayed variable

VAR
```
    Release_Ready = ARRAY[1..3, 1..3, 1..4] OF BOOLEAN;
```

could tell you if the patient has been medically released to go home from the hospital. Thus the element

```
Release_Ready[2,1,3]
```

would be a BOOLEAN variable to indicate if the same patient is ready to have his hospital bill prepared.

Putting the Data In

Program 7–1 loads information into the computer about the patients listed in the hospital. It does this by first declaring array variables in the **VAR** declaration. It names four different data types, each having a fixed number of components. These are the name of the patient, amount owed, days in hospital, and if the patient is release-ready. The indices of each array set limits on the number of components in each of the arrays.

Program 7–1

```
PROGRAM Hospital_Data;
VAR
   Patient_Name : ARRAY[1..3, 1..3, 1..4] OF STRING[79];
   Amount_Owed  : ARRAY[1..3, 1..3, 1..4] OF REAL;
     Days_Here  : ARRAY[1..3, 1..3, 1..4] OF INTEGER;
 Release_Ready  : ARRAY[1..3, 1..3, 1..4] OF BOOLEAN;

 PROCEDURE Patient_Data;
  BEGIN
      {Floor 1, Room 1, Bed 1}
      Patient_Name[1,1,1] := 'Frank Jones';
      Amount_Owed[1,1,1] := 825.43;
      Days_Here[1,1,1] := 3;
      Release_Ready[1,1,1] := FALSE;

      {Floor 1, Room 1, Bed 2}
      Patient_Name[1,1,2] := 'Howard Smith';
      Amount_Owed[1,1,2] := 1679.53;
      Days_Here[1,1,2] := 6;
      Release_Ready[1,1,2] := TRUE;

      {Floor 1, Room 2, Bed 1}
      Patient_Name[1,2,1] := 'Mary Proctor';
      Amount_Owed[1,2,1] := 359.86;
      Days_Here[1,2,1] := 1;
      Release_Ready[1,2,1] := TRUE;
```

Program 7–1 *continued*

```
      {Floor 1, Room 2, Bed 2}
      Patient_Name[1,2,2] := 'Alice McDaniels';
      Amount_Owed[1,2,2] := 3487.75;
      Days_Here[1,2,2] := 5;
      Release_Ready[1,2,2] := FALSE;

      {Floor 1, Room 3, Bed 1}
      Patient_Name[1,3,1] := 'Sammy Adams';
      Amount_Owed[1,3,1] := 589.32;
      Days_Here[1,3,1] := 2;
      Release_Ready[1,3,1] := FALSE;

  END;  {of procedure Patient_Data}
BEGIN
  Patient_Data;
END.
```

The procedure `Patient_Data` simply assigned an index value to a specific data type. Not all of the possible index values were used in order to keep the example program short.

Program 7–1 has developed a tag number which is the index number for each patient. This tag number not only identifies the location of the patient, but the patient's name, amount owed, number of days in the hospital, and whether the patient is ready to be released. If your computer has enough memory, there are many other data types that could carry the same tag, subscript number or index, such as the patient's birth date, name of doctor, home phone, address, insurance policy, or next of kin. At any rate, you now have a method of relating different data types to each other. (You can relate a BOOLEAN to an INTEGER through the same index values; i.e. a certain patient is release-ready and owes a certain amount of money).

Getting the Data Out

Programs can now be developed to display the data in program 7–1 in many different ways. Program 7–2 lists the names of all the patients in room 11. The common index values are the first two (representing the floor number and the room number). The only variable of the index will then be the bed number. Again, note that each component (the patient name in the program) is being accessed by the index.

Program 7–2 had the patient data from the procedure `Patient_Data` entered into it. It then used a **FOR–TO** loop that held the first two indices constant ([1,1,Bed]). Doing this caused only the patients on the first floor, first room (room number 11) to be included in the search. The only variable index used was the bed number. This went from 1 to 4, the minimum and maximum value of this index. Thus, effectively, every bed in the room was searched for a patient.

Program 7–2

```
PROGRAM Hospital_Data;
 VAR
   Patient_Name : ARRAY[1..3, 1..3, 1..4] OF STRING[79];
    Amount_Owed : ARRAY[1..3, 1..3, 1..4] OF REAL;
      Days_Here : ARRAY[1..3, 1..3, 1..4] OF INTEGER;
  Release_Ready : ARRAY[1..3, 1..3, 1..4] OF BOOLEAN;

    PROCEDURE Patient_Data;
      BEGIN
        {Body of procedure goes here.}
      END;  {of procedure Patient_Data}

    PROCEDURE Names_Room_11;
     VAR
     Bed
    :INTEGER;

    BEGIN
       WRITELN('Patients in room 11.');

       FOR Bed := 1 TO 4 DO
         WRITELN(Patient_Name[1,1,Bed]);

    END;  {of procedure Names_Room_11}

 BEGIN
  Patient_Data;
  Names_Room_11;
 END.
```

When executed, program 7–2 would display

```
Patients in Room 11.
Frank Jones
Howard Smith
(No entered data)
(No entered data)
```

Since the third and fourth bed are empty, there is no information entered, and there is no telling what is in these memory locations.

When you set an array in Pascal, such as in the hospital program, Pascal reserves memory locations but doesn't put any data into them until you do. When you ask for output of these memory locations, you will get the data that is there, no matter what it is. This could be very confusing to the program user. Therefore, in Pascal, it is considered good practice to initialize an array before using it.

Initializing an Array

To **initialize an array** simply means setting a known value into each element of the array. One method of doing this in the hospital program would be as shown in program 7–3.

Program 7–3

```
PROGRAM Hospital_Data;
 VAR
   Patient_Name : ARRAY[1..3, 1..3, 1..4] OF STRING[79];
   Amount_Owed : ARRAY[1..3, 1..3, 1..4] OF REAL;
     Days_Here : ARRAY[1..3, 1..3, 1..4] OF INTEGER;
  Release_Ready : ARRAY[1..3, 1..3, 1..4] OF BOOLEAN;

 PROCEDURE Initialize_Data;
    VAR
     Floor,
     Room,
     Bed
   :INTEGER;
      BEGIN
        FOR Floor := 1 TO 3 DO {Three floors in hospital}
          BEGIN
            FOR Room := 1 TO 3 DO {Three rooms per floor}
              BEGIN
                FOR Bed := 1 TO 4 DO {Four beds per room}
                  BEGIN
                    Patient_Name[Floor, Room, Bed] := '';
                    Amount_Owed[Floor, Room, Bed] := 0;
                    Days_Here[Floor, Room, Bed] := 0;
                    Release_Ready[Floor, Room, Bed] := FALSE;
                END;  {Bed loop}
              END;  {Room loop}
            END;  {Floor loop}
        END;  {of procedure Initialize_Data}

  PROCEDURE Patient_Data;
    BEGIN
    {Body of procedure goes here.}
    END;  {of procedure Patient_Data}

 BEGIN
  Initialize_Data;
  Patient_Data;
 END.
```

Program 7–3 is actually using three nested loops to set the value of each component of each data type to a 0 for a number type, null ('') for a string type, and FALSE for a Boolean type. Each of the **FOR-TO** loops causes each index to increment through its full range of values (3 floors, three rooms per floor, and four beds per room), causing every element in the array to be initialized.

However, this initialization procedure is not the most efficient way of initializing an array. Fortunately, Turbo Pascal provides another way of doing this (not all Pascal systems have an easy way of initializing an array). There is a predefined procedure called **FILLCHAR**.

FILLCHAR(Variable, Length, Data**);**

where Variable = variable of any type
Length = the number of bytes to be initialized
Data = an ordinal value to fill the variable(s)

Array variables are usually initialized with the value 0. In order to help find the length (number of bytes) to be initialized, Turbo Pascal has another built-in function called **SIZEOF(**Variable**)**.

SIZEOF(Variable**);**

where **SIZEOF** = the number of bytes filled by the argument
Variable = variable used in the argument

The way these two are used is shown in program 7–4. Program 7–4 does exactly what program 7–3 did; it initializes all the variables. But this time, instead of using all those nested loops, it uses an easy built-in Turbo Pascal procedure and a function to do this task.

Program 7–4

```
PROGRAM Hospital_Data;
 VAR
   Patient_Name : ARRAY[1..3, 1..3, 1..4] OF STRING[79];
    Amount_Owed : ARRAY[1..3, 1..3, 1..4] OF REAL;
      Days_Here : ARRAY[1..3, 1..3, 1..4] OF INTEGER;
  Release_Ready : ARRAY[1..3, 1..3, 1..4] OF BOOLEAN;

PROCEDURE Initialize_Data;
    BEGIN
      FILLCHAR(Patient_Name, SIZEOF(Patient_Name), 0);
      FILLCHAR(Amount_Owed, SIZEOF(Amount_Owed), 0);
      FILLCHAR(Days_Here, SIZEOF(Days_Here), 0);
      FILLCHAR(Release_Ready, SIZEOF(Release_Ready), 0);
    END;   {of procedure Initialize_Data}

  PROCEDURE Patient_Data;
    BEGIN
      {Body of procedure goes here.}
    END;   {of procedure Patient_Data}

BEGIN
  Initialize_Data;
  Patient_Data;
END.
```

You can see in program 7–4 how each data type was initialized by the **FILLCHAR** procedure. In order to determine the number of bytes to be initialized the Turbo function **SIZEOF** is used. Every element in the array will now have a known value (0, ", or FALSE), and there won't be any surprises for array elements that have not yet been assigned (which may be the case if the hospital is not filled). The second procedure, `Patient__Data`, will enter data just as before.

Getting More Data Out

Now that you know how to initialize an array, you are ready to use more of its power. For example, consider program 7–5. It looks for the patients in the hospital who are release-ready; every single bed must be searched and if the release-ready data type is TRUE, data associated with that bed (patient name and amount owed) will be printed to the screen.

Program 7–5

```
PROGRAM Hospital_Data;
 VAR
   Patient_Name : ARRAY[1..3, 1..3, 1..4] OF STRING[79];
   Amount_Owed : ARRAY[1..3, 1..3, 1..4] OF REAL;
      Days_Here : ARRAY[1..3, 1..3, 1..4] OF INTEGER;
  Release_Ready : ARRAY[1..3, 1..3, 1..4] OF BOOLEAN;

 PROCEDURE Initialize_Data;
   BEGIN
     {Initialization process goes here.}
   END;   {of procedure Initialize_Data}

 PROCEDURE Patient_Data;
   BEGIN
     {Body of procedure goes here.}
   END;   {of procedure Patient_Data}

 PROCEDURE Release_Ready_Patients;
   VAR
     Floor,
     Room,
     Bed
  :INTEGER;

   BEGIN
   WRITELN('Patients ready for release:');
   FOR Floor := 1 TO 3 DO {Three floors in hospital}
     BEGIN
     FOR Room := 1 TO 3 DO {Three rooms per floor}
       BEGIN
         FOR Bed := 1 TO 4 DO {Four beds per room}
           BEGIN
             IF Release_Ready[Floor, Room, Bed] = TRUE
```

Program 7–5   *continued*

```
            THEN
              BEGIN
                WRITE('Room = ',Floor,Room,' Bed = ',Bed,' ');
                WRITELN(Patient_Name[Floor, Room, Bed]);
                WRITE('Balance owed = $');
                WRITELN(Amount_Owed[Floor, Room, Bed]:3:3);
              END;
           END;   {Bed loop}
         END;   {Room loop}
    END;   {Floor loop}
  END;   {of procedure Release_Ready_Patients}

  BEGIN
    Initialize_Data;
    Patient_Data;
    Release_Ready_Patients;
  END.
```

Program 7–5 used three nested loops. The output **FOR-TO** loop incremented the index representing each of the three floors. The next loop incremented the index representing each of the three rooms on each floor. The innermost loop incremented the index representing each of the four beds in each room. In the innermost loop, the **IF-THEN** statement checks to see if any of the BOOLEAN types **Release_Ready** is TRUE. If it is, the index that is used for this type will now be the same index used to identify other data associated with that bed—the patient name and balance owed.

When program 7–5 is executed it will now display

```
Patients ready for release:
Room = 11 Bed = 2 Howard Smith
Balance owed = $1679.530
Room = 12 Bed = 1 Mary Proctor
Balance owed = $359.860
```

As you will see, the use of arrays in Pascal provides you with powerful ways of handling data. But first, you need to know a little more about some unique characteristics that determine arrays. That information is in the next section.

Conclusion

This section gave you an idea of what arrays are and how they can be used. In the next section, you will learn more about the structure and use of arrays in Pascal. For now, check your understanding of this section by trying the following section review.

7–1 Section Review

1   Explain the origin of the word array.

2   Give an example of using a subscript.

3   State the advantage of using subscripts.
4   What does it mean to initialize an array?

7–2   Ways of Representing an Array

Some Details

The previous section gave you a good idea of what arrays look like and why to use them. You can think of an array variable as an ordered collection of **elements**, all of the same type. As an example, in the hospital program, the names of the patients were ordered elements of the type STRING[79] and the number of days spent in the hospital were ordered elements of the type INTEGER.

In an array, a particular element is distinguished from other elements by means of an index value enclosed in square brackets. In the hospital program the name of one patient was distinguished from another by the three elements of floor number, room number, and bed number. As an example,

```
PatientName[1,1,1]
```

represented a different person from

```
PatientName[1,1,2]
```

Simple Array

Program 7–6 shows a simple one-dimensional array.

Program 7–6

```
PROGRAM Array_Demo1;

  VAR
  Values

 :ARRAY[0..3] OF INTEGER;
  Number
 :INTEGER;

  BEGIN
    Values[0] := 100;
    Values[1] := 200;
    Values[2] := 300;
    Values[3] := 400;

  FOR Number := 0 TO 3 DO
    WRITELN(Values[Number]);

  END.
```

When program 7–6 is executed, the output will be

```
100
200
300
400
```

Program 7–6 contains four elements, each distinguished by a numerical index. This is an example of a one-dimensional array. Theoretically, there is no limit to the number of dimensions an array can have or to the value for each index in the array. However, there is a practical limitation brought about because any computer has a limited amount of memory.

Any finite ordinal value can be used as an index. An INTEGER cannot be used as a type in an index because it is not finite (the computer will run out of memory space trying to save room for the whole set of integer elements). The values used to index elements of a Pascal array can be a subrange of INTEGER or any scalar or subrange type except INTEGER. This means that you need not declare an array by starting with element 0 or with element 1; it depends upon how you intend to use the elements. In Pascal, an array may have any finite number of dimensions within the memory limitations of your computer.

The fact that any finite ordinal value may be used is illustrated in program 7–7. It uses a CHAR index.

Program 7–7

```
PROGRAM Array_Demo2;

  VAR
   Values

 :ARRAY['A'..'D'] OF INTEGER;

   Number
 :CHAR;

  BEGIN

    Values['A'] := 100;
    Values['B'] := 200;
    Values['C'] := 300;
    Values['D'] := 400;

  FOR Number := 'A' TO 'D' DO

    WRITELN(Values[Number]);

  END.
```

When program 7–7 is executed, the output will be the same as that for program 7–6. This shows that you can use any ordinal value as the index. This is a useful feature because you can now make your programs more descriptive; you're not just limited to whole numbers. You can even use a BOOLEAN type of FALSE and TRUE, as demonstrated in program 7–8. You do need to make sure that FALSE comes before TRUE in the subrange statement (recall that the ordinal value of FALSE is 0). Program 7–8 takes a single-dimension array whose index has two values (FALSE and TRUE), assigns an INTEGER value to each element, and then prints these values on the screen.

Program 7–8

```
PROGRAM Array_Demo3;

  VAR

   Values
 :ARRAY[FALSE..TRUE] OF INTEGER;

   Number
 :CHAR;

  BEGIN

    Values[TRUE] := 100;
    Values[FALSE] := 200;

     WRITELN(Values[TRUE]);
     WRITELN(Values[FALSE]);

  END.
```

When executed, program 7–8 will yield

```
100
200
```

Program 7–8 simply demonstrates that BOOLEAN variables can be used as the index, to reinforce the idea that any finite ordinal value can be used as an index.

Your Own Type

You can even create your own type to be used as an index. Consider program 7–9. Notice that the user-defined type `Color_Code` is used as the index for the arrayed variable. The program will assign the equivalent resistor color code values for the five colors defined in the program. Using your own data type for an array index can help to make the program much more descriptive.

Program 7–9

```
PROGRAM Array_Demo4;
  TYPE
   Color_Code = (Black, Brown, Red, Orange, Yellow);

  VAR

   Values
 :ARRAY[Black..Yellow] OF INTEGER;

   Color
 :Color_Code;

  BEGIN

    Values[Black]  := 0;
    Values[Brown]  := 1;
      Values[Red]  := 2;
   Values[Orange]  := 3;
   Values[Yellow]  := 4;

    FOR Color := Black TO Yellow DO

      WRITELN(Values[Color]);

  END.
```

When program 7–9 is executed, it will cause the screen to display the first five numbers of the color code.

```
0
1
2
3
4
```

Program 7–9 demonstrates that using your own type for an index can make the program itself more descriptive. From the standpoint of the program user this may not be important, but from the standpoint of the programmer it can be very important.

In each of these array programs, the arrayed variable has been defined as **VAR**. This program is somewhat limiting because in all cases it requires you to either use a global variable (which is discouraged by good programming practice) or to define the array **VAR** in each procedure that will use it (which becomes very tedious to program). A better method is to define the array as a type. Once this is done, you can easily declare an array **VAR** in any procedure, keeping your variables local. This is shown in program 7–10 (p. 400). Note that the array used by the procedure is now local to that procedure. Also note how easy it was to make the assignment since the array type was already declared.

Program 7–10

```
PROGRAM Array_Demo5;
  TYPE

   Array_Values  = ARRAY[5..8] OF INTEGER;

PROCEDURE Do_It;

  VAR

   Numbers
:Array_Values;

   Counter
:INTEGER;

  BEGIN

     Numbers[5] := 0;
     Numbers[6] := 1;
     Numbers[7] := 2;
     Numbers[8] := 3;

     FOR Counter := 5 TO 8 DO

        WRITELN(Numbers[Counter]);

  END;    {of procedure Do_It}

BEGIN
  Do_It;
END.
```

When program 7–10 is executed it will simply output the values of each number assigned to each element.

```
0
1
2
3
```

Program 7–10 shows how easily the array was declared as a local variable because its type was declared in the main programming block. Now, any procedure may easily declare a variable of that same type.

There is another important point about program 7–10. The index of the first element in the array is 5 (not 0 or 1). This was done in order to illustrate the fact that you need not start an array with 0 or 1. Now that the array type has been defined, any **VAR** can be assigned to that type.

Passing Arrays as Parameters

There is another advantage of declaring an array type—it gives you the ability to pass arrays easily among different procedures and functions. This is good programming practice because it allows you to keep these variables local. Program 7–11 demonstrates the passing of arrayed components between two procedures. The first procedure is called by the second procedure. Note the parameter list of the first procedure. It uses a value parameter which is defined as the previously-declared array type. The second procedure declares a local variable using the same predefined type. Since both procedures are now using the same type it makes the process of using local variables and the passing of these types between procedures much easier to program.

Program 7–11

```
PROGRAM Array_Demo6;
  TYPE
    String79 = STRING[79];
    Array_Characters = ARRAY[1..4] OF String79;

  PROCEDURE Display_Array(Information : Array_Characters);
    VAR
      Counter
    :INTEGER;
      BEGIN
        FOR Counter := 4 DOWNTO 1 DO
          WRITELN(Information[Counter]);
      END;

  PROCEDURE Load_Them;
    VAR
      These_Characters
    :Array_Characters;

      BEGIN
        These_Characters[1] := 'First sentence';
        These_Characters[2] := 'Second sentence';
        These_Characters[3] := 'Third sentence';
        These_Characters[4] := 'Fourth sentence';

        Display_Array(These_Characters);

      END;

BEGIN
  Load_Them;
END.
```

Observe from program 7–11 that the procedure Load_Them assigns strings to each of the arrayed elements identified as These_Characters. This same procedure then calls another procedure, `Display_Array`, and passes the strings to the called procedure. The procedure `Display_Array` then does a count on the index from 4 **DOWNTO** 1 producing the following results.

```
Fourth sentence
Third sentence
Second sentence
First sentence
```

An array can be passed as a value or variable parameter. Consider program 7–12.

Program 7–12

```
PROGRAM Robot_Arm_Space;

  CONST
    Min_X =  -2;   Max_X = 3;
    Min_Y =  -5;   Max_Y = 5;
    Min_Z =  -2;   Max_Z = 5;

  TYPE
    X_Axis = Min_X..Max_X;
    Y_Axis = Min_Y..Max_Y;
    Z_Axis = Min_Z..Max_Z;

    Space_Movement = ARRAY[X_Axis, Y_Axis, Z_Axis] OF REAL;

    PROCEDURE Initialize_Space(VAR Information : Space_Movement);
      BEGIN
        FILLCHAR(Information, SIZEOF(Information), 0);
      END;

  PROCEDURE Start_With(VAR Observation : Space_Movement;
                         VAR X : X_Axis;
                         VAR Y : Y_Axis;
                         VAR Z : Z_Axis);

    BEGIN
      WRITELN('Enter initial values for X, Y and Z:');
      WRITE('X (-1 to 3) => ');
      READLN(X);
      WRITE('Y (-5 to 5) => ');
      READLN(Y);
      WRITE('Z (-2 to 5) => ');
      READLN(Z);
      Observation[X, Y, Z] := SQRT(X*X + Y*Y + Z*Z);
    END;

  PROCEDURE Get_Information;
```

Program 7–12   *continued*

```
      VAR
         Observe_Space,
         Compute_Space,
         Real_Space
      :Space_Movement;
         X : X_Axis;
         Y : Y_Axis;
         Z : Z_Axis;

      BEGIN
         Initialize_Space(Observe_Space);
         Initialize_Space(Compute_Space);
         Initialize_Space(Real_Space);

         Start_With(Observe_Space, X, Y, Z);

         WRITELN('Starting values: ');
         WRITELN;
         WRITELN('X = ',X,' Y = ',Y,' Z = ',Z);
         WRITELN('Total value = ',Observe_Space[X,Y,Z]);
      END;

BEGIN
  Get_Information;
END.
```

In program 7–12, the procedure `Initialize_Space` is called by the procedure `Get_Information` to initialize all of its arrays and pass them back to the calling procedure. This is an example of using a variable parameter that is an array. In another example, the procedure `Start_With` is called by the procedure `Get_Information` to make use of an array as a variable parameter. This is done by having the coordinate values entered by the program user along with a calculated value for this component to be passed to the calling procedure. In program 7–12, all variables are kept local and the program is broken down into small procedures for each key task. This was achieved by making the array a declared type and then using parameters to pass information between procedures using only local variables.

Arithmetic with Indices

The index of an array can have arithmetic operations performed on it. Program 7–13 (p. 404) does a numerical computation inside the [ ] of the array value.

Execution of program 7–13 yields

```
Numbers[1] = 10
Numbers[2] = 10
Numbers[3] = 10
Numbers[4] = 10
```

Program 7–13

```
PROGRAM Array_Demo_Again;

  TYPE
    Array_Numbers = ARRAY[1..8] OF INTEGER;

  VAR
    Numbers
  :Array_Numbers;
    Counter
  :INTEGER;

    BEGIN

      FILLCHAR(Numbers, SIZEOF(Numbers),0);

      FOR Counter := 1 TO 4 DO
        BEGIN

            Numbers[Counter] := 10;
            Numbers[2*Counter] := Numbers[Counter];

        END;

      FOR Counter := 1 TO 8 DO
        WRITELN('Numbers[',Counter,'] = ',Numbers[Counter]);
    END.
```

```
Numbers[5] = 0
Numbers[6] = 10
Numbers[7] = 0
Numbers[8] = 10
```

In program 7–13, note that there is an arithmetic operation in the index.

```
Numbers[2*Counter]
```

This results in skipping two of the arrayed elements. The only restriction on an expression used as an index value is that the type of the expression's value must be compatible with the index type in the array declaration.

Array Assignments

You can assign the values of one array to another array. In order to do this, the arrays must be **congruent**. Arrays are said to be congruent if they have elements of the same type, the same dimensionality, and the same number of elements in each dimension. As an example, the three following arrays are congruent.

```
TYPE
        Array_1 = ARRAY[0..15, 0..12] OF REAL;
```

```
Array_2 = ARRAY[10..25, 10..22] OF REAL;
Array_3 = ARRAY['a'..'p', 0..12] OF REAL;
```

Even though the index types are different, the three arrays are congruent because they are of the same type (REAL) and they all have 16 by 13 in two dimensions.

The following two arrays are not congruent.

TYPE
```
    This_Array = ARRAY[1..5] OF INTEGER;
    That_Array = ARRAY[1..5, 3..10] OF INTEGER;
```

Even though both of these arrays are of the same type, they do not have the same dimensionality. The first array is one-dimensional, while the second is two-dimensional.

Array Comparisons

Arrays can be compard with each other if they are of congruent types. Only two types of comparisons are allowed: = (equal) or <> (not equal). As an example, if you start with the declaration

VAR
```
    Array_1, Array_2 : ARRAY[1..10] OF CHAR;
```

then each array can be compared. If you have

```
Array_1 = Array_2;
```

the result of the expression is TRUE if every element of **Array_1** has the same value as the corresponding element of **Array_2**. Otherwise, the expression result is FALSE. In a similar fashion the comparison

```
Array_1 <> Array_2
```

will return a TRUE if any element in one array is not exactly the same as an element in the other array.

Packed Arrays

In most Pascal systems you have the option of **packing** an array. Doing this causes each element of the array to try and use as little memory space as possible. When you want to use the array, it must be unpacked.

Turbo Pascal does not use the **PACKED** statement for its arrays, because Turbo automatically tries to conserve memory space every time you declare an array.

Conclusion

This section demonstrated several ways of representing arrays. You saw what was meant by congruent arrays as well as the ability of Pascal to pass arrays as variable and value parameters between procedures. Test your understanding of this section by trying the following section review.

7–2 Section Review

1   State what determines the dimension of an array.
2   State what values may be used as the index of a Pascal array.
3   Explain one of the advantages of declaring an array type.
4   What is meant by a congruent array?
5   State the significance of having congruent arrays.

7–3   Working with Arrayed Data

Discussion

This section illustrates two major programs. One is for getting patient data into a Pascal program and the other is for extracting a particular kind of information from the program.

Putting Data In

The last section presented some necessary details concerning the use of arrays in Pascal. The hospital program can be used again to illustrate the power of arrays when it comes to handling technical data. Program 7–14 illustrates another method for entering patient data into the hospital program. The program first declares each array as a type, so that procedures can now easily declare local variables of the same array type. It's important to note how this is done in the first procedure, `Get_Data`. This procedure contains a nested procedure called `Initialize_Data` that initializes all of the elements in the array. The **REPEAT-UNTIL** loop allows the program user to actively enter information concerning patients. The program user is actually setting the value of each index when entering the floor number, room number, and bed number. Once these are set, data for each component using the same indices is entered by the program user. For example, the patient's name is a component identified as Name[Floor, Room, Bed] where each index has already been assigned a value by the program user.

Program 7–14 (pp. 406–408)

```
PROGRAM Sort_Data;
 USES CRT;

 TYPE
   Patient_Array = ARRAY [1..3, 1..3, 1..4] OF STRING[79];
    Amount_Array = ARRAY [1..3, 1..3, 1..4] OF REAL;
      Days_Array = ARRAY [1..3, 1..3, 1..4] OF INTEGER;
   Release_Array = ARRAY [1..3, 1..3, 1..4] OF BOOLEAN;
```

Program 7–14 *continued*

```
PROCEDURE Get_Data;
 VAR
    Name                {Name of patient.}
 :Patient_Array;

    Amount              {Total amount owed by patient.}
 :Amount_Array;

    Days                {Total days patient in hospital.}
 :Days_Array;

    Release             {Is patient ready for release?}
 :Release_Array;

    Floor,              {Patient floor number.}
    Room,               {Patient room number.}
    Bed                 {Patient bed number.}
 :INTEGER;

    Letter              {User selection variable.}
 :CHAR;

  PROCEDURE Initialize_Data;
   BEGIN
     FILLCHAR(Name, SIZEOF(Name), 0);
     FILLCHAR(Amount, SIZEOF(Amount), 0);
     FILLCHAR(Days, SIZEOF(Days), 0);
     FILLCHAR(Release, SIZEOF(Release), 0);
   END;  {of procedure Initialize_Data}

   BEGIN

     Initialize_Data;
    REPEAT
      ClrScr;
     WRITELN;
     WRITELN('Enter patient information:');
     WRITE('Floor number = ');
        READ(Floor);
     WRITE(' Room number = ');
        READ(Room);
     WRITE(' Bed number = ');
        READLN(Bed);

     WRITELN;
     WRITE('Patient name = ');
        READLN(Name[Floor, Room, Bed]);
     WRITE('Amount owed = ');
        READLN(Amount[Floor, Room, Bed]);
     WRITE('Days in hospital = ');
        READLN(Days[Floor, Room, Bed]);
     WRITE('Select by letter:  Patient ready for release? Y)es N)o: ');
        READLN(Letter);
           IF (Letter = 'Y') OR (Letter = 'y') THEN
              Release[Floor, Room, Bed] := TRUE
              ELSE Release[Floor, Room, Bed] := FALSE;
     WRITELN;
     WRITE('Do you want to enter more data? Y)es N)o: ');
        READLN(Letter);
```

Program 7–14 *continued*

```
   UNTIL (Letter = 'N') OR (Letter = 'n');

 END;   {of procedure Get_Data}

 BEGIN
   Get_Data;
 END.
```

Assume that the following patient data is entered into the program.

```
Enter patient information:
Floor number = 1  Room number = 1  Bed number = 1

Patient name = Frank Jones
Amount owed = 825.43
Days in hospital = 3
Select by letter: Patient ready for release? Y)es N)o: N

Do you want to enter more data? Y)es N)o: Y

Enter patient information:
Floor number = 1  Room number = 1  Bed number = 2

Patient name = Howard Smith
Amount owed = 1679.53
Days in hospital = 6
Select by letter: Patient ready for release? Y)es N)o: Y

Do you want to enter more data? Y)es N)o: Y

Enter patient information:
Floor number = 1  Room number = 2  Bed number = 1

Patient name = Mary Proctor
Amount owed = 359.86
Days in hospital = 1
Select by letter: Patient ready for release? Y)es N)o: Y

Do you want to enter more data? Y)es N)o: Y

Enter patient information:
Floor number = 1  Room number = 2  Bed number = 2

Patient name = Alice McDaniels
Amount owed = 3487.75
```

```
Days in hospital = 5
Select by letter: Patient ready for release? Y)es N)o: N

Do you want to enter more data? Y)es N)o: Y

Enter patient information:
Floor number = 1   Room number = 3   Bed number = 1

Patient name = Sammy Adams
Amount owed = 589.32
Days in hospital = 2
Select by letter: Patient ready for release? Y)es N)o: N

Do you want to enter more data? Y)es N)o: N
```

Program Analysis

All of the arrays in program 7–14 were initialized before the user input any data. The **REPEAT–UNTIL** loop then allowed the program user to enter data about any bed in the hospital, in any order desired. That is, information about patients on the third floor could be entered before information about patients on the first floor; this could be done because the program user was defining the value of each index. As long as the value of each index was kept within the range specified by the type statement originally defining the arrays, any bed in the hospital could have information entered about it in any order chosen by the program user. This enables information to be updated easily and entered in an order that may be dictated by the document from which the program user is receiving the original patient data.

Getting Data Out

Program 7–15 (p. 410) illustrates a method of getting data that represents pieces of data from other elements of the array. This has many useful applications, one of which is demonstrated by program 7–15. It allows the program user to get the total amount owed by all patients on any floor or combination of floors in the hospital. This is accomplished by using three nested **FOR–TO** loops. The statement in the third loop simply keeps a running total of the amount owed by each patient on the selected floor. The outer loop has its limits set by the program user through two variables: `Start` and `Ending`. Note the variable parameter passing between two procedures. Note that all elements in the arrays are initialized before any of the data is used. The initialization process prevents erroneous data from leaping unexpectedly out of the beds.

Program 7–15

```
PROGRAM Hospital_Data;
 VAR
   Patient_Name : ARRAY[1..3, 1..3, 1..4] OF STRING[79];
    Amount_Owed : ARRAY[1..3, 1..3, 1..4] OF REAL;
      Days_Here : ARRAY[1..3, 1..3, 1..4] OF INTEGER;
  Release_Ready : ARRAY[1..3, 1..3, 1..4] OF BOOLEAN;

PROCEDURE Patient_Data;

PROCEDURE Initialize_Data;
     BEGIN
       FILLCHAR(Patient_Name, SIZEOF(Patient_Name), 0);
       FILLCHAR(Amount_Owed, SIZEOF(Amount_Owed), 0);
       FILLCHAR(Days_Here, SIZEOF(Days_Here), 0);
       FILLCHAR(Release_Ready, SIZEOF(Release_Ready), 0);
     END;  {of procedure Initialize_Data}

    BEGIN

       Initialize_Data;
       {Patient data is entered here.}

    END;  {of procedure Patient_Data}

PROCEDURE Total_Bill(VAR Amount : REAL; Start, Ending : INTEGER);

    CONST
     MaxRooms = 3;
     MaxBeds = 4;

    VAR
     Floor,
     Room,
     Bed
    :INTEGER;

    BEGIN

       Amount := 0;

    FOR Floor := Start TO Ending DO

      BEGIN
      FOR Room := 1 TO MaxRooms DO

        BEGIN
        FOR Bed := 1 TO MaxBeds DO

          Amount := Amount + Amount_Owed[Floor, Room, Bed];

        END;  {Bed loop}

      END;  {Room loop}

   END;  {procedure Total_Bill}
```

Program 7–15 *continued*

```
PROCEDURE Get_Money;
 VAR
  Starting_Floor,
  Ending_Floor
:INTEGER;

  Total
:REAL;

  BEGIN

   WRITELN('Compute total bills:');
   WRITELN;

   WRITE('Starting at which floor = ');
   READLN(Starting_Floor);
   WRITE('Ending at which floor = ');
   READLN(Ending_Floor);

   Total_Bill(Total, Starting_Floor, Ending_Floor);

   WRITELN('Total owed = ',Total:5:2);
  END;
BEGIN
 Patient_Data;
 Get_Money;
 END.
```

The first procedure, `Patient_Data`, contained a nested procedure, `Initialize_Data`, that initialized all of the arrays. It also allowed the program user to enter data about any patient in the hospital. The next procedure, `Total_Bill`, did the actual computations of the total using only the floor(s) selected by the program user. Values are passed to and from this procedure via its parameter list. It gets and gives values to its calling procedure, `Get_Money`, which gets the information from the program user as to what floor(s) totals are to be taken. It then calls the procedure `Total_Bill` and passes these values to it. `Total_Bill` uses these values to determine the total owed, and this value is then passed back to the calling procedure where it is displayed on the screen.

Programs such as these are great time savers. Getting data that is the result of many other elements of data is frequently encountered in all fields of technology. You will see more of this as you proceed in your course of study of Pascal.

Conclusion

This section illustrated two major programs: one for an interactive method of getting arrayed data into a program, and another for performing a numerical operation on arrayed data. In the next section, you will discover how to sort information using arrays. Test your understanding of this section by trying the following section review.

7-3 Section Review

1   State what advantage there was in program 7–14 of allowing the program user to enter patient data in any order.
2   Explain how program 7–14 allowed the program user to input patient data in any order.
3   What did program 7–15 illustrate?

7-4   Sorting

Basic Idea

There are many different methods of sorting a list of numbers. The method presented in this section is called a bubble sort, because numbers are bubbled to the top of the list as the sorting is processed.

As an example, consider the following list of numbers.

6
3
6
8

Suppose it was your job to arrange this list in ascending order (smallest number first, largest last). Assume that you are to use the bubble sorting technique to accomplish this. Here are the rules you would use:

Bubble Sorting Rules   (To sort in ascending order.)

1. Test only two numbers at a time, starting with the first two numbers.
2. If the top number is smaller, leave it as is. If the top number is larger, switch the two numbers.
3. Go down one number and compare that number to the number that follows it. These two will be a new pair.
4. Continue this process until no switch has been made in an entire pass through the list.

To sort in descending order, simply change rule 2.

2. If the top number is larger, leave it as is. If the top number is smaller, switch the two numbers.

To sort the given list of numbers, start with the first rule, and test only two numbers at a time, starting with the first two numbers.

6
3

The top number is larger and so, using rule 2, switch the two numbers.

3
6

The list now looks like this.

3
6
6
8

Go one number down and compare it to the number that follows (a new number pair).

6
6

These are both the same. Since they are equal, it makes no difference what you do to them! Now go to the next number pair.

6
8

Since the smaller number is already on the top, leave them as is. You have completed the list, but a switch has been made on this pass through the list; you must, therefore, make another pass through the list. Test the first two numbers.

3
6

No switch is necessary. Now the next two:

6
6

Again, no switch is necessary. The next two:

6
8

Still no required switch. Since there were no switches in this pass through the list, the sorting is completed and the resulting list is

3
6
6
8

A Word about Subscripts

Before you get into the actual sorting of numbers, a discussion about subscripts treated as a single dimension variable will help you understand the sorting program. Program 7–16 (p. 414) demonstrates what happens when you increase or decrease the value of the index by 1. This is done by first loading numerical values for the first seven elements of a single-dimension array. Then a **FOR−TO** loop is used to display the values of each of these elements. For example, the program demonstrates that `Value[2]` is the same as `Value[3−1]` and `Value[1+1]`.

Program 7–16

```
PROGRAM Subscript_Exercise;
 TYPE
   Number_Array = ARRAY[0..6] OF INTEGER;

PROCEDURE Load_Them;
  VAR
   Value
 :Number_Array;

   Index
 :INTEGER;

  BEGIN

    Value[0] := 0;
    Value[1] := 2;
    Value[2] := 4;
    Value[3] := 8;
    Value[4] := 16;
    Value[5] := 32;
    Value[6] := 64;

    FOR Index := 1 TO 5 DO
      BEGIN
        WRITE('  Value[',Index,'] = ',Value[Index]);
        WRITE('  Value[',Index,' + 1] = ',Value[Index + 1]);
        WRITELN(' Value[',Index,'- 1] = ',Value[Index - 1]);
        WRITELN;
      END;
  END; {of procedure Demonstration}

BEGIN
  Load_Them;
END.
```

Note that the value of each of the following numbers depends upon the index of its subscript.

```
Value[0] := 0;
Value[1] := 2;
Value[2] := 4;
Value[3] := 8;
Value[4] := 16;
Value[5] := 32;
Value[6] := 64;
```

Recall that an operation may be performed within the brackets of the subscript notation. Thus, Value[1 + 1] = Value[2] = 4. This is demonstrated by the results of executing program 7–16.

```
Value[1] = 2       Value[1 + 1] = 4       Value[1 - 1] = 0
Value[2] = 4       Value[2 + 1] = 8       Value[2 - 1] = 2
Value[3] = 8       Value[3 + 1] = 16      Value[3 - 1] = 4
Value[4] = 16      Value[4 + 1] = 32      Value[4 - 1] = 8
Value[5] = 32      Value[5 + 1] = 64      Value[5 - 1] = 16
```

Be sure you understand this before going on to the sorting program; if it is not clear to you, you may want to try the interactive exercises that deal with this concept at the end of this chapter.

Sample Program

Program 7–17 illustrates the bubble sorting method. It starts out by defining a single-dimension array **OF REAL** consisting of nine elements and then initializes them. Next, the program user is asked to enter nine numbers in the **FOR-TO** loop. When the program user enters the ninth number the loop is finished and the entered numbers are then sorted in ascending order. The sorting is accomplished in the procedure `Bubble_Sort`. Here two numbers are compared at a time and their location in the array determined by their relative size. The last procedure `Display_Them` then displays the original order of the numbers and their sorted order.

Program 7–17

```
PROGRAM Sort_It;
 CONST
   MaxNumber = 9;

 TYPE
   Array_List = ARRAY[1..MaxNumber] OF REAL;

PROCEDURE Bubble_Sort (NewList, OldList : Array_List);
FORWARD;

PROCEDURE Display_Them (SortedNumbers, UnsortedNumbers : Array_List);
FORWARD;

PROCEDURE Input_Numbers;
 VAR
  Numbers,
  OldOrder
 :Array_List;
   Count
 :INTEGER;

 BEGIN

     {Initialize array}
     FILLCHAR(Numbers, SIZEOF(Numbers), 0);

   WRITELN('Give me nine numbers and I''ll sort them:');
   FOR Count := 1 TO MaxNumber DO
    BEGIN
     WRITE('Number[',Count,'] = ');
     READLN(Numbers[Count]);
    END;
```

Program 7–17 *continued*

```
   OldOrder := Numbers;

   Bubble_Sort(Numbers, OldOrder);

  END;  {of procedure Input_Numbers}
PROCEDURE Bubble_Sort{(NewList, OldList : Array_List)};
  VAR
   Index
 :INTEGER;
   TempValue
 :REAL;
   Switch
 :BOOLEAN;

  BEGIN

  REPEAT
    Switch := FALSE;  {Set sorting flag false.}

    FOR Index := 1 TO MaxNumber DO
      BEGIN
      {Test to see if the number is greater.}

        IF (NewList[Index] > NewList[Index + 1])
              AND (Index <> MaxNumber)
        THEN {A switch is needed}
          BEGIN
           TempValue := NewList[Index];         {Temporarily store value.}
           NewList[Index] := NewList[Index + 1];{Switch order of numbers.}
           NewList[Index + 1] := TempValue;
           Switch := TRUE;  {A switch was made.}
          END;
      END;

  UNTIL Switch = FALSE;

  Display_Them(NewList, OldList);

  END;

 PROCEDURE Display_Them{(SortedNumbers, UnsortedNumbers : Array_List)};
  VAR
    Counter
  :INTEGER;

    BEGIN

      WRITELN(' Original Order         Sorted Order');

      FOR Counter := 1 TO MaxNumber DO
        BEGIN
          WRITE('  ',UnsortedNumbers[Counter]:3:2);
          WRITELN('                          ',SortedNumbers[Counter]:3:2);
        END;
    END;  {of procedure Display_Them}

  BEGIN
    Input_Numbers;
  END.
```

A sample execution of program 7–17 is shown below.

```
Give me nine numbers and I'll sort them:
Number[1] = 6
Number[2] = 7
Number[3] = 5
Number[4] = 8
Number[5] = 4
Number[6] = 9
Number[7] = 3
Number[8] = 0
Number[9] = 2
Original Order                Sorted Order
     6.00                         0.00
     7.00                         2.00
     5.00                         3.00
     8.00                         4.00
     4.00                         5.00
     9.00                         6.00
     3.00                         7.00
     0.00                         8.00
     2.00                         9.00
```

Program Analysis

The procedure `Input_Numbers` initializes the array and then gets nine numbers from the program user. It also has two arrays. Note how all the values of one array can be quickly passed to the values of a congruent array in a single statement

```
OldOrder := Numbers;
```

Then, both of these arrays are passed on to a procedure.

```
Bubble_Sort(Numbers, OldOrder);
```

In the `Bubble_Sort (NewList, OldList)` procedure the numbers are compared as follows.

```
  FOR Index := 1 TO MaxNumber DO
    BEGIN
    {Test to see if the number is greater.}

IF (NewList[Index] > NewList[Index + 1]
   AND (Index <> MaxNumber)
THEN {A switch is needed}
   BEGIN
     TempValue := NewList[Index]; {Temporarily store value.}
     NewList[Index] := NewList[Index + 1];{Switch order of numbers.}
     NewList[Index + 1] := TempValue;
     Switch := TRUE; {A switch was made.}
   END;
END;
```

Two numbers at a time are compared by

```
IF (NewList[Index] > NewList[Index + 1])
   AND (Index <> MaxNumber)
THEN {A switch is needed}
```

This step checks to see if this is the last number in the list. If it is, no comparison will be made (there is nothing to compare it with). If it is not the last number in the list, the numbers are sorted in ascending order. (If the program were to sort in descending order, the $>$ sign would be changed to the $<$ sign.) The switching is accomplished by having a temporary location for the storage of a number. This is equivalent to having a cup of milk and a glass of orange juice and wanting to switch the two liquids between containers. The easiest way to do this is to get a third container, pour the orange juice into it, pour the milk into the glass, and then pour the orange juice into the cup. This switching is done in program 7–17 with the following statements.

```
BEGIN
  TempValue := NewList[Index];            {Temporarily store value.}
  NewList[Index] := NewList[Index + 1];{Switch order of numbers.}
  NewList[Index + 1] := TempValue;
  Switch := TRUE; {A switch was made.}
END;
```

Another important element of the sorting program is the **switch flag**. It is initially set to FALSE. It will only be set to TRUE if a switch is made. Recall that every time a switch is made, you need to go through the list again to make sure that all the numbers are sorted. When a switch is not made, `Switch` will be FALSE (it is set FALSE every time the **REPEAT-UNTIL** loop repeats). When `Switch` = FALSE, the sorting stops. The display routine is then called and both the sorted and unsorted lists of numbers are passed to it.

Conclusion

Bubble sorting was presented here as the most direct method of sorting numbers. In the next section, bubble sorting will be used to sort string variables. Test your understanding of this section by trying the following section review.

7–4 Section Review

1 Briefly describe the process of a bubble sort.
2 For any list of data, what is the minimum number of times a sorting program must go through a list of numbers before the sort is considered completed? Under what circumstances would this happen?
3 What determines if a list of data will be sorted in descending or ascending order?
4 In order to sort the following list in ascending order, how many passes through the list will a bubble sort require?
   8 7 3 1

7–5   Sorting Strings

Basic Idea

Chapter Two introduced the concept of relational operations with string variables. You saw previously how strings can be compared; since the ASCII code for letters of the alphabet is in numerical order, it is easy for the computer to recognize that

```
'A' < 'B'
'AB' < 'AC'
```

and because of this, the program used for sorting numbers requires only a simple modification to sort strings.

Making a Modification

In order to accommodate string variables in a sorting program, the number sorting program of the last section, program 7–17, will be modified to include strings.

The first change necessary will be in the type block. The array type must be changed from REAL to STRING. Thus,

```
TYPE
   Array_List = ARRAY[1..MaxNumber] OF REAL;
```

is changed to

```
TYPE
   Array_List = ARRAY[1..MaxNumber] OF STRING;
```

The other change that is required is to the temporary storage variable `Temp-Value` in the `Bubble_Sort` procedure.

```
PROCEDURE Bubble_Sort{(NewList, OldList : Array_List)};
   VAR
    Index
   :INTEGER;
    TempValue
   :REAL; ← This is changed to STRING;
    Switch
   :BOOLEAN;
```

These changes allow all of the input and sorted data to be handled as strings. Note that the program user could still enter numerical data; it would just be treated as a string (and sorted as a string, not as a number—a big difference).

The last required modification would be in the procedure `Display_Them`.

```
BEGIN
    WRITE(' ',UnsortedNumbers[Counter]:3:2);
    WRITELN('                         ',SortedNumbers[Counter]:3:2);
END;
```

would be changed to

```
BEGIN
    WRITE('  ',UnsortedNumbers[Counter]);
    WRITELN('                          ',SortedNumbers[Counter]);
END;
```

This change is made because STRING data will be displayed, not numerical data. You may also want to make a few other changes for program clarity. For example,

```
WRITELN('Give me nine numbers and I''ll sort them:');
 FOR Count := 1 TO MaxNumber DO
   BEGIN
    WRITE('Number[',Count,'] = ');
    READLN(Numbers[Count]);
   END;
```

should be changed to

```
WRITELN('Give me nine strings and I''ll sort them:');
 FOR Count := 1 TO MaxNumber DO
   BEGIN
    WRITE('String[',Count,'] = ');
    READLN(String[Count]);
   END;
```

Program Execution

When these modifications have been made to program 7–17, you will be able to perform a bubble sort with STRING data. A sample execution of the program is

```
Give me nine strings and I'll sort them:
String[1] = aa
String[2] = AA
String[3] = AB
String[4] = ab
String[5] = Aa
String[6] = aA
String[7] = Ba
String[8] = Ab
String[9] = Zx
Original Order                    Sorted Order
      aa                               AA
      AA                               AB
      AB                               Aa
      ab                               Ab
      Aa                               Ba
      aA                               ZX
      Ba                               aA
      Ab                               aa
      ZX                               ab
```

You can observe from this string sort that capital letters are considered "smaller" than lowercase letters. This is because they are sorted according to their ASCII code values. The ASCII code values of uppercase letters is smaller than the ASCII code values of lowercase letters (refer to the ASCII code chart in the appendix of this book). This also means that the string "AA" will be considered "smaller" than the string "Aa".

Sorting Number Strings

Numerical data can be entered along with string data, and number strings can be sorted, but they will be sorted as string data, not as numerical data. This is illustrated by the following example.

```
Give me nine strings and I'll sort them:
String[1] = 9
String[2] = 90
String[3] = 100
String[4] = 2
String[5] = 5
String[6] = 8
String[7] = 3
String[8] = 0
String[9] = 0
Original Order              Sorted Order
     9                           0
    90                           0
   100                         100
     2                           2
     5                           3
     8                           5
     3                           8
     0                           9
     0                          90
```

Note from this sorted data output that the ordering of the data is correct based upon the ASCII value of the number when treated as a string, although the sorting is not correct based upon the stated numerical value of the string. This is because the data is treated as string data, not numerical data, and the numerical value of the ASCII code is used in the sorting process rather than the actual value of the number itself.

Conclusion

This section demonstrated how to sort strings; it also demonstrated the problems you can run into if you attempt to sort numerical data as string data.

The next section presents some methods that can be used to help you debug your Pascal programs. For now, check your understanding of this section by trying the following section review.

7–5 Section Review

1 Explain how string statements can be compared in Pascal.
2 State the major differences between a numerical and a string sorting program.
3 If the following data were entered into a string sorting program that sorted in descending order, in what order would the sorted data be displayed?
```
Transistor    35    Capacitor    100
```
4 Explain the reason for the resulting string sort in problem 3 above.

7–6 Program Debugging and Implementation— Debugging Methods

Run-Time Errors

You have probably developed programs that seemed to compile, yet when you executed them, you didn't get the answers you expected. This kind of error is called a **run-time error**. This can happen because of an incorrect formula or a mistake in entering the formula. It can also happen because you forgot to initialize certain variables or you didn't structure loops or branches correctly.

These can often be the most difficult bugs to analyze. The compiler is of no help because everything is correct as far as the Pascal coding goes. The problem is embedded somewhere in the program and you are not getting the expected results, or what you are getting is incorrect. There are several techniques for finding these run-time errors. One way is to carefully read through your program (or have someone else read through it with you). This is a fine technique for smaller programs. However, in larger programs consisting of many procedures with complex loops and branches, taking the time to read through every one may not be the most efficient method. One technique referred to as **tracing** is used by many professional programmers. It's actually an easy addition to any program and you may want to consider this method for your own use.

A Sample Problem

First, a program that contains a problem will be illustrated and the problem explained. Then a **debug routine** will be imbedded within the program. You will then see how the debug routine can be evoked to give you insight into what is happening within the program. The sample program is program 7–18.

The procedure `Loop_It;` will loop forever because the condition for the loop is never met. The entire program will compile successfully because there are no Pascal programming errors, but you will experience a run-time error in program 7–18.

You need to see what is actually going on while this part of the program is executing. A debug routine usually consists of two parts:

1. Visual
2. Stop/Start

Program 7–18

```
PROGRAM Problem_Demo;

  PROCEDURE Loop_It;
   VAR
    Counter
  :INTEGER;

     BEGIN

      Counter := 0;    {Initialize counter}

        REPEAT

            Counter := Counter + 2;

            {Body of counting loop...}

        UNTIL Counter = 9;

      END;  {of procedure Loop_It}

  BEGIN
    {Other procedures and functions}
    Loop_It;

  END.
```

The visual part causes the values of one or more variables to be displayed while the stop/start part allows you to "single-step" through the program. A debug routine for the procedure Loop_It from program 7–18 is shown in program 7–19.

Program 7–19

```
PROGRAM Problem_Demo_2;

  PROCEDURE Loop_It;
   VAR
    Counter
  :INTEGER;

     BEGIN

      Counter := 0;  {Initialize counter}

          REPEAT
```

Program 7–19   *continued*

```
{Debug Routine}   WRITELN('Counter = ',Counter);
                  READLN;

            Counter := Counter + 2;

            {Body of counting loop...}

        UNTIL Counter = 9;

    END;  {of procedure Loop_It}

  BEGIN
    {Other procedures and functions}
    Loop_It;
  END.
```

When program 7–19 is executed, the output will be

```
Counter = 0
 [Return]
Counter = 2
 [Return]
Counter = 4
 [Return]
Counter = 6
 [Return]
Counter = 8
 [Return]
Counter = 10
 [Return]
```

Now you can single-step through the program and observe the action of the loop counter. Its continued increase confirms what you should have suspected; the debug routine quickly shows you that it never achieves the loop exit requirements.

Auto Debug

If you are in the process of developing a large program and you are testing various partially-completed procedures within the program you may consider using an **auto-debug routine.** This is similar to the debug routine you have seen with the added advantage that you can easily turn it on or off with one simple command. This is especially helpful if you have several procedures with debug routines in them. Program 7–20 demonstrates the auto-debug method.

Program 7–20

```
PROGRAM Problem_Demo_3;
  VAR
   Debug
 :BOOLEAN;

  PROCEDURE Loop_It;
   VAR
    Counter
 :INTEGER;

    BEGIN

      Debug := TRUE;

      Counter := 0;   {Initialize counter}

        REPEAT
{Debug Routine} IF Debug THEN
                BEGIN
                  WRITELN('Counter = ',Counter);
                  READLN;
                END;

           Counter := Counter + 2;

           {Body of counting loop...}

         UNTIL Counter = 9;

    END;   {of procedure Loop_It}

  BEGIN
    {Other procedures and functions}
    Loop_It;
  END.
```

If the BOOLEAN variable `Debug` is made global, you only need to set it TRUE or FALSE depending on whether you want it active in all the procedures.

Auto-Debug Procedure

An **auto-debug procedure** is simply a Pascal procedure that is included in your developing program. This is not feasible with all types of programs, but for certain programs where the variable types are the same, this is particularly useful. Program 7–21 (p. 426) illustrates a debug procedure.

Program 7–21

```
PROGRAM Problem_Demo_4;

  PROCEDURE Debug (Active   :BOOLEAN;
                   Display  :INTEGER;
                   StartStop :BOOLEAN);
    BEGIN
      IF Active
        THEN
          BEGIN
            WRITELN(Display);
            IF StartStop
              THEN READLN
            ELSE
          END;
      END; {of procedure Debug}

  PROCEDURE Loop_It;
   VAR
    Counter
  :INTEGER;

    BEGIN

      Counter := 0;   {Initialize counter}

        REPEAT

          Debug(TRUE, Counter, FALSE);

            Counter := Counter + 2;

            {Body of counting loop...}

        UNTIL Counter = 10;

    END;  {of procedure Loop_It}

  BEGIN
    {Other procedures and functions}
    Loop_It;
  END.
```

Now this procedure can be called by any part of the program; it will output the value of a given variable and offers a stop/start option as well as an on/off option. This can be a very useful method for quickly debugging large programs. It is suggested that if such auto-debug procedures are used, that they be highlighted so they can be easily removed from the final program. Many programmers will keep two source copies of their programs—one with the auto debug in it and the other without it. You modify the version with the debug routines in it, make a backup copy, remove the debug commands, and you now have a new copy of the modified version without the debug routines.

Conclusion

This section presented some of the debugging methods used by professional programmers for debugging large Pascal programs. You saw how to include a basic debug command, yet make it easy to deactivate. You also saw how to create an automatic debugging feature that could be used for certain types of programs. Try some of these methods in the next program you develop. For now, test your understanding of this section by trying the following section review.

7–6 Section Review

1 Explain what is meant by a run-time error.
2 Does the compiler catch run-time errors? Explain.
3 What is usually contained in a debug routine?
4 What is meant by an auto-debug routine?

7–7   Case Study

Discussion

This case study utilizes many of the features of Pascal that you have learned up to this point. During the presentation of this case study you will see the need for another feature in Pascal, paving the way into the next chapter on records. Don't worry about this for now; just know that there is an easier way of developing the program you will see.

Problem Statement

Develop a Pascal program for a stock broker that would perform the following.

1. Allow the program user to enter
   Name of the stock.
   Purchase price of the stock.
   Number of shares purchased.
   Identify if the stock is blue chip.
2. Allow the program user to modify
   The price of any stock.
   Remove an existing stock from the list.
3. Allow the program user to observe a list in alphabetical order of
   Name of the stock.
   Number of shares.
   Original purchase price.
   If stock is blue chip.
   Net worth of any one stock group.
   Gain/loss in stock group.

Design Process

Once the problem is stated in writing, use the Problem Solution Guide to help identify the steps needed to solve the problem.

PROBLEM SOLUTION GUIDE   `PROCEDURE Explain__Program`

Formula: **Explain operation of program to user.**
Values Used: **None.**

Step	Computation	What did you actually do?
1	Explain program to the user.	Use at the start of the program. Do not repeat this part of the program.

PROBLEM SOLUTION GUIDE   `PROCEDURE Display__Main__Menu`

Formula: **Check to see if option selected is valid.**
Values Used: **A, B, C, D, E**

Step	Computation	What did you actually do?
1	Decide what feature of the program to use.	Think about all the options the program user would have for this program.
2	Options are: A. Enter new stock. B. Update price. C. Sell stock. D. List stock. E. Exit program.	Decide how many options the program user will need for this program.

PROBLEM SOLUTION GUIDE   `PROCEDURE Input__New__Stock`

Formula: **Allow user to continually enter data until an exit is desired.**
Values Used: **NameOfStock, NumberOfStock, CostOfStock, TypeOfStock, MoreInput**

Step	Computation	What did you actually do?
1	Ask user to input: Stock name Number of shares Cost per share Is it Blue Chip? Another entry?	Consider all of the information that would actually have to be entered by the program user.

PROBLEM SOLUTION GUIDE   **PROCEDURE** `Update_Stock_Price`

Formula: Display stocks by name and have user select which stock (by number). Compute gain/loss:
`Difference := NewValue * Quantity - Cost * Quantity`
Values Used: NameOfStock, NumberOfStock, CostOfStock, NewValue, Difference

Step	Computation	What did you actually do?
1	Ask user to input: Stock name New Price	Consider all of the information that would actually have to be entered by the program user.

PROBLEM SOLUTION GUIDE   **PROCEDURE** `Sell_Stock`

Formula: Display stocks by name and ask user to select stock to be deleted (by number).
Values Used: NameOfStock, NumberOfStock, CostOfStock, TypeOfStock, TotalValue, GainLoss

Step	Computation	What did you actually do?
1	Ask user to input: Stock name to be deleted.	Consider all of the information that would actually have to be entered by the program user. Realize that the stock and all associated data will have to be "pushed up" to the space left by the deleted stock.
2	Stock must be sorted so it fills the gap left by the omitted stock.	

PROBLEM SOLUTION GUIDE   **PROCEDURE** `List_Stock`

Formula: This presents an alphabetized display of the stock.
Values Used: NameOfStock, NumberOfStock, CostOfStock, TypeOfStock, Value, GainLoss

Step	Computation	What did you actually do?
1	This must be an alphabetized listing of the stock and the following information: Name of stock Number of shares Cost per share Type (Blue chip) Value of stock Gain/Loss	Stock will have to be alphabetized before this procedure is used.

PROBLEM SOLUTION GUIDE   PROCEDURE Alphabetize_Stock

Formula: **This presents an alphabetized display of the stock.**
Values Used: **NameOfStock, NumberOfStock, CostOfStock, TypeOfStock, Value, GainLoss**

Step	Computation	What did you actually do?
1	List stock in alphabetical order. All other items associated with that stock must also follow the process of alphabetization	Stock must be alphabetized every time a new stock is entered.

PROBLEM SOLUTION GUIDE   PROCEDURE Program_Repeat

Formula: **Asks program user if program exit is desired.**
Values Used: **UserChoice**

Step	Computation	What did you actually do?
1	Ask user if they are sure they want to exit the program.	Cause a message to appear at the bottom of the screen.

PROBLEM SOLUTION GUIDE   PROCEDURE Continue

Formula: **Presents "Press RETURN/ENTER to continue . . ."**
Values Used: **None**

Step	Computation	What did you actually do?
1	Have RETURN/ENTER message appear at bottom of screen.	Use this in any procedure to stop the program execution and wait for user response.

Analysis of these Problem Solution Guides shows that this program involves very little actual computation. There is, however, a lot of arranging of information, and the "formula" sections of the Problem Solution Guides do not contain mathematical computations as much as how program information will be formulated. The Problem Solution Guides indicate that there will be nine procedures called

```
Explain_Program;
Display_Main_Menu;
Input_New_Stock;
Update_Stock_Price;
Sell_Stock;
List_Stock;
```

```
Alphabetize_Stock;
Program_Repeat;
Continue;
```

Algorithm Development Guide

The next step in the design process is to develop an algorithm. Once you have completed the thinking process required by the Problem Solution Guides, you have a good idea of what needs to be done, and algorithm development becomes much easier.

ALGORITHM DEVELOPMENT GUIDE

 I. Explain Program to User
   A. State what the program will do
   B. State what the program user can do

 II. Display Main Menu
   A. Have program user select from the following:
     1. Enter new stock
     2. Update stock price
     3. Sell stock
     4. List stock
     5. End the program

 III. Have Stock Alphabetized
   A. Every time a new entry is made

 IV. Program Repeat
   A. Repeat the above sequence starting at II

The algorithm gives an overview of the main parts of the program. The details of the program have already been studied in the Problem Solution Guides. The next step is to elaborate on each of the procedures.

Control Blocks Guide

Using the Control Blocks Guide helps you think through what is actually going into the procedure. This will get you thinking about the three types of program blocks: action, loop, and branch. Once this is done, you will have excellent documentation to fall back on as you develop the actual program.

CONTROL BLOCKS GUIDE (PROCEDURES)

Action Blocks

Number	Action Taken	
1	Program title.	`[PROCEDURE Program_Title]`
2	Explain program to user.	`[PROCEDURE Explain_Program]`
3	Update a stock price.	`[PROCEDURE Update_Stock_Price]`
4	Confirm program exit.	`[PROCEDURE Program_Repeat]`
5	Stops program for user.	`[PROCEDURE Continue]`

Counting Loop Blocks

No. 6 What is repeated: The number of stocks alphabetically below the stock that has been deleted.
Starting Value: Index value of the deleted stock.
Increment: One
Ending Value: Maximum stock entries. `[PROCEDURE Sell_Stock]`

No. 7 What is repeated: Name, Quantity, Cost, Type, Value, Gain/Loss
Starting Value: One
Increment: One
Ending Value: Maximum stock entries. `[PROCEDURE List_Stock]`

Sentinel Loop Blocks

No. 8 What is repeated: Information about a new stock.
Under what conditions: As long as the user wants another stock entry or the maximum number of stocks have been entered. `[PROCEDURE Input_New_Stock]`

No. 9 What is repeated: All information about the stock.
Under what conditions: Until list is in alphabetical order.
`[PROCEDURE Alphabetize_Stock]`

No. 10 What is repeated: Entire program except `Explain_Program`.
Under what conditions: Until user exits program.
`[PROCEDURE Main_Programming_Sequence]`

Branch Blocks

No. 11 Condition for branch: User input—one of five choices.
If condition is met: Branch to required procedure.
If condition is not met: Repeat the selection. `[PROCEDURE Display_Main_Menu]`

As you can see from the Control Blocks Guides, two new procedures have been added. One contains the sentinel loop that allows the entire program to be repeated (with the exception of the program explanation) until the program user exits the program. The other is a title for the program. Of course the program title along with the program instructions will not be repeated.

Next, you should determine what kind of program constants and variables you will need.

Program Constants and Variables

The completed Constants and Variables Guide is shown below.

CONSTANTS AND VARIABLES GUIDE

Formulas	Constant or Variable	Meaning	Type R	I	C	B	S	Block(s) Used
	MaxStocks	Maximum number of stock entries.		X				Global CONST
ARRAY [1..MaxStocks]	Stock_Name_ Array	Name of stock					X	Global TYPE
ARRAY [1..MaxStocks]	Quantity_ Array	Number of shares		X				Global TYPE
ARRAY [1..MaxStocks]	Money_ Array	Cost of share	X					Global TYPE
ARRAY [1..MaxStocks]	Blue_Chip_ Array	If stock is Blue chip				X		Global TYPE
	Decision	User choice			X			All blocks where a decision is required
	StockName	Name of stock	Stock_Name_Array					Use as actual parameter
	Quantity	Number of shares	Quantity_Array					Use as actual parameter
	Cost	Cost of share	Money_Array					Use as actual parameter
	Value	Present stock value	Money_Array					Use as actual parameter
	Difference	Gain/Loss	Money_Array					Use as actual parameter
	BlueChip	If stock is Blue chip	Blue_Chip_Array					Use as actual parameter
	NameOfStock	Name of stock	Stock_Name_Array					Use as formal parameter
	NumberOfStock	Number of shares	Quantity_Array					Use as formal parameter
	CostOfStock	Cost of share	Money_Array					Use as formal parameter
	Values	Present stock value	Money_Array					Use as formal parameter
	GainLoss	Gain/Loss	Money_Array					Use as formal parameter
	TypeOfStock	If stock is Blue chip.	Blue_Chip_Array					Use as formal parameter
	Index	Array index		X				Use for counter in loops
	IndexMax	Maximum entries made by user		X				To end index loops

Using the Constants and Variables Guide, you have identified twenty constants, types, and variables. For a large programming project, different members of a programming group would now be given the assignment of producing individual procedures. You, for example, could be assigned the Input_New_Stock procedure. Everyone now knows how many procedures there will be in the program and the agreed-upon identifiers that will be used for program consistency.

An important point about this program is that there are no global variables; you know it is considered poor programming practice to use global variables. If you had a team of programmers working on different procedures and there were global variables, there is no telling what values could be left in them. Thus, there are no global variables.

Pascal Design Guide

Already a tremendous amount of information has been developed concerning this complex programming assignment, but you now have a consistent reference point— something to fall back on once you begin program coding. If this were being done by a programming team, each team member would have the same reference document.

When you use the Pascal Design Guide, you are beginning to create the actual program structure. The Design Guide is illustrated in program 7–22. Note that even though programmers were assigned different procedures and didn't use the exact identifier names listed in the Constants and Variable Guide, it doesn't make any difference since all of them are local.

Program 7–22 (pp. 434–441)

```
    PROGRAM Stock__Broker;
{ ================================================================
            Developed by:  Your Name
                  Date:

    The program will allow the user to do the following:

    Enter:
          Name of stock.
          Purchase price per share.
          Number of shares.
          Identify if stock is "blue chip."

    Program maintenance:
          Current price of stock.
          Stock sold.

    Process:
          List stock in alphabetical order and show the following:
          Purchase price.
          Number of shares owned.
          Net worth of stock.
          Total price gain or loss.
          If stock is "blue chip."

**********************************************************************}
{Library UNITS}
 USES
 CRT;
```

Program 7–22   *continued*

```
CONST
MaxStocks = 10;

TYPE
Stock_Name_Array = ARRAY[1..MaxStocks] OF STRING[79];
   Quantity_Array = ARRAY[1..MaxStocks] OF INTEGER;
      Money_Array = ARRAY[1..MaxStocks] OF REAL;
 Blue_Chip_Array = ARRAY[1..MaxStocks] OF BOOLEAN;

  String79 = STRING[79];

  {There are no other global quantities.}

{*************************************************************************
                  Procedures Used:
-------------------------------------------------------------------------
 PROCEDURE Main_Programming_Sequence;

  DESCRIPTION:
     This procedure calls all of the primary procedures used in this
     program.  It is the first procedure given in order to make the
     structure of the program easy to follow.

  PARAMETERS:
     None.

  SAMPLE CALL:
     BEGIN
        Main_Programming_Sequence;
-------------------------------------------------------------------------}

   PROCEDURE Program_Title;
   FORWARD;
   {
    DESCRIPTION:
        This procedure displays the program title and waits for the
        program user to press the RETURN/ENTER key.

    PARAMETERS:
        None.

    SAMPLE CALL:
        Program_Title;
-------------------------------------------------------------------------}
   PROCEDURE Explain_Program;
   FORWARD;
   {
    DESCRIPTION:
        This procedure explains the purpose of the program to the program
        user.

    PARAMETERS:
        None.

    SAMPLE CALL:
        Explain_Program;
-------------------------------------------------------------------------}
   PROCEDURE Display_Main_Menu(VAR UserChoice :CHAR);
   FORWARD;
   { .
```

Program 7–22 *continued*

```
        DESCRIPTION:
            This procedure displays the main menu to the program user
            and asks to make one of the following selections:
                    Select by letter:
                        A] New stock purchase.
                        B] Update stock price.
                        C] Sell stock.
                        D] List stock.

        PARAMETERS:
            UserChoice [output] = Returns user selection.

        SAMPLE CALL:
            Display_Main_Menu(UserSelection);
----------------------------------------------------------------------}
    PROCEDURE Alphabetize_Stock(VAR StockName :Stock_Name_Array;
                                VAR Quantity :Quantity_Array;
                                VAR     Cost :Money_Array;
                                VAR BlueChip :Blue_Chip_Array;
                              VAR TotalStock :INTEGER;
                                VAR Value,
                                    Difference :Money_Array);
    FORWARD;
    {
        DESCRIPTION:
            This procedure alphabetizes the list of entered stocks.

        PARAMETERS:
            StockName [input/output] = Name of the stock.
            Quantity [input/output]  = Number of shares.
            Cost [input/output]      = Cost per share.
            BlueChip [input/output]  = If stock is blue chip.
            TotalStock [input/output] = Total number of a stock.
               Value [input/output] = Value of a stock.
            Difference [input/output] = Net gain or loss of a stock.

        SAMPLE CALL:
            Alphabetize_Stock(Name, Number, Amount, FALSE, Total,
                              Value, GainLoss);
----------------------------------------------------------------------}
    PROCEDURE Input_New_Stock (VAR NameOfStock : Stock_Name_Array;
                               VAR NumberOfStock : Quantity_Array;
                               VAR CostOfStock : Money_Array;
                               VAR TypeOfStock : Blue_Chip_Array;
                               VAR IndexMax : INTEGER;
                                 VAR Value,
                                     GainLoss :Money_Array);
    FORWARD;
    {
      DESCRIPTION:
          This procedure allows the program user to input the following:
                  Name of stock.
                  Purchase price of the stock.
                  Number of shares of the stock.
                  Identify if the stock is "blue chip."

        PARAMETERS:
            NameOfStock [Output] = Stock name.
          NumberOfStock [Output] = Amount of shares.
            CostOfStock [Output] = Cost per share.
            TypeOfStock [Output] = If stock is blue chip.
               IndexMax [Output] = Number of entries.
                  Value[Output] = Value of stock.
               GainLoss[Output] = Gain or loss in value of a stock.
```

Program 7–22 *continued*

```
        SAMPLE CALL:
            Input_New_Stock(Name, Shares, Cost, Stock, MaxNumber, Amount,
                            Value, Difference);
--------------------------------------------------------------------------}
    PROCEDURE Update_Stock_Price(StockName :Stock_Name_Array;
                                  Quantity :Quantity_Array;
                                      Cost :Money_Array;
                                  BlueChip :Blue_Chip_Array;
                                  IndexMax :INTEGER;
                                  VAR Value,
                                  Difference :Money_Array);
    FORWARD;
    {
      DESCRIPTION:
          This procedure allows the program user to update the price
          of previously entered stock.  An alphabetized list of stocks
          is presented and the program user is asked to select one
          by number.

      PARAMETERS:
          StockName [Input] = Name of stock.
          Quantity [Input] = Number of shares.
          Cost [Input] = Cost per share.
          BlueChip [Input] = If stock is blue chip.
          IndexMax [Input] = Maximum number of different stocks.
          Value [Output] = New value of the stock.
          Difference [Output] = Price gain or loss.

      SAMPLE CALL:
          Update_Stock_Price(Name, Number, Amount, TRUE, Maximum, Total,
                             GainLoss);
--------------------------------------------------------------------------}
    PROCEDURE Sell_Stock(VAR StockName :Stock_Name_Array;
                         VAR Quantity :Quantity_Array;
                         VAR Cost :Money_Array;
                         VAR BlueChip :Blue_Chip_Array;
                         VAR MaxIndex :INTEGER;
                         VAR TotalValue,
                             GainLoss :Money_Array);
    FORWARD;
    {
      DESCRIPTION:
          This allows the program user to delete a stock from the list
          of previously entered stock.  An alphabetized list of stocks
          is presented and the program user is asked to select one
          by number.

      PARAMETERS:
          StockName [Input/Output] = Name of stock.
          Quantity [Input/Output] = Number of shares.
          Cost [Input/Output] = Price per share.
          BlueChip [Input/Output] = If stock is blue chip.
          MaxIndex [Input/Output] = Maximum number of different stocks.
          TotalValue [Input/Output] = Total value of a stock.
          GainLoss [Input/Output] = Amount of gain or loss from a stock.

      SAMPLE CALL:
          Sell_Stock(Name, Number, Amount, TRUE, 15, Totals, Difference);
--------------------------------------------------------------------------}
    PROCEDURE List_Stock(StockName :Stock_Name_Array;
                          Quantity :Quantity_Array;
                              Cost :Money_Array;
                        TypeOfStock :Blue_Chip_Array;
                          IndexMax :INTEGER;
                          Value,
                          GainLoss :Money_Array);
```

Program 7–22   *continued*

```
   FORWARD;
   {
     DESCRIPTION:
         This program will display an alphabetized list of previously
         entered stocks along with the following information:
            Purchase price.
            Number of shares owned.
            Net worth of stock.
            Total price gain or loss.
            If stock is "blue chip."

     PARAMETERS:
         StockName [Input] = Name of stock.
          Quantity [Input] = Number of shares.
              Cost [Input] = Cost per share.
       TypeOfStock [Input] = If stock is blue chip.
          IndexMax [Input] = Number of stock entries.
             Value [Input] = Value of stock.
          GainLoss [Input] = Gain or loss for a stock.

     SAMPLE CALL:
         List_Stock(Name, Number, Amount, FALSE, Entries,
                    Dollars, Totals);
   --------------------------------------------------------------------------}
PROCEDURE Program_Repeat(VAR UserChoice : CHAR);
FORWARD;
{
     DESCRIPTION:
         This procedure asks the program user if the program is to
         be repeated.

     PARAMETERS:
         UserChoice (input) = User choice to repeat program.

     SAMPLE CALL:
         Program_Repeat(Decision);
   --------------------------------------------------------------------------}
PROCEDURE Continue;
FORWARD;
{
   DESCRIPTION:
       This procedure displays the message:
           Press RETURN/ENTER to continue...
       and waits for the user to respond.

   PARAMETERS:
       None.

   SAMPLE CALL:
       Continue;
   --------------------------------------------------------------------------

   ***********************************************************************}
PROCEDURE Main_Programming_Sequence;
{========================================================================}
 VAR
  Decision
:CHAR;

  IndexMax,
  Index
:INTEGER;

  StockName                {Name of purchased stock}
:Stock_Name_Array;
```

Program 7–22   *continued*

```
    Quantity               {Amount of stock purchased}
  :Quantity_Array;

    Value,                 {Total value of a stock}
    Difference,            {Gain or loss in dollars for a stock}
    Cost                   {Cost of purchased stock}
  :Money_Array;

    BlueChip               {If stock is blue chip}
  :Blue_Chip_Array;

    BEGIN

      IndexMax := 0;
      Program_Title;
      Explain_Program;

    REPEAT

      Display_Main_Menu(Decision);

        CASE Decision OF

          'A','a' : Input_New_Stock(StockName, Quantity, Cost,
                                    BlueChip, IndexMax, Value, Difference);
          'B','b' : Update_Stock_Price(StockName, Quantity, Cost,
                                    BlueChip,IndexMax, Value, Difference);
          'C','c' : Sell_Stock(StockName, Quantity, Cost, BlueChip,
                                    IndexMax, Value, Difference);
          'D','d' : List_Stock(StockName, Quantity, Cost, BlueChip,
                                    IndexMax, Value, Difference);
          'E','e' : Program_Repeat(Decision);

        END;  {of case}

      {Alphabetize if new stock entered}
      IF (Decision = 'A') OR (Decision = 'a') THEN
          Alphabetize_Stock(StockName, Quantity, Cost, BlueChip, IndexMax,
                            Value, Difference);

    UNTIL (Decision = 'Y') OR (Decision = 'y');
{========================================================================}
  END;  {of procedure Main_Programming_Sequence.}

  PROCEDURE Program_Title;
{----------------------------------------------------------------------}
    BEGIN

{----------------------------------------------------------------------}
  END;  {of procedure Program_Title}

  PROCEDURE Explain_Program;
{----------------------------------------------------------------------}
    BEGIN

{----------------------------------------------------------------------}
  END;  {of procedure Explain_Program.}

 PROCEDURE Display_Main_Menu{(VAR UserChoice :CHAR)};
{----------------------------------------------------------------------}
    BEGIN
```

Program 7–22   *continued*

```
{--------------------------------------------------------------------}
  END;   {of procedure Display_Main_Menu}

  PROCEDURE Input_New_Stock{(VAR NameOfStock : Stock_Name_Array;
                             VAR NumberOfStock : Quantity_Array;
                             VAR CostOfStock : Money_Array;
                             VAR TypeOfStock : Blue_Chip_Array;
                             VAR IndexMax : INTEGER
                             VAR Value,
                                 GainLoss :Money_Array)};
{--------------------------------------------------------------------}
  BEGIN

{--------------------------------------------------------------------}
  END;   {of procedure Input_New_Stock.}

  PROCEDURE Update_Stock_Price{(StockName :Stock_Name_Array;
                               Quantity :Quantity_Array;
                                   Cost :Money_Array;
                               BlueChip :Blue_Chip_Array;
                               IndexMax :INTEGER
                               VAR Value,
                               Difference :Money_Array)};
{--------------------------------------------------------------------}
  BEGIN

{--------------------------------------------------------------------}
  END;   {of procedure Update_Stock_Price}

  PROCEDURE Sell_Stock{(VAR StockName :Stock_Name_Array;
                        VAR Quantity :Quantity_Array;
                        VAR Cost :Money_Array;
                        VAR BlueChip :Blue_Chip_Array;
                        VAR MaxIndex :INTEGER
                        VAR TotalValue,
                            GainLoss :Money_Array)};
{--------------------------------------------------------------------}
  BEGIN

{--------------------------------------------------------------------}
  END;   {of procedure Sell_Stock}

  PROCEDURE List_Stock{(StockName :Stock_Name_Array;
                        Quantity :Quantity_Array;
                            Cost :Money_Array;
                      TypeOfStock :Blue_Chip_Array;
                        IndexMax :INTEGER;
                        VAR Value,
                            GainLoss :Money_Array)};
{--------------------------------------------------------------------}
  BEGIN

{--------------------------------------------------------------------}
  END;   {of procedure List_Stock}

  PROCEDURE Program_Repeat{(VAR UserChoice : CHAR)};
{--------------------------------------------------------------------}
  BEGIN

{--------------------------------------------------------------------}
  END;    {or procedure Program_Repeat}

  PROCEDURE Alphabetize_Stock{(VAR StockName :Stock_Name_Array;
                               VAR Quantity :Quantity_Array;
                               VAR      Cost :Money_Array;
```

Program 7–22 *continued*

```
                        VAR BlueChip    :Blue_Chip_Array;
                        VAR TotalStock  :INTEGER;
                        VAR Value,
                            Difference :Money_Array)};
{ -------------------------------------------------------------------}
   BEGIN

{ -------------------------------------------------------------------}
   END;   {of procedure Alphabetize_Stock}

   PROCEDURE Continue;
   { -----------------------------------------------------------------}
   BEGIN

{ -------------------------------------------------------------------}
   END;    {of procedure Continue}

   BEGIN
   Main_Programming_Sequence;
   END.
```

Conclusion

This section presented an actual case study of a programming assignment. You can begin to appreciate all of the work that is involved in a complex program—in many cases it is a team effort, as was the case here. The self-test for this chapter contains the completed program procedures. Test your understanding of this material by trying the following section review.

7–7 Section Review

1 What is the maximum number of shares that can be entered into this program?
2 How many procedures were used in the final program design?
3 Explain why array types were used.
4 What can the program user do with this program?

Summary

1 An array can be thought of as an arrangement of information.
2 An array is an easy way of keeping track of things.
3 Arrays can be used to assign different values to similar variables.
4 The dimension of an array is determined by the number of indices.
5 Arrays may be of any type.
6 An array index may be of any ordinal type except INTEGER. However, the index may be a finite set of INTEGER values as long as the amount of available computer memory is not exceeded.
7 You must make sure that an array contains a known value before working with it.
8 Initializing an array is the process of setting all elements in the array to zero.

9   Turbo Pascal provides an easy method of initializing arrays.

10   You may define an array type and then define variables as this defined type.

11   The index of an array may have arithmetic operations performed in it.

12   Most Pascal systems have the option of packing an array, but Turbo Pascal has already done this for you.

13   There are many ways of sorting data. One of the simplest methods is called a bubble sort.

14   A bubble sort compares two numbers at a time and makes a decision to switch their order depending upon the relative size of the numbers.

15   A bubble sort may be used with STRING data.

16   Sorting strings works because each letter of the alphabet is kept inside the computer as a sequential number code.

17   When sorting STRING data, numbers will not be sorted according to their numerical value but according to their ASCII value.

18   Run-time errors are errors that are not detected by the Pascal compiler; the program is coded correctly, but does not perform as desired.

19   Debug routines are used to assist you in finding run-time errors.

20   An auto-debug routine is a debug routine that can be easily activated or deactivated depending on the needs of the programmer.

Interactive Exercises

Directions

These exercises require that you have access to a computer and software that supports Pascal, specifically, the Turbo Pascal Development System, version 4.0, from Borland International. They are provided here to give you valuable experience and immediate feedback on what the concepts and commands introduced in the chapter will do. They are also fun.

Exercises

1   Program 7–23 will display uninitialized variables. What kind of output will you get?

2   Redo program 7–23, but this time change the array variable type to REAL, then run the

Program 7–23

```
PROGRAM Ie_1;
 VAR
  ArrayVariable
 :ARRAY[0..5] OF INTEGER;
  Index
 :INTEGER;

    BEGIN
      FOR Index := 0 TO 5 DO
       WRITELN(ArrayVariable[Index]);
    END.
```

program again. Record what you get. Do the same thing again, but this time with the array variable as type CHAR; record your results. Try it again with type STRING[79]. Be sure to record your results each time so you can share them in class to see if other students are getting the same results.

3 Enter program 7–24 and compile it. Now determine the largest value that the constant MaxValue can be assigned. What happens when you exceed this value?

Program 7–24

```
PROGRAM Ie_2;
 CONST
  MaxValue = 1000;

 VAR
  ArrayVariable
 :ARRAY[0..MaxValue] OF INTEGER;
  Index
 :INTEGER;

    BEGIN
      FOR Index := 0 TO 6 DO
        WRITELN(ArrayVariable[Index]);
    END.
```

4 Determine if program 7–25 will compile and execute. Record what happens.

Program 7–25

```
PROGRAM Ie_3;
 CONST
  MaxValue = 1000;

 VAR
  ArrayVariable
 :ARRAY[0..MaxValue] OF INTEGER;
  Index
 :INTEGER;

    BEGIN
      FOR Index := 2 DOWNTO -5 DO
        WRITELN(ArrayVariable[Index]);
    END.
```

5   Program 7–26 is similar to program 7–25. The difference is that the constant `MaxValue` has been changed to a new constant `MinValue`. Predict what the program will do and then try it.

Program 7–26

```
PROGRAM Ie_4;
 CONST
  MinValue = -100;

 VAR
  ArrayVariable
 :ARRAY[0..MinValue] OF INTEGER;
  Index
 :INTEGER;

    BEGIN
      FOR Index := 2 DOWNTO MinValue DO
       WRITELN(ArrayVariable[Index]);
     END.
```

6   Predict whether program 7–27 will compile, then try it. What results did you get?

Program 7–27

```
PROGRAM Ie_5;

 VAR
  ArrayVariable
 :ARRAY[0..1,0..1,0..1,0..1,0..1,0..1,0..1,0..1,0..1,0..1]
   OF INTEGER;
  Index
 :INTEGER;

    BEGIN
       WRITELN(ArrayVariable[1,1,1,1,1,1,1,1,1,1]);
     END.
```

7   Predict whether program 7–28 will compile. What were your results? How many elements does this array contain?

Program 7–28

```
PROGRAM Ie__5;
VAR
 ArrayVariable
:ARRAY[0..3,0..3,0..3,0..3,0..3,0..3,0..3,0..3,0..3,0..3]
  OF INTEGER;
 Index
:INTEGER;

  BEGIN
    WRITELN(ArrayVariable[1,1,1,1,1,1,1,1,1,1]);
  END.
```

    8   Look at program 7–29, predict what will happen, then try it. What results did you get?

Program 7–29

```
PROGRAM Ie__6;
 TYPE
  Words = STRING[79];
  MySystem = (This, That, TheOther);
  FirstArray = ARRAY[0..2, 'A'..'C'] OF Words;
  SecondArray = ARRAY[This..TheOther, 5..7] OF Words;

 VAR
  Variables__1
 :FirstArray;
  Variables__2
 :SecondArray;

  BEGIN
    Variables__1[0,'A'] := 'Hello';
    Variables__1[2,'C'] := 'Goodbye';

    Variables__2 := Variables__1;
    WRITELN(Variables__2[This, 5]);
    WRITELN(Variables__2[TheOther, 7]);
  END.
```

Pascal Commands

Identifier : **ARRAY**[$index_1$, $index_2$, . . . $index_N$] **OF** TYPE

where Identifier = any legal Pascal variable identifier
          Index = any scalar type or subrange except INTEGER

FILLCHAR(Variable, Length, Data);

where Variable = variable of any type
           Length = the number of bytes to be initialized
               Data = an ordinal value to fill the variable(s)

SIZEOF(Variable);

where SIZEOF = returns the number of bytes filled by the argument
           Variable = variable used in the argument

Self-Test

Directions

The program for this self-test contains all the procedures for the case study presented in section 7–7. Note that the first part of the program presented in the case study must be used with what follows in order for the program to compile and be executed. Answer the questions for this self-test by referring to the procedures shown in program 7–30 and by making reference to the first part of the program contained in program 7–22 in section 7–7.

Program 7–30 (pp. 446–452)

```
  PROCEDURE Program_Title;
 {----------------------------------------------------------------------}
    BEGIN

     ClrScr;  {Clear the screen}

     WINDOW(20,5,79,25);
     WRITELN('*******************************************');
     WRITELN('*                                         *');
     WRITELN('*          STOCK MARKET ANALYZER          *');
     WRITELN('*                                         *');
     WRITELN('*                   by                    *');
     WRITELN('*                                         *');
     WRITELN('*             A. Stock Broker             *');
     WRITELN('*                                         *');
     WRITELN('*******************************************');
     WINDOW(1,1,80,25);

     Continue;

 {----------------------------------------------------------------------}
 END;  {of procedure Program_Title}

  PROCEDURE Explain_Program;
 {----------------------------------------------------------------------}
    BEGIN

     ClrScr;  {Clear the screen}
```

Program 7–30 *continued*

```
      WRITELN;
      WRITELN('This program is used to simulate a stock market');
      WRITELN('analysis program.');
      WRITELN;
      WRITELN('You may enter the following:');
      WRITELN;
      WRITELN('    * Name of stock ');
      WRITELN('    * Purchase price per share ');
      WRITELN('    * Number of shares ');
      WRITELN('    * Identify if stock is Blue Chip');

         Continue;

      ClrScr;    {Clear the screen}
      WRITELN;
      WRITELN('Maintain the following information:');
      WRITELN;
      WRITELN('    * Current price of stock');
      WRITELN('    * Delete stock from list (Sell stock)');

         Continue;

      ClrScr;       {Clear the screen}
      WRITELN;
      WRITELN('Process the following information:');
      WRITELN;
      WRITELN('    * List stock in alphabetical order with');
      WRITELN('      -Purchase price');
      WRITELN('      -Number of shares owned');
      WRITELN('      -Net worth of stock');
      WRITELN('      -Total price gain or loss');
      WRITELN('      -If stock is blue chip');

         Continue;

{-----------------------------------------------------------------------}
  END;  {of procedure Explain_Program.}

 PROCEDURE Display_Main_Menu{(VAR UserChoice :CHAR)};
{-----------------------------------------------------------------------}
  VAR

   CorrectChoice
 :BOOLEAN;

    BEGIN

      ClrScr;  {Clear the screen.}

    REPEAT

      CorrectChoice := TRUE;

      WRITELN;
      WRITELN;
      WRITELN('              Select by letter:');
      WRITELN;
      WRITELN('A] Enter new stock.    B] Update stock price.');
      WRITELN('C] Sell stock.         D] List stock.');
      WRITELN('E] End this program.');
      WRITELN;
      WRITE('Your selection => ');
      READLN(UserChoice);
```

Program 7–30   *continued*

```
        {Error checking:}
        IF NOT ((UserChoice IN ['A'..'E']) OR (UserChoice IN ['a'..'e']))
            THEN
                BEGIN
                    WRITELN;
                    WRITELN('Please enter a letter from A to D');
                    CorrectChoice := FALSE;
                END;
        {end error checking}

    UNTIL CorrectChoice = TRUE;

{-----------------------------------------------------------------------}
    END;   {of procedure Display__Main__Menu}

    PROCEDURE Input__New__Stock{(VAR NameOfStock : Stock__Name__Array;
                                VAR NumberOfStock : Quantity__Array;
                                VAR CostOfStock : Money__Array;
                                VAR TypeOfStock : Blue__Chip__Array;
                                VAR IndexMax : INTEGER
                                VAR Value,
                                    GainLoss :Money__Array)};
{-----------------------------------------------------------------------}
    VAR

    Index                  {Array index}
  :INTEGER;

    MoreInput              {Variable to indicate more input}
    CHAR;                  { and type of stock}

        BEGIN

            Index := IndexMax;   {Initialize index}
            ClrScr;

            WINDOW(5,3,25,79);

        REPEAT

            Index := Index + 1;

            WRITELN;
            WRITE(Index,'] Name of stock => ');
            READLN(NameOfStock[Index]);

            WRITELN;
            WRITE(Index,'] Number of shares => ');
            READLN(NumberOfStock[Index]);

            WRITELN;
            WRITE(Index,'] Price of stock => ');
            READLN(CostOfStock[Index]);

            WRITELN;
            WRITE('Is stock blue chip Y/N => ');
            READLN(MoreInput);

                IF (MoreInput = 'Y') OR (MoreInput = 'y')
                    THEN TypeOfStock[Index] := TRUE
                    ELSE
                    TypeOfStock[Index] := FALSE;

            GainLoss[Index] := 0;
            Value[Index] := NumberOfStock[Index] * CostOfStock[Index];
```

Program 7–30 *continued*

```
          WRITELN;
          WRITE('Do you have another entry Y/N => ');
          READLN(MoreInput);

     UNTIL  (Index = 10) OR (MoreInput = 'N') OR (MoreInput = 'n');
     IF Index >= 10 THEN WRITELN('You already have 10 entries.');
     IndexMax := Index;
{---------------------------------------------------------------------}
  END;   {of procedure Input_New_Stock.}

 PROCEDURE Update_Stock_Price{(StockName :Stock_Name_Array;
                               Quantity :Quantity_Array;
                                   Cost :Money_Array;
                               BlueChip :Blue_Chip_Array;
                               IndexMax :INTEGER
                               VAR Value,
                               Difference :Money_Array)};
{---------------------------------------------------------------------}
     VAR

   NewValue
 :REAL;

   Selection,
   Index                {Array index}
 :INTEGER;

   MoreInput            {Variable to indicate more input}
 :CHAR;                 { and type of stock}
     BEGIN

       ClrScr;   {Clear the screen}
       WRITELN;

       FOR Index := 1 TO IndexMax DO
         BEGIN
          WRITELN(Index,'] ',StockName[Index]);
         END;

        WRITELN;
        WRITELN('Select stock you wish to update by number = ');
        READLN(Selection);
        Index := Selection;

        WRITELN;
        WRITE('Enter new share value of ',StockName[Index],' = ');
        READLN(NewValue);

       Value[Index] := NewValue * Quantity[Index];
  Difference[Index] := Value[Index]-Cost[Index] * Quantity[Index];
{---------------------------------------------------------------------}
  END;   {of procedure Update_Stock_Price}

  PROCEDURE Sell_Stock{(VAR StockName :Stock_Name_Array;
                        VAR Quantity :Quantity_Array;
                        VAR Cost :Money_Array;
                        VAR BlueChip :Blue_Chip_Array);
                        VAR MaxIndex :INTEGER
                        VAR TotalValue,
                            GainLoss :Money_Array)};
{---------------------------------------------------------------------}
```

Program 7–30 *continued*

```
    VAR

    Index,
    Selection
   :INTEGER;

      BEGIN

          ClrScr;    {Clear the screen}
          WRITELN;

          FOR Index := 1 TO MaxIndex DO
            BEGIN
             WRITELN(Index,'] ',StockName[Index]);
            END;

           WRITELN;
           WRITELN('Select stock you wish to sell by number = ');
           READLN(Selection);

             FOR Index := Selection TO MaxIndex DO
               BEGIN
                IF Index <> MaxIndex THEN
                  BEGIN
                    StockName[Index] := StockName[Index+1];
                    Quantity[Index] := Quantity[Index+1];
                    Cost[Index] := Cost[Index+1];
                    BlueChip[Index] := BlueChip[Index+1];
                    TotalValue[Index] := TotalValue[Index+1];
                    GainLoss[Index] := GainLoss[Index+1];
                  END;
               END;

           MaxIndex := MaxIndex -1;
{--------------------------------------------------------------------------}
   END;    {of procedure Sell_Stock}

   PROCEDURE List_Stock{(StockName :Stock_Name_Array;
                          Quantity :Quantity_Array;
                              Cost :Money_Array;
                        TypeOfStock :Blue_Chip_Array;
                          IndexMax :INTEGER;
                          VAR Value,
                          GainLoss :Money_Array)};
{--------------------------------------------------------------------------}
   VAR
    Index
   :INTEGER;

     BEGIN

        ClrScr;    {Clear the screen}
        WRITELN;
        WRITELN(' Stock Name    Quantity    Cost    Type    Value    Gain/Loss');
        WRITELN('---------------------------------------------------------------');

        WINDOW(1,3,79,24);
        FOR Index := 1 TO IndexMax DO
          BEGIN
           GOTOXY(3,1+Index);
           WRITE(StockName[Index]);
           GOTOXY(20,1+Index);
           WRITE(Quantity[Index]);
           GOTOXY(30, 1+Index);
           WRITE(Cost[Index]:3:2);
```

Program 7–30   *continued*

```
            GOTOXY(40, 1+Index);
             IF TypeOfStock[Index] = TRUE
              THEN WRITELN('Blue')
             ELSE WRITELN(' ');

             GOTOXY(50, 1+Index);
             WRITE(Value[Index]:3:2);

             GOTOXY(60, 1+Index);
             WRITE(GainLoss[Index]:3:2);

           END;

         WINDOW(1,1,79,25);

         Continue;

{------------------------------------------------------------------}
  END;   {of procedure List_Stock}

  PROCEDURE Program_Repeat{(VAR UserChoice : CHAR)};
{------------------------------------------------------------------}
    BEGIN

       GOTOXY(5,25);
       WRITE('Are you sure you wish to exit this program (Y/N) => ');
       READLN(UserChoice);

{------------------------------------------------------------------}
  END;     {or procedure Program_Repeat}

  PROCEDURE Alphabetize_Stock{(VAR StockName :Stock_Name_Array;
                               VAR Quantity :Quantity_Array;
                               VAR      Cost :Money_Array;
                               VAR BlueChip :Blue_Chip_Array;
                               VAR TotalStock :INTEGER;
                               VAR Value,
                                   Difference :Money_Array)};
{------------------------------------------------------------------}
      VAR

    TempReal               {Temporary location for real variable}
   :REAL;

    Index,                 {Index number}
    TempInteger            {Temporary location for integer variable}
   :INTEGER;

   MoreInput               {Variable to indicate more input}
  :CHAR;                   { and type of stock}

   Switch,                 {Switch flag, indicates if a swap was made}
   TempBoolean             {Temporary location for Boolean variable}
  :BOOLEAN;

   TempString              {Temporary location for string variable}
  :String79;

    BEGIN

      REPEAT
        Switch := FALSE;

        FOR Index := 1 TO TotalStock DO
          BEGIN
            IF (StockName[Index] > StockName[Index + 1])
```

Program 7–30   *continued*

```
                           AND
                           (Index <> TotalStock)
                           THEN    {A switch is needed}
                             BEGIN
                                TempString := StockName[Index];
                                StockName[Index] := StockName[Index + 1];
                                StockName[Index + 1] := TempString;

                                {Swap everything else}
                                  TempInteger := Quantity[Index];
                                  Quantity[Index] := Quantity[Index + 1];
                                  Quantity[Index + 1] := TempInteger;

                                TempReal := Cost[Index];
                                Cost[Index] := Cost[Index + 1];
                                Cost[Index + 1] := TempReal;

                                TempBoolean := BlueChip[Index];
                                BlueChip[Index] := BlueChip[Index + 1];
                                BlueChip[Index + 1] := TempBoolean;

                                TempReal := Value[Index];
                                Value[Index] := Value[Index + 1];
                                Value[Index + 1] := TempReal;

                                TempReal := Difference[Index];
                                Difference[Index] := Difference[Index + 1];
                                Difference[Index + 1] := TempReal;

                                  Switch := TRUE;    {A switch was made}
                                END;
                           END;

                       UNTIL Switch = FALSE;
{---------------------------------------------------------------------}
   END;   {of procedure Alphabetize_Stock}

   PROCEDURE Continue;
   {---------------------------------------------------------------------}
     BEGIN

         GOTOXY(5,25);
         WRITE('Press RETURN/ENTER to continue...');
         READLN;

   {---------------------------------------------------------------------}
     END;    {of procedure Continue}

   BEGIN
   Main_Programming_Sequence;
   END.
```

Questions

1 In the procedure `Display_Main_Menu`, what is the reason for the statement

```
IF NOT ((UserChoice IN ['A'..'E']) OR
         (UserChoice IN ['a'..'e']));
```

2 Why is it necessary, in procedure `Input_New_Stock`, to set `Index := IndexMax` in the first part of the procedure?

3 In the procedure `Update_Stock_Price` explain why it is necessary to have two variable parameters.

4 Referring to the procedure `List_Stock`, state the purpose of the **GOTOXY** commands. Explain what purpose the `1 + Index` serves within the **GOTOXY** command.

5 In the procedure `Alphabetize_Stock`, why is it necessary to swap everything else if the name of the stock needs to be swapped?

6 What is the purpose of the procedure `Continue`?

7 Explain the purpose of the procedure `Program_Repeat`.

8 What is the maximum number of stocks that can be entered into this program? How did you determine this?

9 Referring to the procedure `Main_Programming_Sequence`, when is the list of stocks alphabetized?

10 In the procedure `Input_New_Stock`, why is the arrayed variable `Gain-Loss[Index]` set to 0?

Problems

General Concepts

Section 7–1

1 What is an array?

2 Name some of the things that can be done when programming with arrays.

3 Give an example of the use of a subscript.

4 What are some of the advantages of using subscripts?

5 Give the general form for a Pascal array.

6 State what the number of indices in an array indicate.

7 What is a three-dimensional array?

Section 7–2

8 Describe an array variable.

9 State what values may be used as an index for a Pascal array.

10 Can an array index be a subrange of INTEGERs? Explain.

11 Will Pascal accept an array index that is a user-defined type? Explain.

12 Can arrays be passed as value parameters? As variable parameters?

13 Are arithmetic operations allowed with an array index?

14 Explain what is meant by congruent arrays.

15 What is the significance of having congruent arrays?

Section 7–3

16 State the advantage of declaring array types.

17 Explain what is meant by entering data interactively.

18 What is the advantage of using arrayed data?

Section 7–4

19 State the rules for bubble sorting numbers.

20 What is the minimum number of times a program must go through a list of data in a bubble sort?

21 What numerical operation must be performed on a array index when sorting? Explain.

22 What determines if a list will be sorted in an ascending or descending order?

23 Why is a temporary value used when switching data in a sorting program?

Section 7–5

24 Explain how string data is compared in Pascal when it comes to sorting.

25 What is the major difference between string sorting and numerical sorting in Pascal?

26 Describe the potential problem you may encounter if you try to sort numbers as strings.

Section 7–6

27 State what is meant by a run-time error.

28 Will the compiler catch run-time errors?

29 What is a debug routine?

30 State what is usually contained in a debug routine.

31 What is an auto-debug routine?

Program Analysis

32 Will program 7–31 compile? If not, why not?

Program 7–31

```
PROGRAM Problem_1;
 VAR
  AnArray : ARRAY[5..1] OF INTEGER;
  BEGIN
   WRITELN(AnArray[3]);
  END.
```

33 Do you think program 7–32 will compile? If not, where is the problem?

Program 7–32

```
PROGRAM Problem_2;
  TYPE
   MyType = (n5, n4, n3, n2, n1);

  VAR
   ArrayType : ARRAY[n5..n1] OF REAL;

  BEGIN
   ArrayType[n3] := 55;
  END.
```

34 Program 7–33 has a user-defined index in the array. Will this compile? If not, why not?

Program 7–33

```
PROGRAM Problem_3;
  TYPE
   MyType = (n5, n4, n3, n2, n1);
   ArrayType = ARRAY[MyType] OF REAL;

  VAR
   NewArray
   :ArrayType;

  BEGIN
   NewArray[n3] := 55;
  END.
```

35 Program 7–34 uses type INTEGER in its index. Does it compile? If not, what is the trouble?

Program 7–34

```
PROGRAM Problem_4;
  TYPE
   MyType = (n5, n4, n3, n2, n1);
   ArrayType = ARRAY[INTEGER] OF MyType;

  VAR
   NewArray
   :ArrayType;

  BEGIN
   NewArray[n3] := n1;
  END.
```

36 Look at program 7–35. It uses a type BOOLEAN as an index. Will this program compile? Why or why not?

Program 7–35

```
PROGRAM Problem_5;
  TYPE
   MyType = (n5, n4, n3, n2, n1);
   ArrayType = ARRAY[BOOLEAN] OF MyType;
```

Program 7–35   *continued*

```
VAR
 NewArray
:ArrayType;
BEGIN
 NewArray[FALSE] := n1;
END.
```

37   Program 7–36 uses a different kind of index. Will it compile?

Program 7–36

```
PROGRAM Problem__6;
TYPE
 IntegerRange = (N1, N2, N3, N4, N5);
 ArrayType = ARRAY[IntegerRange] OF INTEGER;

VAR
 NewArray
:ArrayType;

BEGIN
 NewArray[N1] := n1;
END.
```

38   Program 7–37 uses an arrayed variable that is the same user-defined type as its index. Is there any problem here? Will the program compile?

Program 7–37

```
PROGRAM Problem__7;
  TYPE
   IntegerRange = (N1, N2, N3, N4, N5);
   ArrayType = ARRAY[IntegerRange] OF IntegerRange;

  VAR
   NewArray
  :ArrayType;

  BEGIN
   NewArray[N1] := N5;
  END.
```

Program Design

For all of the following programs use the structure assigned to you by your instructor.

Electronics Technology

39 The circuit in figure 7–3 is a parallel-series circuit. Develop a Pascal program that will compute the total resistance of any branch selected by the program user. The total resistance of any one branch is

$$R_T = R_1 + R_2 + R_3$$

where R_T = total branch resistance in ohms
R_1, R_2, R_3 = value of each resistor in ohms

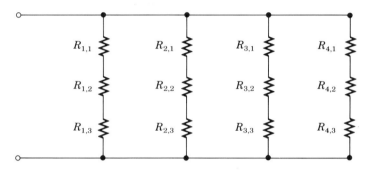

Figure 7–3   Circuit for Problems 39, 40, and 41

40 Modify the program for problem 39 so that the total resistance of any combination of branches may be found by the program user. The total resistance of any parallel branches is found by first determining the total resistance of that branch, then using the parallel resistance formula for finding the total resistance. The parallel resistance formula is

$$R_T = 1/(1/R_{T1} + 1/R_{T2} \ldots + 1/R_{TN})$$

where R_T = total resistance in ohms
$R_{T1}, R_{T2}, \ldots R_{TN}$ = total branch resistance in ohms

41 Expand the program for problem 40 so that the program user can enter the value of each resistor and the resistors will be displayed in numerical order.

42 Develop a Pascal program that computes the power dissipation of each resistor in a series circuit. The program user may select how many resistors there will be in the circuit. The program will sort the resistors by their power dissipation and display their value and subscript number (the subscript number represents the order in which they appear in the circuit). The program user enters the value of the voltage source in volts. Power dissipation in a resistor may be determined by

$$P = I^2R$$

where P = power dissipation in watts
     I = current in the resistor in amps
     R = value of resistor in ohms

The current in each resistor in a series circuit is the same. It may be determined from

$$I = V_S/R_T$$

where I = circuit current in amps
     V_S = source voltage in volts
     R_T = total circuit resistance in ohms

Business Applications

43 Develop a Pascal program that will display the amount of money in any safety deposit box that is contained in a wall which has 10 rows by 8 columns of these boxes.

Computer Science

44 Create a Pascal program that will take a 4 × 5 array and multiply the first column by any other column and display the sum of the products.

Drafting Technology

45 Create a Pascal program that will give the program user the color of an area of the grid system shown in figure 7–4. The program user must enter the row and column number.

Column

	1	2	3	4
1	Red	Green	Blue	White
2	Violet	Amber	Brown	Black
3	Orange	Pink	Magenta	Yellow
4	Silver	Gold	Slate	Pink

Row

Figure 7–4 Grid System for Problem 45

Agriculture Technology

46   Develop a Pascal program that will rate dairy cows according to their age, milk-producing ability (quarts per day), and cost of feed per day. The program user must be able to access the name of any dairy cow by milk-producing ability, age, or cost of feed per day.

Health Technology

47   A pharmacist needs a Pascal program that will allow her to find any prescription by (1) name of medication, (2) name of patient, or (3) cost of medication. The program user must be able to locate all of this information by entering any one of the above three items.

Manufacturing Technology

48   A robotics company requires a Pascal program that will produce the following information about factory workers on an assembly line: (1) name of worker, (2) hourly wages, (3) hours worked per week, and (4) total weekly pay. The program user must be able to locate all of the above information by entering any one of the above items.

Business Applications

49   Modify the program of problem 43 so the program user may enter the amount of money in any vault. Have the program display the amounts in each vault by displaying a matrix on the screen replicating the structure of the vault (use the **GOTOXY(X,Y)**) command.

Computer Science

50   Change the program in problem 44 so that the program user may select any two columns to be multiplied.

Drafting Technology

51   Modify the program in problem 45 so that the coordinates of all the areas of the same color will be displayed by the program user. Make the program so the program user may input the color of each area. The program must be able to sort the colors in order according to their frequency on the matrix.

Agriculture Technology

52   Modify the program in problem 46 so that the program will display the names of cows sorted according to any of the program parameters selected by the program user.

Health Technology

53   Change the program in problem 47 so that the program user may have all the medications for any one patient sorted alphabetically and the total price displayed.

Manufacturing Technology

54   Modify the program in problem 48 so the program user may find the total wages paid for the week to any factory worker. The program must allow the program user to locate any factory worker by his weekly income earnings, or to locate those who have made less than a specified amount, or those who have made more than a specified amount.

8 Records and Files

Objectives

This chapter will give you a chance to learn:

1 The meaning of a Pascal record.
2 The basic concepts and reasons for Pascal records.
3 The development of a Pascal program using records.
4 What is meant by a variant part of a record.
5 A method of sorting records.
6 The meaning of a Pascal file.
7 The basic concepts and reasons for Pascal files.
8 The difference between text files and random-access files.
9 What compiler directives are, and some examples.
10 A case study in modifying an existing program.

Key Terms

Record
Field
Variant Records
Files
I/O
Text File

End-of-Line Marker
End-of-File Marker
Sequential Files
Random-Access Files
Compiler Directive

Outline

Chapter seven showed you the power of arrays, as well as the weakness of arrays. Recall the stock program used for the self-test. During the sorting process, each element of information concerning the stock had to be switched when the name of the stock was being alphabetized. The same thing would have been true with the hospital program. The only way patient information was linked together was by the use of the array index. If you had to change one of the indices on any single patient item (as an example, for sorting), then all the other arrayed data had to have their indices changed accordingly.

The other problem with the hospital and stock market programs of the last chapter is that when you turn the computer off, you lose all of your entered data! It is the purpose of this chapter to remedy both of these situations. Here, you will see a new and powerful way of keeping track of data—and here you will also see how to easily preserve the program data you have so painstakingly entered. This is a very useful chapter. It shows you how to use Pascal in order to let the computer do some of the tasks it's good at: arranging, retrieving, and storing information.

8–1   What is a Record?

Introduction

Chapter seven dealt with the arrangement of data in a unique way—using arrays. However, you saw a need to group various pieces of data together (such as information about a patient or a particular type of stock) that required the grouping of information of different data types. For example, in the hospital program, the patient's name was a STRING, while the bill was a REAL, days in the hospital was an INTEGER, and whether or not the patient was ready for release was a BOOLEAN type.

This is such a common occurrence when working with data that the Pascal language has developed a much easier way of doing this, called a record. This section will introduce you to this important and powerful concept.

Basic Idea

Look at figure 8–1. It illustrates two different ways of grouping different data types together. The first method, called the arrayed method, identifies data of the same

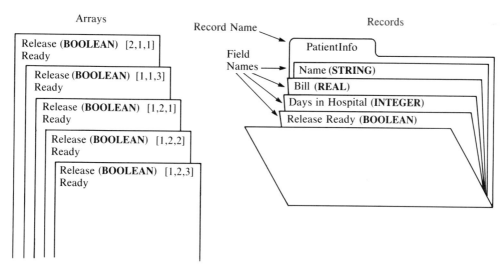

Figure 8-1 Two Ways of Handling Data

group by use of subscripts. The second method, called the record method, keeps track of different types of data belonging to the same group by a folder called a record.

Program 8-1 presents the general idea of a Pascal record. The program inputs data about a single patient, then outputs the same data. The difference is the type of structure used to handle this information; it is done in the context of a record.

Program 8-1

```
PROGRAM Record_Demo_1;
  VAR
    Patient_Info : RECORD
                     Name : STRING[79];
                     AmountOwed : REAL;
                     DaysHere : INTEGER;
                     ReleaseReady : BOOLEAN;
                   END;

  PROCEDURE Input_Information;
    BEGIN
      Patient_Info.Name := 'Frank Jones';
      Patient_Info.AmountOwed := 825.43;
      Patient_Info.DaysHere := 3;
      Patient_Info.ReleaseReady := FALSE;
    END;
```

Program 8–1   *continued*

```
PROCEDURE Output_Data;
  BEGIN
    WRITELN(Patient_Info.Name);
    WRITELN(Patient_Info.AmountOwed:3:2);
    WRITELN(Patient_Info.DaysHere);
    WRITELN(Patient_Info.ReleaseReady);
  END;

BEGIN
 Input_Information;
 Output_Data;
END.
```

When program 8–1 is executed the output is

```
Frank Jones
825.43
3
FALSE
```

The syntax for a variable of **RECORD** is

```
Identifier : RECORD
                FieldName₁:TYPE;
                FieldName₂:TYPE;
                      •
                      •
                      •
                FieldNameₙ:TYPE;
              END;
```

Look at an example from program 8–1.

```
Patient_Info : RECORD
                  Name : STRING[79];
                  AmountOwed : REAL;
                  DaysHere : INTEGER;
                  ReleaseReady : BOOLEAN;
                END;
```

`Patient_Info` is the identifier for the record variable. What follows is a list of the **fields** for that record. As you can see, each of the fields may be of different types (they don't have to be different; it's up to you). The **RECORD** block terminates with an **END;**.

Thus a record identifier is nothing more than a collection of fields. Each field may be referenced individually as follows.

```
Identifierᵣ.Identifierᵩ
```

where $Identifier_R$ = the record identifier

$Identifier_F$ = the field identifier

As an example, when data was entered about the patient in program 8–1, the following syntax was used.

```
Patient_Info.Name := 'Frank Jones';
Patient_Info.AmountOwed := 825.43;
Patient_Info.DaysHere := 3;
Patient_Info.ReleaseReady := FALSE;
```

Note that the syntax uses the record identifier, then the required period, followed by the field identifier. The same syntax was used to output the information to the screen.

```
WRITELN(Patient_Info.Name);
WRITELN(Patient_Info.AmountOwed:3:2);
WRITELN(Patient_Info.DaysHere);
WRITELN(Patient_Info.ReleaseReady);
```

Record Arrays

Going to all this trouble to create a record in program 8–1 may seem like a lot of work for nothing. There seems to be more coding involved than when using arrays. However, the real power of using records is when you combine them with arrays and have an array of records. Look at program 8–2. It does exactly the same thing as program 8–1 except that the structure is now made into an array of a record. Nothing different will happen; the purpose of the program is to simply demonstrate a method of creating an arrayed record. Note that **RECORD** is now a type rather than a variable.

Program 8–2

```
PROGRAM Record_Demo_2;
  TYPE
    PatientRecord = RECORD
                      Name : STRING[79];
                      AmountOwed : REAL;
                      DaysHere : INTEGER;
                      ReleaseReady : BOOLEAN;
                    END;
  VAR
    PatientData : ARRAY[1..3, 1..3, 1..4] OF PatientRecord;

  PROCEDURE Input_Information;
    BEGIN
      PatientData[1,1,1].Name := 'Frank Jones';
      PatientData[1,1,1].AmountOwed := 825.43;
      PatientData[1,1,1].DaysHere := 3;
      PatientData[1,1,1].ReleaseReady := FALSE;
    END;
```

Program 8–2   *continued*

```
PROCEDURE Output_Data;
  BEGIN
    WRITELN(PatientData[1,1,1].Name);
    WRITELN(PatientData[1,1,1].AmountOwed:3:2);
    WRITELN(PatientData[1,1,1].DaysHere);
    WRITELN(PatientData[1,1,1].ReleaseReady);
  END;

BEGIN
  Input_Information;
  Output_Data;
END.
```

Program 8–2 has now made a type of **RECORD**; it is no longer a variable. This was done so there could be an array of records. Recall that for the hospital program, there are three floors, three rooms per floor, and four beds to a room. Now, each patient bed is being assigned just one variable that is a **RECORD** of other variables called fields.

```
TYPE
  PatientRecord = RECORD
                    Name : STRING[79];
                    AmountOwed : REAL;
                    DaysHere : INTEGER;
                    ReleaseReady : BOOLEAN;
                  END;
VAR
  PatientData : ARRAY[1..3, 1..3, 1..4] OF PatientRecord;
```

Note that the identifier **PatientData** is a three-dimensional array of records. There are advantages to this. You can now access an individual field within the array. Another advantage is that you can access the record itself. When you alphabetize, you do not have to swap individual pieces of data—you only need to swap the record. Once the swap is made, you can then access the individual fields within the record.

The stock market program presented as a case study in chapter seven could have benefited from the use of this kind of structure. You will have an opportunity to see this for the case study in this chapter.

Using the WITH

Before proceeding further with the applications of records, there are some other helpful details to learn about their structure. For example, it still takes a lot of typing to access individual fields of the record. This is where the **WITH** statement becomes helpful. For example, instead of accessing each field using

```
PROCEDURE Input_Information;
  BEGIN
    PatientData[1,1,1].Name := 'Frank Jones';
    PatientData[1,1,1].AmountOwed := 825.43;
    PatientData[1,1,1[.DaysHere := 3;
    PatientData[1,1,1].ReleaseReady := FALSE;
  END;
```

the same procedure could be constructed using the Pascal reserved word **WITH**.

```
PROCEDURE Input_Information;
  BEGIN

    WITH PatientData[1,1,1] DO
      BEGIN
        Name := 'Frank Jones';
        Amount Owed := 825.43;
        DaysHere := 3;
        ReleaseReady := FALSE;
      END;
END; {of procedure Input_Information.}
```

The **WITH** statement is a shorthand method for referencing any individual field of the record. The syntax of the **WITH** statement is

WITH RecordIdentifier **DO** Statement

The use of the **WITH** is illustrated in program 8–3. What the program does is similar to what was done in program 8–2; the difference is in the structure.

Program 8–3

```
PROGRAM Record_Demo_3;
 TYPE
   PatientRecord = RECORD
                     Name : STRING[79];
                     AmountOwed : REAL;
                     DaysHere : INTEGER;
                     ReleaseReady : BOOLEAN;
                   END;
 VAR
   PatientData : ARRAY[1..3, 1..3, 1..4] OF PatientRecord;

  PROCEDURE Input_Information;
    BEGIN
```

Program 8–3 *continued*

```
    WITH PatientData[1,1,1] DO
     BEGIN
      Name := 'Frank Jones';
      AmountOwed := 825.43;
      DaysHere := 3;
      ReleaseReady := FALSE;
     END;

 END;   {of procedure Input_Information.}

PROCEDURE Output_Data;
  BEGIN

    WITH PatientData[1,1,1] DO
      BEGIN
        WRITELN(Name);
        WRITELN(AmountOwed:3:2);
        WRITELN(DaysHere);
        WRITELN(ReleaseReady);
      END;

  END;    {of procedure Output_Data.}

BEGIN
  Input_Information;
  Output_Data;
END.
```

You may want to compare program 8–3 with program 8–2. Note the amount of coding the **WITH** statement saves. You now have two methods of accessing any field in a record.

Passing Records

Program 8–4 demonstrates the accessing of individual record fields or the accessing of a whole record. The program has three procedures. In the procedure **Main_ Programming_Sequence**, the records are passed to the other procedures as either a variable or value parameter; this is the example of accessing the entire record. The other two procedures then access the individual record fields.

In program 8–4, the procedure **Main_Sequence** calls two other procedures. One of these (procedure **Input_Information**) demonstrates the use of value and variable parameter passing. For example, the Floor, Room, and Bed numbers are passed as value parameters to the procedure and the data about the patient is passed back to the calling procedure as a value parameter.

Program 8–4

```
PROGRAM Record_Demo_4;
 TYPE
   PatientRecord = RECORD
                     Name : STRING[79];
                     AmountOwed : REAL;
                     DaysHere : INTEGER;
                     ReleaseReady : BOOLEAN;
                   END;

   PatientData = ARRAY[1..3, 1..3, 1..4] OF PatientRecord;

  PROCEDURE Output_Data(Floor, Room, Bed : INTEGER;
                        PatientInfo : PatientData);

    BEGIN

      WITH PatientInfo[Floor, Room, Bed] DO
        BEGIN
          WRITELN(Name);
          WRITELN(AmountOwed:3:2);
          WRITELN(DaysHere);
          WRITELN(ReleaseReady);
        END;

    END;   {of procedure Output_Data.}

  PROCEDURE Input_Information(Floor, Room, Bed : INTEGER;
                              VAR PatientInput : PatientData);
  VAR
    Release
  :CHAR;

    BEGIN

     WITH PatientInput[Floor, Room, Bed] DO
      BEGIN
       WRITE('Patient Name ');
       READLN(Name);
       WRITE('Amount Owed ');
       READLN(AmountOwed);
       WRITE('Days Here ');
       READLN(DaysHere);
       WRITE('Release Ready (Y/N) ');
       READLN(Release);
         IF (Release = 'Y') OR (Release = 'y') THEN
           ReleaseReady := TRUE
            ELSE
           ReleaseReady := FALSE;
      END;

    END;   {of procedure Input_Information.}
```

Program 8–4   *continued*

```
PROCEDURE Main_Sequence;
 VAR
  Floor,
  Room,
  Bed
 :INTEGER;
  Choice
 :CHAR;
  HospitalRecords
 :PatientData;

  BEGIN

   REPEAT

     WRITELN('Give information for input/output: ');
     WRITELN;
      WRITE('Floor number = ');
      READLN(Floor);
      WRITE('Room number = ');
      READLN(Room);
      WRITE('Bed Number = ');
      READLN(Bed);
     WRITELN;
   WRITELN('Do you want to (E)nter or (R)etrieve information?');
   WRITE('Select by letter (E) or (R) => ');
     READLN(Choice);
      IF (Choice = 'E') OR (Choice = 'e') THEN
       Input_Information(Floor, Room, Bed, HospitalRecords)
      ELSE
       Output_Data(Floor, Room, Bed, HospitalRecords);

      WRITE('Do you wish another entry (Y/N) ');
      READLN(Choice);

   UNTIL (Choice = 'N') OR (Choice = 'n');

 END;  {of procedure Main_Sequence}

  BEGIN
   Main_Sequence;
  END.
```

Note in program 8–4 that in order to pass all of the data about each record field (that is, the name of the patient, amount owed, etc.) only one operation had to be performed using the name of the record. This was done twice, once as a variable parameter,

```
PROCEDURE Input_Information(Floor, Room, Bed : INTEGER;
                           VAR PatientInput : PatientData);
```

and once as a value parameter.

```
PROCEDURE Output_Data(Floor, Room, Bed : INTEGER;
                      PatientInfo : PatientData);
```

This results in a great saving of programming code. It also helps in easily keeping track of many different items of different types that have some common relationship. In this case, it is varied information about a patient.

Conclusion

This section presented the concept of a Pascal record; you saw how to construct a record and access each of its fields. You also saw how to use the reserved word **WITH** to simplify accessing individual fields. The real time-saving feature of a record is that you can treat it as a single variable and all of its fields will follow. The other major advantage is that the fields may be of different types.

In the next section, you will see different kinds of records along with some important applications. For now, test your understanding of this section by trying the following section review.

8–1 Section Review

1   State the difference between a Pascal array and a Pascal record in terms of how they can keep track of grouped data.
2   What are the different elements of a record called?
3   State the two methods by which an individual element of a record may be accessed.
4   Can a Pascal record be passed between procedures? Explain.

8–2   Ways of Representing Records

Introduction

Pascal records are rich in the variety of ways they may be used; this section will demonstrate some of them. You will find many potential technical applications for this powerful feature of Pascal.

Records Within Records

You can have a Pascal record within another record. Consider program 8–5 on page 472; it shows the first record treated as a field in the second record.

Program 8–5

```
PROGRAM Example_1;
 TYPE
     First_Record = RECORD
                         Field1 : INTEGER;
                         Field2 : REAL;
                         Field3 : BOOLEAN;
                     END;

 VAR
     Second_Record : RECORD
                         SecondField1 : First_Record;
                         SecondField2 : INTEGER;
                         SecondField3 : REAL;
                     END;

 BEGIN
 END.
```

As you can see from program 8–5, the variable `Second_Record` contains a record (`First_Record`) in the field called `SecondField1`. This is a case of a record containing another record. The hospital program could have a record contain a record of the patient's medical history. This **sub-record** can be treated as a single variable or its individual fields can be accessed.

Arrays Within Records

You can also have Pascal arrays within a record. For example, consider program 8–6. Here, one field of the record is an array.

Program 8–6

```
PROGRAM Example_2;
 TYPE
     Array_List = ARRAY[1..10, 1..10] OF REAL;

     First_Record = RECORD
                         Field1 : Array_List;
                         Field2 : REAL;
                         Field3 : BOOLEAN;
                     END;

 BEGIN
 END.
```

Program 8–6 contains a record for which one of its fields is an array. You could use this for a list of prices of medical items used by the patient.

Arrays of Records with Arrays

You can have an array of records in which one or more of the fields can be an array. Consider, for example, program 8–7; it contains an array of records in which one of the record fields contains an array. This situation could be used for patient data where one of the data items is itself an array (such as different types of illnesses a patient has had).

Program 8–7

```
PROGRAM Example_3;
  TYPE
      Array_List = ARRAY[1..10, 1..10] OF REAL;

      First_Record = RECORD
                      Field1 : Array_List;
                      Field2 : REAL;
                      Field3 : BOOLEAN;
                    END;

      Second_Array = ARRAY[1..5, 1..10] OF First_Record;

  BEGIN
  END.
```

In program 8–7, `First_Record` is a record and `Field1` is an array within the record. `Second_Array` is then made to be an array of that record.

Records Within Array Records

You can have an even more complex arrangement of records and arrays. For example, program 8–8 demonstrates a record as a field of another record, both of which are arrays.

Program 8–8

```
PROGRAM Example_4;
  TYPE
      Array_List = ARRAY[1..10, 1..10] OF REAL;
```

Program 8–8 *continued*

```
   First_Record = RECORD
                      Field1 : Array_List;
                      Field2 : REAL;
                      Field3 : BOOLEAN;
                   END;

   Second_Array = ARRAY[1..5, 1..10] OF First_Record;

   Second_Record = RECORD
                      SecondField1 : Second_Array;
                      SecondField2 : First_Record;
                      SecondField3 : Array_List;
                   END;
BEGIN
END.
```

As shown in program 8–8, `Second_Record` contains a field called `SecondField1` which is an array of more records called `First_Record`. `First_Record` also contains a field called `Second_Array`. The point of all this is that Pascal allows you to structure your programs with increasing complexity. Thus you could have a situation where many different data items were structured in various ways within records and arrays. For example, you could have a record for each hospital patient that contained his or her financial history (credit rating) and within that could be an array of records that contained a history of the payment of past debts. This could be further extended to include an array of records about the medical history of the patient's parents, brothers, and sisters. The ability of structuring data into arrays and records gives you tremendous power in the use of information.

Variant Records

There may be times when you need to store different kinds of data in a record field depending upon a previous condition. For example, consider this electronic parts inventory program. Program 8–9 does little more than illustrate the structure of a variant.

Program 8–9

```
PROGRAM Variation_1;
  TYPE

    Item = (Resistor, Capacitor, Inductor);
```

Program 8–9   *continued*

```
    Parts_Inventory = RECORD
                        StockNumber : ARRAY[1..100] OF INTEGER;
                        Supplier : STRING[79];
                        Quantity : INTEGER;
                          CASE ItemType : Item OF
                            Resistor : (Wattage, Ohms : REAL);
                            Capacitor : (VoltageRating, Farads : REAL);
                            Inductor : (Q, Henrys : REAL);
                        END;

  VAR
      StockItem : Parts_Inventory;

BEGIN
END.
```

There are three different components in the inventory program in program 8–9: resistors, capacitors, and inductors. Each of these items requires different information. This is achieved by the Pascal feature called a **variant record**. This allows mutually exclusive fields (fields that will never be used at the same time) to share the same storage space within the record, resulting in a saving of source code since you don't have to make a separate record for the variant.

The program excerpt in program 8–10 demonstrates how the variants may be accessed. Note that the procedure `Input_Items` uses the filed identifiers (`Wattage`, `Ohms`, etc.) as declared in the variant part of the record; these identifiers are passed by simply using the record identifier `StockItem`.

Program 8–10

```
PROGRAM Variation_2;
  TYPE

    Item = (Resistor, Capacitor, Inductor);

    Parts_Inventory = RECORD
                        StockNumber : ARRAY[1..100] OF INTEGER;
                        Supplier : STRING[79];
                        Quantity : INTEGER;
                          CASE ItemType : Item OF
                            Resistor : (Wattage, Ohms : REAL);
                            Capacitor : (VoltageRating, Farads : REAL);
                            Inductor : (Q, Henrys : REAL);
                        END;

  VAR
      StockItem : Parts_Inventory;

  PROCEDURE Input_Items(Selection : Item);
```

Program 8–10   *continued*

```
   BEGIN
    WITH StockItem DO
     BEGIN
      CASE Selection OF
       Resistor : BEGIN
                    WRITE('Resistor value in ohms => ');
                    READLN(Ohms);
                    WRITE('Resistor wattage in watts => ');
                    READLN(Wattage);
                  END;
        Capacitor : BEGIN
                    WRITE('Capacitor value in farads => ');
                    READLN(Farads);
                    WRITE('Capacitor voltage rating in volts => ');
                    READLN(VoltageRating);
                  END;
         Inductor : BEGIN
                    WRITE('Inductor value in henrys => ');
                    READLN(Henrys);
                    WRITE('Inductor Q (no units) => ');
                    READLN(Q);
                  END;
       END;  {of CASE}
    END;
  END;  {of procedure Input_Items}

 PROCEDURE User_Selection;
  VAR
   Choice,
   Selection
  :CHAR;

    BEGIN

REPEAT
     WRITELN;
     WRITELN('Select (R)esistor, (C)apacitor, (I)nductor by letter: ');
     WRITE('Your selection => ');
     READLN(Choice);
      Selection := UpCase(Choice); {Converts input to upper case letter}
     UNTIL Selection IN ['R', 'C', 'I'];
        CASE Selection OF
         'R' : Input_Items(Resistor);
         'C' : Input_Items(Capacitor);
         'I' : Input_Items(Inductor);
        END;   {Case}

    END;   {of procedure User_Selection}

BEGIN
  User_Selection;
END.
```

Observe from program 8–10, that the procedure User_Selection calls
the procedure Input_Items and passes one of three predeclared types
(Resistor, Capacitor, or Inductor) to the called procedure. The called
procedure Input_Items then selects the variant to be used by calling the
record StockItem and using a **CASE** statement to make the selection. Once this
is done, the identifiers for each **READLN** input are pulled from the variant part of the

record. Again this illustrates the ability of Pascal to pass all of the record fields, including selected variants, to procedures within the program.

There is a new feature in program 8–10. It is the error-trapping contained in the procedure `User__Selection`.

```
REPEAT
     WRITELN;
     WRITELN('Select (R)esistor, (C)apacitor, (I)nductor by letter: ');
     WRITE('Your selection => ');
     READLN(Choice);
      Selection := UpCase(Choice); {Converts input to upper case letter}
     UNTIL Selection IN ['R', 'C', 'I'];
```

Note the use of the Turbo Pascal built-in function **UpCase**. This function converts a CHAR variable to its uppercase. All characters that are not in the range of a..z are ignored. The Pascal statement

```
IN ['R', 'C', 'I',];
```

is a BOOLEAN that will only be TRUE if the user selects R, C, or I. If this isn't TRUE the **REPEAT** causes the whole selection process to be done over again.

Conclusion

This section presented some of the many ways of expressing a Pascal record. You saw how records could be nested inside arrays and other records. In the next section you will see how to use records in various different ways. Check your understanding of this section by trying the following section review.

8–2 Section Review

1   Can a record be an array? Can an array be a record? Explain.
2   Can a record contain a record? Explain.
3   Explain what is meant by a variant record.
4   What are mutually exclusive fields?

8–3   Introduction to Files

Introduction

You have come a long way in the development of your programming skills. Many of the Pascal programs that may now seem quite familiar to you represent powerful programming concepts. However, none of these programs were able to save information for later use.

Consider the hospital program. You could save the program itself to the disk (using the Turbo Pascal System), but any data about a specific patient input while the program was active could not be saved to the disk for use again at another time. The same thing was true about the stock market program presented in chapter seven—any information entered about stocks was lost when the program was terminated.

This section will show you how to create Pascal programs that will allow data to be stored on and retrieved from the disk. Armed with this knowledge, you will be able to create technical programs that can not only allow you to input data but also automatically save that data to the disk (or any other data to the disk) for use again.

Basic Idea

Pascal has a very special kind of structured variable. It resembles an array in that it consists of a sequence of distinct variable components all of the same type. It differs from an array because the number of components is unknown; you will soon learn its use.

If you had special Pascal commands that would take all the information about patients that you entered while the program was active and save it on the disk, the next time you used the program you could use another Pascal command that would get the old data from the disk. This would have been a nice feature to have had with the stock market program presented in chapter seven.

Inputting and Outputting

Pascal has two commands that are used for getting data into a program and for getting data out. Consider program 8–11. It uses the built-in Pascal procedures **WRITE** and **WRITELN** to input data from the keyboard and the built-in Pascal procedure **READLN** to output data to the screen.

Program 8–11

```
PROGRAM Input__Output;
  VAR
    InOut;
  :INTEGER;

  BEGIN
    WRITE('Give me a whole number => ');
    READLN(InOut);
    WRITE('The number is: ');
    WRITELN(InOut);
  END.
```

Consider what you already know about the **READLN** procedure. It causes the Pascal program to input data from the keyboard (in this case a whole number). Also consider the **WRITELN** procedure; it will cause Pascal to output data to the console screen. Another way of looking at the **READLN** and **WRITELN** procedures (as well as the **READ** and **WRITE** procedures) is that these are **input/output** procedures: the **READLN** (and **READ**) are input and the **WRITELN** (and **WRITE**) are output. These built-in Pascal procedures are used to exchange information with the keyboard and the screen. This input/output process is called I/O. Thus, Pascal already has methods of exchanging data between the active program and peripheral devices such as the keyboard and the monitor screen. The question is how to make these built-in I/O procedures cause the program to exchange data between the floppy disk and data entered by the program user from the keyboard while the program is active.

What's in a Name?

The developers of Pascal decided to call any external source to output and input information (such as the keyboard, console monitor, printer, or location on a floppy disk) a file. This means that as far as Pascal is concerned, when you input data into your program with a **READLN** procedure, that information is coming from a file (the keyboard). Hence, you can see how a file is like an array but its size is unknown. For example, if you ask the program user to input the name of an item, you have no way of knowing how many characters will be input by the program user into the program through the **keyboard file**.

The **WRITELN** procedure causes the program to output data to the **console file**. This means, as you know, that the data will appear on your screen. Again, this is similar to an array but the **WRITELN** procedure has no way of knowing how many characters will be output to the console file.

A Secret Revealed

There is a little secret about the **WRITE** and **WRITELN** built-in procedures. The real syntax for these procedures is

WRITE([Filevariable], variable);

and

WRITELN([Filevariable], variable);

where [**Filevariable**] specifies the file to output data to.

For example, recall chapter three, where you learned how to output data to the printer (which you now know to be an output file). You used

WRITE(LST, 'This goes to the printer.');

where **LST** is the file variable you use to let Pascal know that you want the output file to be the printer, not the console monitor.

For convenience, the developers of Pascal allow you to omit **Filevariable** in the **WRITE** or **WRITELN** procedure and when you do, the output automatically goes to the screen. This is why you may not have been aware of Pascal's little "secret".

Consider program 8–12 on page 480. It allows the program user to select where the data will be output—either the screen or the printer.

Program 8–12

```
PROGRAM Your_Choice;
  USES CRT;
    VAR
     FileType
    :TEXT;

    BEGIN
     WRITE('Output to (S)creen or (P)rinter => ');
      IF UPCASE(ReadKey) = 'P' THEN
        ASSIGN(FileType, 'PRN')
      ELSE
        ASSIGN(FileType, 'CON');

        REWRITE(FileType);

        WRITELN(FileType, 'This goes where you choose!');
    END.
```

There is a lot of new information presented in program 8–12. For now, just note that the **WRITELN** procedure is used to actually output the data. The other new Pascal statements you see (**ASSIGN** and **REWRITE**) are used to help get the variable **FileType** set up. Don't worry about these for now; you'll learn about them shortly. One other new item is the variable type TEXT; you'll learn about that shortly too. First, look at getting data in from a file.

About Reading

The **READ** and **READLN** built-in Pascal procedures also have the same little secret as the **WRITE** and **WRITELN** procedures. Their syntax is

READ([FileVariable], Variable);

and

READLN([FileVariable], Variable);

where **FileVariable** will indicate from what peripheral device (file) the data is to be input.

Again, as a convenience, the developers of Pascal allow you to omit **File-variable** from the **READ** and **READLN** procedures; Pascal defaults to inputting information from the keyboard file.

Conclusion

This section presented the basic concept of what is meant in Pascal by a file. You saw that the **READLN**, **READ**, **WRITELN**, and **WRITE** are built-in Pascal procedures that actually direct data into or out of the computer.

In the next section, you will discover the actual programming details of saving data on your disk and getting data from it. For now, test your understanding of this section by trying the following section review.

8–3 Section Review

1   Explain what is meant by a Pascal file.
2   What does I/O mean?
3   Name some Pascal input files and output files.
4   State the Pascal commands that actually output data to a file.
5   State the Pascal commands that actually input data from a file.

8–4   File Details

Introduction

The last section presented the idea of what was meant by a file in Pascal. In this section, you will see the actual programming details required in a Pascal program to have the program create a disk file, store information into it, and retrieve information from it.

A Text File

A **text file** is nothing more than a string of ASCII characters, usually designed to hold information that people can read.

Turbo Pascal provides a predefined file type called TEXT. Specifically, a variable type TEXT means a file containing characters organized into lines. This type of variable will be used to hold the identification of the file you want to create and use on the disk.

Pascal needs to know the name of the disk file for I/O operations involving the disk. This name will actually represent a physical location on the disk. Once Pascal knows the name of the file, it can tell DOS (the Disk Operating System) the exact spot on the disk to physically place the READ/WRITE head of the disk drive so that I/O can be performed.

Disk I/O is done cooperatively with DOS and as such the name of disk file must be a legal DOS filename (refer to the Appendix for an introduction to MS-DOS).

A DOS Review

As a summary of legal MS-DOS file names, recall that a legal file name has the form

`[Drive:]FileName[.EXT]`

where `Drive:` = the name of the disk drive that contains the disk to which you want to I/O (Drive `A:` is on the left, drive `B:` on the right; for stack drives, `A:` is on top, `B:` is on the bottom)

FileName = the name of the file (up to eight characters)
.EXT = an extension to the file name (up to three characters)

Both the **Drive:** and **.EXT** are optional. If **Drive:** is not specified, then the default drive will be used. **A:** is the left (or top) drive, **B:** the right (or bottom) drive—the colon is necessary.) For example,

B:AFILE.01

is a disk file name on a disk in drive **B:**, with the name **AFILE**, and an extension of **.01**. Your Pascal program must now somehow take a DOS file name and convert it into a type TEXT.

Assigning Things

In order to assign the actual DOS name of the file to a TEXT variable, Pascal uses the **ASSIGN** procedure.

ASSIGN([FileVariable], 'FileName');

where **FileVariable** = a variable of type TEXT
FileName = a legal DOS file name

The application of this procedure is illustrated in program 8–13.

Program 8–13

```
PROGRAM Example_8_13;
  VAR
   TextName
  :TEXT;

  BEGIN
     ASSIGN(TextName, 'THISFILE');
  END.
```

Program 8–13 does nothing more than assign the name of the DOS file **THISFILE** to the TEXT variable **TextName**. However, this is an important first step. The next step depends upon you, the creator of the program. There are some choices you need to consider.

Disk File Conditions

The next thing you need to tell your Pascal program is which one of the following four conditions will you want to use.

Table 8–1   Disk File Conditions

Condition	Meaning
1	The disk file does not exist and you want to create it on the disk, and add some information.
2	The disk file already exists and you want to get information from it.
3	The disk file already exists and you want to add more information into it while preserving the old information that was already there.
4	The disk file already exists and you want to get rid of all of the old information and add new information.

The Pascal program needs to know which one of the four options you want to use. A detailed programming description for each of the above four options now follows.

File Does Not Exist

Program 8–14 creates a new file, named **NEWFILE**. This is done by first assigning the name of the file, then creating the file, and lastly, closing the file.

Program 8–14

```
PROGRAM Example_8_14;
  VAR
   TextName
  :TEXT;

  BEGIN

     ASSIGN(TextName, 'NEWFILE');
     REWRITE(TextName);
     CLOSE(TextName);

  END.
```

When program 8–14 is executed, the program will not cause any output to the monitor. However, a new file will be created on the active disk drive. If your Turbo Pascal disk was in drive **A:** and the program disk where you save your programs in drive **B:**, the new file would appear on the disk in drive **B:**. This is because the default drive is always the active drive. When using the Turbo Pascal environment in a two-drive system, you normally have the Pascal programming disk in drive **A:** and the disk with your programs in drive **B:**.

Activation of program 8–14 has caused a DOS file to be created on the disk called **NEWFILE** (DOS will automatically capitalize all letters no matter what combination of upper and lower case letters you use for the name of the disk file). The built-in

Pascal procedure that causes this to happen is **REWRITE(TextName)**. The form of this procedure is

REWRITE(FileName**);**

where **FileName** = the TEXT variable that has already had the legal DOS file
                    name assigned to it

The Pascal **REWRITE** procedure causes a new file to be created. If a file by that name had already existed, it would have been erased and a new file by the same name would have been created. There is another important command in program 8–14. The **CLOSE(NewFile);** is another built-in Turbo Pascal procedure. It closes a previously-opened file. The **REWRITE** procedure will not only create a new disk file, it will also open it. Opened files must be closed before terminating the program or opening other files. If you forget to do this, information in your disk files could become lost. Program 8–14 created a new file. Now see the same program with some extra code that puts some information into the file.

Putting Data Into the File

Program 8–15 creates a new file and puts some data into it. The name of the file created is **NEWFILE**. Note that some data goes to the screen while other data goes to the file. See if you can tell the difference.

Program 8–15

```
PROGRAM Example_8_15;
  VAR
   TextName
  :TEXT;

  BEGIN

     ASSIGN(TextName, 'NEWFILE');
     REWRITE(TextName);
     WRITELN('This goes to the screen.');
     WRITELN(TextName, 'This goes to the disk file.');
     CLOSE(TextName);

  END.
```

The only difference between program 8–14 and program 8–15 is that program 8–15 uses two **WRITELN** procedures. One directs data to the console file (and thus appears on your screen); the other directs data to the disk file that was already opened by the **REWRITE** procedure. Again, note that the opened file was closed.

Note that in order to get the data to the disk file, the **WRITELN** had to have

```
WRITELN(TextName, 'This goes to the file.');
```

The variable `TextName` was necessary in order to direct the data to the disk file. Note that when the **WRITELN** procedure omits `TextName`, Pascal defaults the data to the console screen.

```
WRITELN('This goes to the screen.');
```

The next step is to be able to get data that has been put into a disk file.

An Existing File

Program 8–16 creates a disk file, puts data into it, and then gets the data back.

Program 8–16

```
PROGRAM Example_8_16;
  VAR
   TextName
  :TEXT;

  PROCEDURE Create_Disk_File;

   BEGIN

      ASSIGN(TextName, 'NEWFILE');
      REWRITE(TextName);
      WRITELN('This goes to the screen.');
      WRITELN(TextName, 'This goes to the disk file.');
      CLOSE(TextName);

   END;

  PROCEDURE Read_Disk_File;
   VAR
    Information
   :STRING[79];

    BEGIN

      ASSIGN(TextName, 'NEWFILE');
      RESET(TextName);
      READLN(TextName, Information);
      WRITELN(Information);
      CLOSE(TextName);

    END;

BEGIN
  Create_Disk_File;
  Read_Disk_File;
END.
```

As you can see, program 8–16 contains two procedures. The first procedure, `Create_Disk_File;`, creates a new disk file called `NEWFILE`. Then some information is put into the file and it is closed. The second procedure, `Read-_Disk_File;`, opens the existing file by using the built-in Turbo Pascal procedure **RESET**. Once the disk file is opened, the **READLN** procedure is used to get a line of information from the disk file and the **WRITELN** is used to direct the information to the console screen.

The Pascal procedure for opening a disk file is

RESET(FileName);

where `FileName` = the TEXT variable that has already had the legal DOS file name assigned to it

If a disk file by the assigned name does not exist, a run-time error will result.

Observe in program 8–16 that the **READLN** procedure contains the TEXT variable that has already had the legal DOS file name assigned to it.

READLN(TextName, Information);

This causes a line of text to be placed into the **STRING**[79]; variable `Information`. This is what you have been using the **READLN** procedure for in the past, except that all of your input to the program has been through the keyboard file rather than from a disk file.

Adding More Information

Program 8–17 considers the case of having an existing file that already contains data and needing to add more data to it.

Program 8–17

```
PROGRAM Make_a_Disk_File;
  TYPE
   Data = STRING[255];

  VAR
   FileName
 :TEXT;

PROCEDURE Create_Disk_File;

  BEGIN

    ASSIGN(FileName, 'NEWFILE');
    REWRITE(FileName);
    WRITELN('This data goes to the screen.');
    WRITELN(FileName, 'This data goes to the file.');
    CLOSE(FileName);
```

Program 8–17   *continued*

```
    END;
PROCEDURE Read_Disk_File;
 VAR
  Information,
  MoreStuff
  :Data;

  BEGIN
     ASSIGN(FileName, 'NEWFILE');
     RESET(FileName);
     READLN(FileName, Information);
     WRITELN(Information);
     READLN(FileName, Information);
     WRITELN(Information);
     CLOSE(FileName);

  END;

PROCEDURE Add_to_File;

  BEGIN
     ASSIGN(FileName, 'NEWFILE');
     APPEND(FileName);
     WRITELN(FileName, 'This is more information to be added.');
     CLOSE(FileName);
  END;

BEGIN
  Create_Disk_File;   {Create a new file and add information.}
  Read_Disk_File;     {Read information from an existing file.}
  Add_to_File;        {Add more information to an existing file.}
  Read_Disk_File;     {Read information from an existing file.}
END.
```

Program 8–17 contains three different procedures. The first, Create_
Disk_File, opens a new disk file called NEWFILE and puts some information
into it. The second procedure, Read_Disk_File, reads information from
the existing file. The third procedure, Add_to_File, adds new information to
an existing disk file while preserving the old information. Again the procedure
Read_Disk_File is used to read information from the existing disk file.

Execution of program 8–17 produces the following output

```
This data goes to the disk file.

This data goes to the disk file.
This is more information to be added.
```

The new built-in Turbo Pascal procedure **APPEND** was used to open an existing
disk file and add more information to it while leaving existing information intact. If a
disk file by the given name did not exist, a run-time error would occur. The syntax is

```
APPEND([FileName]);
```

where **FileName** **=** the text variable that has already had the legal DOS file name
assigned to it

Revisiting the Ways

Look now at the four conditions for working with text files that were first presented
earlier in this section. You have seen a program example for each one. The built-in
Turbo Pascal procedure that should be used for each can now be identified.

Table 8–2   Disk File Conditions and Commands

Condition	Meaning	Command
1	The disk file does not exist and you want to create it on the disk, and add some information.	**REWRITE**
2	The disk file already exists and you want to get information from it.	**RESET**
3	The disk file already exists and you want to add more information into it while preserving the old information that is already there.	**APPEND**
4	The disk file already exists and you want to get rid of all of the old information and add new information.	**REWRITE**

Conclusion

You are now able to create, add to, and read from a disk text file. However, the
example programs presented here were understandably kept simple to illustrate the
concept. You need information about what to do if the disk text file you are working
with contains many lines of information. How do you get all of these lines? How do
you get the line you want and ignore the other lines? These important and practical
considerations are presented in the next section. Test your understanding of this
section by trying the following section review.

8–4 Section Review

1   Explain the meaning of a text file.
2   State what the name of a disk file actually means to the Pascal program.
3   Describe a legal DOS file name.
4   State the four possible conditions you may encounter when working with a disk file.
5   Name the built-in Turbo Pascal procedures you would use to implement each of the four
    conditions you described in question 4 above.

8–5   Inside Files

Introduction

This section will show you that text files are stored on the disk in much the same manner as they are stored on the monitor screen. Keep this in mind as you cover the following material; it will help you visualize what is happening with disk files.

Using WRITELN and READLN

To set the stage for the following discussion, consider program 8–18. It creates a new file and enters three separate lines of text into it, and then the file is closed. Next, a second procedure opens the existing file and reads the three lines of text from it. Observe that the actual writing to the file is done with **WRITELN** statements and the actual reading from the FILE is done with **READLN** statements. This is handled in much the same way you would write text to the screen and read information from the keyboard. The difference is that the disk is now being used as both the output and the input.

Program 8–18

```
PROGRAM Text_Files_1;

VAR
 FileName
:TEXT;

 PROCEDURE Create_and_Enter;
  BEGIN

   ASSIGN(FileName, 'NEWFILE');
   REWRITE(FileName);

   WRITELN(FileName, 'Line One.');
   WRITELN(FileName, 'Line Two.');
   WRITELN(FileName, 'Line Three.');

   CLOSE(FileName);

  END;

 PROCEDURE Read_File;
  VAR
   FileLine
:STRING[79];
```

Program 8–18   *continued*

```
   BEGIN

      RESET(FileName);

      READLN(FileName, FileLine);
      WRITELN('First reading of file => ',FileLine);

      READLN(FileName, FileLine);
      WRITELN('Second reading of file => ',FileLine);

      READLN(FileName, FileLine);
      WRITELN('Third reading of file => ',FileLine);

      CLOSE(FileName);

   END;
BEGIN
 Create_and_Enter;
 Read_File;
END.
```

When program 8–18 is executed the output is

```
First reading of file => Line One.
Second reading of file => Line Two.
Third reading of file => Line Three.
```

Program 8–18 created a new file on the disk and then entered three lines into that file with **WRITELN** statements. Recall that a **WRITELN** statement automatically places a carriage return at the end of the line. This is exactly what has happened to each line entered into the disk file; because of the **WRITELN** statement, there is a carriage return at the end of each of the three lines of text in the file.

In program 8–18 a **READLN** was used to read each line of the disk file. Recall that a **READLN** will automatically return to the next line once it has completed its task. Program 8–19 will illustrate this process.

Looking at READ

Program 8–19 does not use **READLN** to read the disk file; instead it uses **READ**. This is the only difference between program 8–18 and program 8–19. What do you think the results will be?

Program 8–19

```
PROGRAM Text_Files_2;

VAR
 FileName
:TEXT;

 PROCEDURE Create_and_Enter;
  BEGIN

    ASSIGN(FileName, 'NEWFILE');
    REWRITE(FileName);
    WRITELN(FileName, 'Line One.');
    WRITELN(FileName, 'Line Two.');
    WRITELN(FileName, 'Line Three.');

    CLOSE(FileName);

  END;

  PROCEDURE Read_File;
   VAR
    FileLine
   :STRING[79];

  BEGIN

    RESET(FileName);

    READ(FileName, FileLine);
    WRITELN('First reading of file => ',FileLine);

    READ(FileName, FileLine);
    WRITELN('Second reading of file => ',FileLine);

    READ(FileName, FileLine);
    WRITELN('Third reading of file => ',FileLine);

    CLOSE(FileName);

  END;
BEGIN
 Create_and_Enter;
 Read_File;
END.
```

Program 8–19 is identical to program 8–18 with the important exception that **READ** has replaced **READLN** in the procedure that reads data from the disk file. The output of program 8–19 is

```
First reading of file => Line One.
Second reading of file =>
Third reading of file =>
```

Notice that the last two readings are blanks; the first **READ** never allowed a return to the next line for another **READ**. Thus, the last two **READ**s did nothing more than read the end of the first line of text where there is nothing located. This is also what the **READ** does when you are inputting information from the program user rather than from the disk.

Looking at WRITE

Program 8–20 is the same as program 8–18 except that the **WRITELN** will be replaced with **WRITE**. Notice the difference between using a **WRITE** and a **WRITELN** in a text file.

Program 8–20

```
PROGRAM Text_Files_3;

VAR
 FileName
:TEXT;

 PROCEDURE Create_and_Enter;
  BEGIN

    ASSIGN(FileName, 'NEWFILE');
    REWRITE(FileName);

    WRITE(FileName, 'Line One.');
    WRITE(FileName, 'Line Two.');
    WRITE(FileName, 'Line Three.');

    CLOSE(FileName);

  END;

  PROCEDURE Read_File;
   VAR
     FileLine
  :STRING[79];

  BEGIN

    RESET(FileName);

    READLN(FileName, FileLine);
    WRITELN('First reading of file => ',FileLine);
```

Program 8–20   *continued*

```
      READLN(FileName, FileLine);
      WRITELN('Second reading of file => ',FileLine);

      READLN(FileName, FileLine);
      WRITELN('Third reading of file => ',FileLine);

      CLOSE(FileName);

   END;

BEGIN
  Create_and_Enter;
  Read_File;
END.
```

When program 8–20 is executed the output is

```
First reading of file => Line One.Line Two.Line Three.
Second reading of file =>
Third reading of file =>
```

You can see the effect of using the **WRITE** instead of the **WRITELN** when inputting information into the disk file; since the **WRITE** does not place a carriage return at the end of each line, one statement was written after the other when the file was created.

These demonstrations show that you may use **WRITE**, **WRITELN**, **READ**, or **READLN** on these files provided you are aware of the results you will get. Understanding that the differences in these commands have essentially the same effect on disk files as they do on the screen may help reduce some run-time bugs in your programs.

What You Have Seen

You have seen that the basic components of a text file are characters. They are structured into **lines**, each line terminated by an **end-of-line marker** (the same as a RETURN/ENTER). You will see that this kind of file is also terminated by an **end-of-file marker**, done automatically whenever you close an opened file. Since the length of each line may vary, the position of a given line cannot be calculated, so these types of files are called **sequential**.

Working with Numbers

You may be wondering if Turbo Pascal accepts numbers in text files. It does, but all of the data is stored as a text file. For example, consider program 8–21 (p. 494). It opens a file and enters a set of numbers into it. The second procedure then reads these numbers back from the disk file, but no longer as numbers!

Program 8–21

```pascal
PROGRAM Text_Files_4;

VAR
 FileName
:TEXT;

 PROCEDURE Create_and_Enter;

  BEGIN

   ASSIGN(FileName, 'NEWFILE');
   REWRITE(FileName);

   WRITELN(FileName, 123);
   WRITELN(FileName, 231);
   WRITELN(FileName, 321);

   CLOSE(FileName);

  END;

  PROCEDURE Read_File;
   VAR
    FileLine
  :STRING[79];

  BEGIN

   RESET(FileName);

   READLN(FileName, FileLine);
   WRITELN('First reading of file => ',FileLine);

   READLN(FileName, FileLine);
   WRITELN('Second reading of file => ',FileLine);

   READLN(FileName, FileLine);
   WRITELN('Third reading of file => ',FileLine);

   CLOSE(FileName);

  END;

BEGIN
 Create_and_Enter;
 Read_File
END.
```

The main feature of program 8–21 is that whole numbers are input into the disk file. Note that the **READLN** that reads the file uses a STRING variable to read the numbers! Execution of program 8–21 reveals

```
First reading of file => 123
Second reading of file => 231
Third reading of file => 321
```

This verifies that INTEGERs are stored in the disk text file as CHARacters. Does Turbo Pascal store REAL numbers on the disk file as CHARacters as well? Yes, it does; if the first number in program 8–21 were of type REAL, it would have been stored in the disk file as a CHARacter in the form

```
1.230000000E+2
```

and when you did a **READLN** to get it back, that is how it would appear.

Converting Back and Forth

Turbo Pascal will automatically convert a number back to an INTEGER or REAL from a disk text file, as demonstrated by program 8–22. Note that the first procedure simply creates a new disk file and stores some numbers in it. The second procedure opens the file and reads the numbers from it. Then the first two numbers read from the disk file are added to each other to get a sum. Look at the program, then at the results.

Program 8–22

```pascal
PROGRAM Text_Files_With_Numbers_1;

VAR
 FileName
:TEXT;

  PROCEDURE Create_and_Enter;
    VAR
     UserInput
    :INTEGER;

  BEGIN

    ASSIGN(FileName, 'NEWFILE');
    REWRITE(FileName);

    WRITE('Give me a number => ');
    READLN(UserInput);
    WRITELN(FileName, UserInput);
    WRITELN(FileName, 231);
    WRITELN(FileName, 321);

    CLOSE(FileName);
```

Program 8–22 *continued*

```
    END;

    PROCEDURE Read_File;
     VAR
       FileNumbers
     :ARRAY [1..3] OF INTEGER;

    BEGIN

      RESET(FileName);

      READLN(FileName, FileNumbers[1]);
      READLN(FileName, FileNumbers[2]);
      READLN(FileName, FileNumbers[3]);

      CLOSE(FileName);

      WRITELN('The first number =  ',FileNumbers[1]);
      WRITELN('The second number = ',FileNumbers[2]);
      WRITELN('The third number = ',FileNumbers[3]);

      WRITELN('The sum of the first two numbers is:');
      WRITELN(FileNumbers[1] + FileNumbers[2]);

    END;

 BEGIN
  Create_and_Enter;
  Read_File;
 END.
```

Execution of program 8–22 yields

```
Give me a Number => 12
The first number = 12
The second number = 231
The third number = 321
The sum of the first two numbers is:
243
```

As you can see, the program produced a correct answer for the sum of the first two numbers (12 + 231 = 243) that were originally stored as CHARacters. However, Turbo Pascal easily converted them back to numbers for you.

As you can see, Turbo will convert the character string stored in the disk file back to the required number type, allowing the arithmetic operation to be performed.

Different Types

Since you can write and read these sequential text files to the disk as if they were
going to the screen, you should be able to **WRITELN** a STRING, INTEGER, REAL, or
BOOLEAN to the disk (since this is what you can do to the screen). You should also
be able to **READLN** a STRING, INTEGER, and REAL, but not a BOOLEAN (again this
is what you can do from the keyboard). This is demonstrated by program 8–23.

Program 8–23

```
PROGRAM Text_Files_With_Types;
VAR
 FileName
:TEXT;

 PROCEDURE Create_and_Enter;
   VAR
    StringLine
   :STRING[79];

    IntegerInput
   :INTEGER;

    RealInput
   :REAL;

    BooleanInput
   :BOOLEAN;

  BEGIN

   ASSIGN(FileName, 'NEWFILE');
   REWRITE(FileName);

   StringLine := 'This is a string.';
   WRITELN(FileName, StringLine);

   IntegerInput := 123;
   WRITELN(FileName, IntegerInput);

   RealInput := 12.3;
   WRITELN(FileName, RealInput);

   BooleanInput := TRUE;
   WRITELN(FileName, BooleanInput);

   CLOSE(FileName);

  END;
```

Program 8–23   *continued*

```
   PROCEDURE Read_File;
   VAR
    StringLine
   :STRING[79];

    WholeNumber
   :INTEGER;

    AnyNumber
   :REAL;

    TrueFalse
   :BOOLEAN;

  BEGIN

    RESET(FileName);
    READLN(FileName, StringLine);
    READLN(FileName, WholeNumber);
    READLN(FileName, AnyNumber);
    CLOSE(FileName);

    WRITELN('The stored string is =  ',StringLine);
    WRITELN('The stored whole number is = ',WholeNumber);
    WRITELN('The stored real number is = ',AnyNumber);

  END;

BEGIN
 Create_and_Enter;
 Read_File;
END.
```

Note that program 8–23 does with a disk file exactly what it does when outputting to the screen or inputting from the keyboard. When program 8–23 is executed, the results are

```
The stored string is = This is a string.
The stored whole number is = 123
The stored real number is = 1.2300000000E+01
```

Using the EOF Marker

Turbo Pascal recognizes the command **EOF**. It means End Of File. This command returns a Boolean value of TRUE or FALSE. It is useful for reading disk text files when you don't know the length of the text file. Program 8–24 demonstrates its use.

Program 8–24

```pascal
PROGRAM Using_the_End_of_File_Marker;

VAR
 FileName
:TEXT;

 PROCEDURE Create_and_Enter;
  VAR
   LineNumber
 :INTEGER;

  BEGIN

    ASSIGN(FileName, 'NEWFILE');
    REWRITE(FileName);

     FOR LineNumber := 1 TO 10 DO
       BEGIN
        WRITELN(FileName, 'Line ',LineNumber,' of text.');
       END;

   CLOSE(FileName);

  END;

  PROCEDURE Read_File;
   VAR
    StringLine
    :STRING[79];

  BEGIN

    RESET(FileName);

    WHILE NOT EOF(FileName) DO
      BEGIN
        READLN(FileName, StringLine);
        WRITELN(StringLine);
      END;

    CLOSE(FileName);

  END;
BEGIN
 Create_and_Enter;
 Read_File;
END.
```

The first procedure, `Create__and__Enter`, uses a **FOR-TO** loop to enter lines of data into the disk file. You could also have used a **FOR-TO** loop in the second procedure, `Read__File`, because you would know where to stop the loop (at the count of 10). Suppose, however, you didn't know how many lines of text were in the disk file. This is where the **EOF** comes in handy. This structure uses a **WHILE-DO**; note the syntax.

WHILE NOT EOF(FileName) DO

When program 8–24 is executed, the output is

```
Line 1 of text.
Line 2 of text.
Line 3 of text.
Line 4 of text.
Line 5 of text.
Line 6 of text.
Line 7 of text.
Line 8 of text.
Line 9 of text.
Line 10 of text.
```

Conclusion

This section completes your study of sequential files called text files. In the next section, you will be introduced to another important and practical kind of file, the random-access FILE. Now you can store information on a disk for later use. You will soon learn how to retrieve information anywhere in the disk file you choose. For now, check your understanding of this section by trying the following section review.

8–5 Section Review

1   State the difference between **WRITELN** and **WRITE** as far as inputting information to a disk file is concerned.
2   State the difference between **READLN** and **READ** as far as outputting information from a disk file is concerned.
3   Describe the basic structure of a text file.
4   Explain what is meant by a sequential file. What is the significance of this?
5   What data types may be placed into a disk text file? What data types may be read from the same file?
6   Explain the significance of the **EOF** marker in a text file.

8–6   Random-Access Files

Introduction

This section introduces you to the power of a **random-access file**. You will see how to enter a variety of data of different types, store it on a disk file, then retrieve any piece of the data you please. This type of disk file works well with inventory programs such as the hospital or stock market program.

Basic Idea

A random-access file in Pascal consists of a sequence of components, for which the number of the components is not determined by the definition of the file. Instead, the Turbo Pascal system keeps track of each component by use of a **file pointer**. Each time a component is written to or read from the file, the file pointer of that file is advanced to the next component.

In a random-access file, all components are of equal length and thus the position of a specific component can be calculated. This means that the file pointer can be moved to any component of the file, thus providing random access to any element of the file.

Sample Program

To give you an idea of what all of this means, consider program 8–25.

Program 8–25

```
PROGRAM Random_Access_1;

  TYPE
    PatientRecord = RECORD
                      Name : STRING[79];
                      AmountOwed : REAL;
                      DaysHere : INTEGER;
                      ReleaseReady : BOOLEAN;
                    END;

    VAR
      PatientFiles : FILE OF PatientRecord;
      PatientInfo : PatientRecord;

  PROCEDURE Create_File;

    BEGIN

      ASSIGN(PatientFiles, 'PATIENT.DTA');
      REWRITE(PatientFiles);

      WITH PatientInfo DO
        BEGIN
          Name := 'Frank Jones';
          AmountOwed := 825.43;
          DaysHere := 3;
          ReleaseReady := FALSE;
        END;
```

Program 8–25 *continued*

```
               WRITE(PatientFiles, PatientInfo);

       CLOSE(PatientFiles);

    END;

  PROCEDURE Read_the_File;

   BEGIN
     ASSIGN(PatientFiles, 'PATIENT.DTA');
     RESET(PatientFiles);

     SEEK(PatientFiles, 0);
     READ(PatientFiles, PatientInfo);

       WITH PatientInfo DO
         BEGIN
           WRITELN('Patient name => ',Name);
           WRITELN('Amount owed => ',AmountOwed:3:2);
           WRITELN('Days in hospital => ',DaysHere);
           WRITELN('Ready for for release => ',ReleaseReady);
         END;

   END;

BEGIN
 Create_File;
 Read_the_File;
END.
```

You can see from program 8–25 that it contains several features that have already been introduced. First, it contains a type record.

```
TYPE
  PatientRecord = RECORD
                    Name : STRING[79];
                    AmountOwed : REAL;
                    DaysHere : INTEGER;
                    ReleaseReady : BOOLEAN;
                  END;
```

This is exactly the same kind of record that was introduced in the first section of this chapter. You use this because you will now define a file as a record type. In program 8–25, this was done as follows.

```
VAR
  PatientFiles : FILE OF PatientRecord;
  PatientInfo : PatientRecord;
```

`PatientFiles` will now be a file consisting of the previously defined `PatientRecord`. This is similar to defining an array except that the number of elements is not known. Note also the defining of another variable, `PatientInfo`, which is not a file but simply a single record type of the previously defined record. Keep in mind that there are two separate record variables in the program: one is a file of records and the other is a single record (both of the same type).

The program has two procedures; the Pascal file code in the first procedure should be familiar.

```
ASSIGN(PatientFiles, 'PATIENT.DTA');
REWRITE(PatientFiles);
```

The **ASSIGN** procedure does exactly the same thing here as it did before. It is telling the Pascal program to identify a special location on the disk called **PATIENT.DTA** as the file record variable **PatientFiles**. The **REWRITE** procedure is doing the same thing it did before—causing a new disk file to be created by the name of **PATIENT.DAT** (and destroying an old one by that name if it already exists).

The next part of the procedure is entering data into the single record variable `PatientInfo`.

```
WITH PatientInfo DO
  BEGIN
    Name := 'Frank Jones';
    AmountOwed := 825.43;
    DaysHere := 3;
    ReleaseReady := FALSE;
  END;
```

This is no different from what was presented in the first section of this chapter. Next, the data that has been placed into the single record variable of `Patient-Info` is now placed in the file record variable `PatientFiles`:

```
WRITE(PatientFiles, PatientInfo);
```

This is really no different from what was happening before with sequential files. You are assigning the variable that you just defined to the actual disk file just as you did before; all of the data entered about the patient is now transferred to the actual disk file. It should be pointed out that this is stored as element 0 (the first element of a random-access file has the number 0).

As before, the **CLOSE** procedure is used to close the file. This built-in Turbo Pascal procedure is just as important here as it was for sequential files. So far, except for **FILE OF**, there have been no new Pascal commands; they have all been the same as for the sequential text files just presented. The only difference is that the type RECORD has been used rather than the type TEXT.

Getting Stuff Out

Data has now been put into the file; the next step is to access it. In program 8–25, the second procedure `Read_the_File` uses the same Pascal instructions to

ASSIGN, **RESET**, and **READ** the file. The difference is the built-in Turbo Pascal procedure **SEEK**. Note how it is used in program 8–25.

```
ASSIGN(PatientFiles, 'PATIENT.DTA');
RESET(PatientFiles);
SEEK(PatientFiles, 0);
READ(PatientFiles, PatientInfo);
```

Recall that you could think of a random-access file as having a number of components that could be accessed by a pointer. This means that each component has a number, the first of which is 0. The information that was put into the disk file by program 8–25 has the position 0. The **SEEK** procedure allows you to position the file pointer at the beginning of any record so that the next **READ** or **WRITE** operation is done on that record. The syntax is

```
SEEK(FileName, RecordNumber);
```

where **FileName** = the variable that has had the name of the disk file previously assigned to it

RecordNumber = an integer that represents the number of the record starting with 0

Using READ and WRITE

In random-access files, only **READ** and **WRITE** have meaning since these files are not arranged into lines and thus there are no end-of-line markers. The **READ** and **WRITE** procedures should only be used with one or more values of the component type. Unlike sequential FILEs, therefore, you don't need to worry about when to use **READLN** or **WRITELN** since they are never used with random-access files.

Application Program

Consider program 8–26.

Program 8–26

```
PROGRAM Random_Access_Files;
USES CRT;

  TYPE
    PatientRecord = RECORD
                      Name : STRING[79];
                      Floor, Room, Bed : INTEGER;
                      AmountOwed : REAL;
                      DaysHere : INTEGER;
                      ReleaseReady : BOOLEAN;
                    END;
```

Program 8–26   *continued*

```
VAR
  PatientFiles : FILE OF PatientRecord;
  PatientInfo : PatientRecord;
PROCEDURE Enter_Information(VAR NewData : PatientRecord);
 VAR
  Floor,
  Room,
  Bed
 :INTEGER;
  Choice
 :CHAR;

  BEGIN

    WITH NewData DO
       BEGIN
          WRITE('Floor = ');
          READLN(Floor);
          WRITE('Room number = ');
          READLN(Room);
          WRITE('Bed number = ');
          READLN(Bed);

          WRITE('Patient name = ');
          READLN(Name);
          WRITE('Amount owed => ');
          READLN(AmountOwed);
          WRITE('Days in hospital => ');
          READLN(DaysHere);
          WRITE('Release Ready (Y/N) => ');
          READLN(Choice);
            IF UPCASE(Choice) = 'Y'
              THEN ReleaseReady := TRUE
            ELSE
               ReleaseReady := FALSE;
          END;
  END;
PROCEDURE Make_New_File;
 VAR
  FileName
:STRING[79];
  Choice
:CHAR;

BEGIN

   WRITELN('What is the name of the new file: ');
   READLN(FileName);

   ASSIGN(PatientFiles, FileName);
   REWRITE(PatientFiles);

REPEAT

   Enter_Information(PatientInfo);
   WRITE(PatientFiles, PatientInfo);

   WRITE('Do you wish to enter another record (Y/N) => ');
   READLN(Choice);
```

Program 8–26 *continued*

```
   UNTIL UPCASE(Choice) = 'N';

       CLOSE(PatientFiles);

   END;
 PROCEDURE Open_Existing_File;
  VAR
  FileName
 :STRING[79];
  Choice
 :CHAR;

   BEGIN

       WRITELN('What is the name of the existing file: ');
       READLN(FileName);

       ASSIGN(PatientFiles, FileName);
       RESET(PatientFiles);

       SEEK(PatientFiles, FILESIZE(PatientFiles));

   REPEAT

       Enter_Information(PatientInfo);
       WRITE(PatientFiles, PatientInfo);

       WRITE('Do you wish to enter another record (Y/N) => ');
       READLN(Choice);

   UNTIL UPCASE(Choice) = 'N';

       CLOSE(PatientFiles);

   END;
 PROCEDURE Get_Information;
  VAR
  FileName
 :STRING[79];
  RecordNumber
 :INTEGER;

   BEGIN

       WRITELN('What is the name of the file: ');
       READLN(FileName);
       WRITE('Patient Number => ');
       READLN(RecordNumber);

       ASSIGN(PatientFiles, FileName);
       RESET(PatientFiles);

       SEEK(PatientFiles, RecordNumber);

       READ(PatientFiles, PatientInfo);
```

Program 8–26   *continued*

```
        WITH PatientInfo DO
          BEGIN
          WRITELN('Patient name => ',Name);
          WRITELN('Floor = ',Floor,' Room = ',Room,' Bed = ',Bed);
          WRITELN('Amount owed => ',AmountOwed:3:2);
          WRITELN('Days in hospital => ',DaysHere);
          WRITELN('Ready for release => ',ReleaseReady);
          WRITELN('Floor number => ',Floor);
          WRITELN('Room number => ',Room);
          WRITELN('Bed number => ',Bed);
          END;
        CLOSE(PatientFiles);

END;

PROCEDURE Make_Selection;
 VAR
  Selection
 :CHAR;
   Again
 :BOOLEAN;

  BEGIN

     CLRSCR;

   REPEAT

      WRITELN;
      WRITELN('Select by number:');
      WRITELN('1] Create a new file.  2] Add to an existing file.');
      WRITELN('3] Read records.');
      WRITELN;
      WRITE('Your selection => ');
      READLN(Selection);
       IF Selection IN ['1','2','3'] THEN Again := TRUE
        ELSE
        Again := FALSE;

    UNTIL Again = TRUE;

    CASE Selection OF

      '1' : Make_New_File;
      '2' : Open_Existing_File;
      '3' : Get_Information;

    END;  {of case}

END; {of procedure Make_Selection}

BEGIN
  Make_Selection;
END.
```

Program 8–26 allows you to

1. Create a new disk file and add patient information.
2. Add more patient information to an existing disk file.
3. Read information from the disk file.

The program has five procedures. The last procedure, `Make_Selection`, asks the program user to make one of three selections. Depending upon the user selection, one of three procedures will be evoked through the action of the **CASE** statement. If the program user has decided to make a new disk file, the procedure `Make_New_File` will be executed.

The procedure `Make_New_File` asks the program user to enter a name to be used for the disk file. This is then put into the **VAR FileName** and the **ASSIGN** and **REWRITE** procedures are then evoked. The procedure `Enter_Information(PatientInfo)`, for which `PatientInfo` is a variable parameter, is implemented. `Enter_Information` asks for information pertaining to the patient and this data is passed back to the calling procedure, `Make_New_File`. The disk file is created with the **REWRITE** procedure and the record is entered into the disk file with the **WRITE(PatientFiles, PatientInfo);** statement.

If the program user chooses to add information to an existing disk file, the procedure `Open_Existing_File` is evoked. Again the program user is asked for the name of the disk file to be opened. But now, the new information must be added past the existing records that already exist in the disk file, using the statement

SEEK(PatientFiles, FILESIZE(PatientFiles));

This moves the file pointer to the end of the last record within the disk file. The built-in Turbo Pascal function **FILESIZE** has the following syntax

FILESIZE(FileName);

will return an INTEGER that represents the number of records within the disk file.

If the program user had selected to read records, the procedure `Get_Information` would have been evoked. Here the **RESET** procedure is used to open the user-named file. The program user is asked for the patient record number, and this record number sets the file pointer to the corresponding record using

SEEK(PatientFiles, RecordNumber);

You can thus see that these files are randomly accessible.

Programming Dangers

You will see that there are some problems with program 8–26. One of the great dangers in this program is asking the program user to input the name of the disk file to be created. This could result in the unintentional destruction of an existing disk file that just happened to have the same name the program user selected. Another problem is asking the program user for the name of the disk file when information is to be added or the file is to be read. A run-time error could occur here if the file name selected by the program user did not exist. The program debugging and implementation section of this chapter will show you a method of making this kind of program more user-friendly.

Conclusion

This concludes your introduction to Pascal records and files. In the next section you will be shown how to avoid some of the pitfalls that can be encountered with user selection of disk files. Test your understanding of this section by trying the following section review.

8–6 Section Review

1   Briefly describe a Pascal random-access file.
2   What Turbo Pascal procedure is used to locate the position of a record within a random-access file?
3   Is there any difference in the use of the **ASSIGN**, **REWRITE**, or **RESET** procedures in sequential text files and random-access files?
4   Name the built-in Turbo Pascal procedure that can be used to determine the number of files in a random-access file.

8–7   Program Debugging and Implementation— Compiler Directives

Compiler Directives

A **compiler directive** is nothing more than an instruction to the Turbo Pascal compiler. These directives may be placed anywhere in the program a comment { } is allowed. The syntax for a compiler directive is

`{$D-}`

It's important to note that no spaces are allowed in such a directive. The command must start with a left bracket {, be immediately followed by a dollar sign $ and the letter representing the directive, then a plus + or minus sign −, followed immediately by a right bracket }.

An Application

In the last section you were told about a potential problem if the program user was allowed to select a name for a disk file; the program user could mistakenly select the name of an existing disk file and cause it to be erased when creating a new file of the same name. The user could also select a disk file name that did not exist when trying to add information to an existing file; a run-time error would occur and the program would terminate. For example, consider program 8–27.

Program 8–27

```
PROGRAM Run_Time_Error;

   TYPE
     AnyRecord = RECORD
                   OneItem : CHAR;
                 END;

   VAR
     AnyFile : FILE OF AnyRecord;
     ThisRecord : AnyRecord;

   PROCEDURE Read_File;
    VAR
     FileName
    :STRING[79];

      BEGIN
        WRITELN('Enter the name of the file you wish to read from:');
        READLN(FileName);

        ASSIGN(AnyFile, FileName);
        RESET(AnyFile);

        SEEK(AnyFile, 0);
        READ(AnyFile, ThisRecord);

        WRITE('The character is: ');
        WRITELN(ThisRecord.OneItem);
      END;

BEGIN
 Read_File;
END.
```

Program 8–27 asks the program user to select the name of a disk file for opening and reading out some data. If the file name selected by the program user does not exist, a run-time error will occur and the program will be aborted. In order to prevent this the following program modification could be made.

Program 8–28

```
PROGRAM Better_Approach;

   TYPE
     AnyRecord = RECORD
                   OneItem : CHAR;
                 END;

   VAR
     AnyFile : FILE OF AnyRecord;
     ThisRecord : AnyRecord;
```

Program 8–28 *continued*

```
PROCEDURE Open_Existing_File;
VAR
 FileName
:STRING[79];
 FileExists
:BOOLEAN;

 BEGIN
   REPEAT
     WRITE('Enter the name of the file you wish to read from: ');
     READLN(FileName);

     ASSIGN(AnyFile, FileName);

     {$I-}       {Turn I/O error checking off.}
     RESET(AnyFile)
     {$I+};      {Turn I/O error checking back on.}

     FileExists := (IORESULT = 0);

      IF NOT FileExists THEN
        WRITELN('A file by that name does not exist.');

     UNTIL FileExists;
   END;
BEGIN
 Open_Existing_File;
END.
```

Program 8–28 will stay in a continuous loop until the program user selects a disk file name that already exists. The key to this program is the compiler directive that turns off I/O error checking.

```
{$I-} {Turn I/O error checking off.}
RESET(AnyFile)
{$I+}; {Turn I/O error checking back on.}
```

The letter I followed by a minus sign tells the Turbo Pascal compiler to turn off its automatic I/O error checking. Thus, when **RESET(AnyFile)** is evoked and a run-time error occurs because there is no existing disk file by the required name, the program will not announce a run-time error and stop the program. Instead, it is the programmer's responsibility to develop the program so it will handle any type of I/O errors. This is accomplished by the statement

```
FileExists := (IORESULT = 0);
```

FileExists is a BOOLEAN variable. The built-in Turbo Pascal procedure **IORESULT** will return a 0 if there is no I/O error and a non-zero value if there is.

The compiler directive **{$I+}** used in program 8–28 turns Turbo Pascal's automatic I/O error checking back on.

Making an I/O Error Function

It is more convenient to develop a standard function that can be used to check if a disk file already exists. This function can then be used in any of your programs that deal with disk files. The use of such a function is illustrated in program 8–29.

Program 8–29

```
PROGRAM Another_Way;
   TYPE
     Words = STRING[79];

     AnyRecord = RECORD
                   OneItem : CHAR;
                 END;

   VAR
     AnyFile : FILE OF AnyRecord;
     ThisRecord : AnyRecord;

   FUNCTION FileExists(FileName : Words) : BOOLEAN;
     VAR
      AnyFile
    :FILE;

       BEGIN

          ASSIGN(AnyFile, FileName);

          {$I-}      {Turn I/O checking off.}
          RESET(AnyFile);
          CLOSE(AnyFile);
          {$I+}

          FileExists := (IORESULT = 0);

       END;

   PROCEDURE Read_File;
    VAR
     FileName
    :STRING[79];

     BEGIN
       WRITELN('Enter the name of the file you wish to read from:');
       READLN(FileName);

     IF FileExists(FileName) THEN
       BEGIN
         ASSIGN(AnyFile, FileName);
         RESET(AnyFile);

         SEEK(AnyFile, 0);
         READ(AnyFile, ThisRecord);

         WRITE('The character is: ');
         WRITELN(ThisRecord.OneItem);
       END
     ELSE
       WRITELN('A file by that name does not exist.');
```

Program 8–29   *continued*

```
   END;

BEGIN
 Read_File;
END.
```

Program 8–29 uses a user-developed BOOLEAN function that returns a TRUE or FALSE condition depending upon the existence of a disk file name selected by the program user. This function may be used to indicate to the program user that a selected file name does not exist, or as shown in program 8–30, that a selected disk file name does exist.

Program 8–30

```
PROGRAM Check_for_File;

   TYPE
     Words = STRING[79];

     AnyRecord = RECORD
                   OneItem : CHAR;
                 END;

   VAR
     AnyFile : FILE OF AnyRecord;
     ThisRecord : AnyRecord;

   FUNCTION FileExists(FileName : Words) : BOOLEAN;
     VAR
      AnyFile
     :FILE;

       BEGIN

         ASSIGN(AnyFile, FileName);

         {$I-}      {Turn I/O checking off.}
         RESET(AnyFile);
         CLOSE(AnyFile);
         {$I+}

         FileExists := (IORESULT = 0);

       END;
```

Program 8–30 *continued*

```
PROCEDURE Create_File;
  VAR
   FileName
  :STRING[79];
   Choice
  :CHAR;

    BEGIN
      WRITELN('Enter the name of the file you wish to create:');
      READLN(FileName);

    IF FileExists(FileName) THEN
    BEGIN
      WRITELN('A file by that name already exists.');
      WRITELN('If you create another file by that name');
      WRITELN('the existing one will be destroyed.');
      WRITELN;
      WRITELN('Are you sure you want to destroy the');
      WRITELN('existing file?');
      WRITELN;
      WRITELN('Enter (Y) to destroy existing file.');
      WRITELN('Enter (N) to choose a different file name.');
      WRITELN;
      WRITE('Your choice => ');
      READLN(Choice);
        IF UPCASE(Choice) = 'N' THEN EXIT
    END;

    ASSIGN(AnyFile, FileName);
    REWRITE(AnyFile);

    {Information entered in file.}

    CLOSE(AnyFile);
    END;
BEGIN
 Create_File;
END.
```

Program 8–30 warns the program user that the chosen disk file name already exists. Recall that when a new disk file is created, any other disk file by the same name is destroyed first. Program 8–30 alerts an unwary program user to this fact before proceeding. Program 8–30 uses a built-in Turbo Pascal procedure called **EXIT.** When it is evoked the program will exit the procedure and continue with the rest of the program as if the procedure had been completed in a normal fashion.

Conclusion

This section introduced you to some of the tricks of the trade in dealing with disk files. Your ability to create a program that has the capability to store and retrieve data from disk files gives you great programming power. However, you need to temper that power with a balance of user friendliness to make the program useful to the non-programmer. Check your understanding of this section by trying the following section review.

8–7 Section Review

1 What is a compiler directive?
2 State the syntax for a compiler directive.
3 Explain what would happen if the program user attempted to cause a Pascal program to open a disk file that did not exist.
4 What is the Turbo Pascal compiler directive that turns off automatic I/O error checking? What is the directive that turns it back on?
5 What is used in a program to check if the disk file does exist when automatic I/O checking is disabled?

8–8   Case Study

Discussion

The case study for this chapter is a modification of an existing Pascal program—the program used for the case study and self-test in chapter seven.

Recall that the program was developed for a stock broker. It allowed information about stocks to be entered, the information was then alphabetized, and stock prices could be changed as well as new ones added and existing ones deleted. The program was essentially a case study in developing a classical data base management system that did not use records or disk files for the storing and retrieving of data.

The case study in this chapter demonstrates the advantage of using records in a Pascal program of this type. Here you will also see how the data can be saved to and retrieved from the disk.

Problem Statement

Modify the stock market program (programs 7–22 and 7–30) so that it will make use of the Pascal record structure. The program must also be modified so that user-input data may be saved to a disk file and automatically retrieved each time the program is used. Otherwise, the program is to retain all of the features of the original program.

Design Process

The design process here is somewhat simplified in that it is a modification of an existing program. There are several approaches that can be taken here, but it is good programming practice to make one program modification at a time and check that the program will work under the new conditions. After you are sure that the program performs as required with its new change, go back and make the second change. This process will help you focus attention on one specific change at a time, which will be a great aid in the program-modification process.

Making it a Record

The first step is to define the record type.

```
TYPE
    StockRecord = RECORD
                    StockName : STRING[79];
                    Quantity  : INTEGER;
                        Cost  : REAL;
                    BlueChip  : BOOLEAN;
                       Value  : REAL;
                    GainLoss  : REAL;
                  END;

    StockArray = ARRAY[1..MaxStocks] OF StockRecord;
```

This program excerpt shows that once the record has been defined, an array of records can then be defined.

RECORD Advantages

The advantages of using a record become apparent when passing parameters between procedures. For example, the alphabetizing procedure parameters are much simpler than before.

```
PROCEDURE Alphabetize_Stock(VAR StockInfo : StockArray
                               IndexMax : INTEGER);
```

Another obvious advantage to the use of Pascal records is simplifying the alphabetizing procedure itself. If the name of a stock needs to be swapped it is only necessary to then swap the whole record and not each of the individual stocks, as illustrated in this program excerpt.

```
VAR
 TempRecord {Temporary storage for record}
:StockArray;
 Index
:INTEGER;
 Switch
:BOOLEAN;

 BEGIN
```

```
REPEAT
  Switch := FALSE;

FOR Index := 1 TO IndexMax DO
  BEGIN
   IF (StockInfo[Index].StockName > StockInfo[Index+1].StockName)
    AND (Index < IndexMax)
     THEN {Switch the records}
      BEGIN
       TempRecord[Index] := StockInfo[Index];
       StockInfo[Index] := StockInfo[Index+1];
       StockInfo[Index+1] := TempRecord[Index];
       Switch := TRUE;
      END;

   END;
  UNTIL Switch = FALSE;

END;
```

You may want to compare the simplicity of swapping records as opposed to swapping individual array items as was done in the original stock market program.

Adding File Capabilities

The program must check to see if a disk data file has already been created. If not, it must do so. If a data file already exists, however, creating a new one would destroy all previously entered stock market data. If a disk file does exist, the data must be read into the program so that it can be merged, alphabetized, and displayed along with any new user data or data modifications.

The following program excerpt performs the required process (**FileExists** is the programmer-defined function already discussed).

```
IF NOT(FileExists('STOCKS.DTA'))
  THEN {File does not exist so create it}

   BEGIN
    ASSIGN (StockFile, 'STOCKS.DTA');
    REWRITE(StockFile);
    CLOSE(StockFile);
   END

  ELSE {Count the number of records in the file.}

   BEGIN
    ASSIGN(StockFiles, 'STOCKS.DTA');
    RESET(StockFiles); {Opens the file}
    IndexMax := FILESIZE(StockFiles);
    SEEK(StockFiles, 0);
```

```
FOR Index := 1 TO IndexMax DO
  BEGIN
    READ(StockFiles, MoreStuff);
    Stocks[Index] := MoreStuff;
  END;
CLOSE(StockFiles);

END;
```

Program Suggestion

The final suggested changes for new program modifications are presented in the self-test section of this chapter.

Conclusion

In this case study you saw a modification of an existing program. In doing this exercise, the advantages of using a record structure were presented along with suggested techniques for saving the material to a disk file. Check your progress by trying the following section review.

8–8 Section Review

1   State the main points of the problem statement for the case study presented in this section.
2   Describe a recommended procedure to use when modifying an existing program.
3   What was the first modification of the program presented in this section?
4   State at least two advantages that a record structure has in this program modification.
5   What is needed in the program to automatically interact with a disk file to store and retrieve data?

Summary

1   Pascal offers another method for organizing information called a record.
2   A Pascal record contains a record identifier along with a collection of record elements called fields.
3   In a record, a field is an individual data element of a specific type.
4   A Pascal record may be an array.
5   A record field may be of any type, including another record.
6   The **WITH** statement is used to make programming with record fields easier.
7   Records may be treated as a single variable and are a convenient way of passing parameters between procedures.
8   Variant records are a method of storing different kinds of information between different fields.
9   The built-in Turbo Pascal function **UPCASE** converts all lower case letters to upper case.

10   A Pascal file is similar to that of an array except that the number of elements is not known.

11   I/O is the process of inputting and outputting information.

12   The **READ** and **READLN** procedures are Pascal's input commands.

13   The **WRITE** and **WRITELN** procedures are Pascal's output commands.

14   Pascal treats any input or output device as a file.

15   Examples of output files are the console screen, printer, and disk.

16   Examples of input files are the keyboard and disk.

17   A test file consists of a string of ASCII characters.

18   Disk files must use legal DOS names because of the interaction of Pascal and the system DOS when working with disk files.

19   There are four possible conditions that can be encountered when working with disk files: 1] The file doesn't exist. 2] The file does exist and information is to be taken from it. 3] The file does exist and information needs to be added to it. 4] The file does exist and its old data must be replaced with new data.

20   Turbo Pascal provides methods for dealing with all four of the possible conditions encountered when working with disk files.

21   A disk file must always be closed before the program is terminated or before opening another disk file.

22   The end-of-file (**EOF**) is a convenient way of creating a loop that will insure that all the information of a text file is retrieved. It can also be used to put the file pointer at the end of the file in order to add new information.

23   Text files are sequential files.

24   Random-access files contain elements that are individually accessible.

25   Turbo Pascal provides the **SEEK** procedure to move the file pointer to the desired file element.

26   Pascal's **READ** and **WRITE** procedures are the only ones to use for I/O with random-access files since there is no concept of a line within a file.

27   A Turbo Pascal compiler directive is an instruction to the Turbo Pascal compiler.

28   The Turbo Pascal compiler directive that turns I/O error checking on or off is useful for making programs with more user-friendly disk file operations.

29   It's useful to make an I/O error function in disk file programs that is of type BOOLEAN. The function simply identifies if a file exists.

Interactive Exercises

Directions

These exercises require that you have access to a computer and software that supports Pascal, specifically, the Turbo Pascal Development System, version 4.0, from Borland International. They are provided here to give you valuable experience and immediate feedback on what the concepts and commands introduced in the chapter will do—they are also fun.

Exercises

1   Program 8–31 (p. 520) has about the simplest record structure possible. Predict what you think will be displayed on the screen, then try the program.

Program 8–31

```
PROGRAM IE_1;
 VAR
     ThisThing : RECORD
                      ThatThing : CHAR;
                 END;
 BEGIN
    ThisThing.ThatThing := '?';
    WRITELN(ThisThing.ThatThing);
 END.
```

2 Program 8–32 is a simple example of using the **WITH**. Compare this structure to that in program 8–30. Do you get the same results? What conclusions can you make about the necessity of the **WITH**?

Program 8–32

```
PROGRAM IE_2;
 VAR
     ThisThing : RECORD
                      ThatThing : CHAR;
                 END;
 BEGIN
    ThisThing.ThatThing := '?';
    WITH ThisThing DO
    WRITELN(ThatThing);
 END.
```

3 Program 8–33 shows a variable now defined as the record type. Other than the structure of the program, is there any difference between program 8–32 and program 8–33 in terms of output?

Program 8–33

```
PROGRAM IE_3;
 TYPE
     ThisThing = RECORD
                      ThatThing : CHAR;
                 END;
 VAR
    AnotherThing : ThisThing;
```

Program 8–33 *continued*

```
BEGIN
    AnotherThing.ThatThing := '?';
    WITH AnotherThing DO
    WRITELN(ThatThing);
END.
```

4   Program 8–34 illustrates a slightly different use of the **WITH** statement. Compare program 8–34 to program 8–33. Do you expect any differences in the output? What happens when you try it?

Program 8–34

```
PROGRAM IE_4;
  TYPE
      ThisThing = RECORD
                      ThatThing : CHAR;
                  END;
  VAR
      AnotherThing : ThisThing;

  BEGIN
      WITH AnotherThing DO
      ThatThing := '?';
      WITH AnotherThing DO
      WRITELN(ThatThing);
  END.
```

5   Program 8–35 illustrates yet another method of using the **WITH**. This time it is followed by a **BEGIN–END**. Compare this method to the previous methods. Any preference?

Program 8–35

```
PROGRAM IE_5;
  TYPE
      ThisThing = RECORD
                      ThatThing : CHAR;
                  END;
  VAR
      AnotherThing : ThisThing;
```

Program 8–35   *continued*

```
BEGIN
   WITH AnotherThing DO
    BEGIN
      ThatThing := '?';
      WRITELN(ThatThing);
    END;
END.
```

6 Program 8–36 is an illustration of using an array as one of the elements of a record. Note also the use of the **WITH** statement. What do you predict will be displayed on the screen? Try it and see.

Program 8–36

```
PROGRAM IE_6;
 TYPE
    ThisThing = RECORD
                   ThatThing : CHAR;
                   MoreThings : ARRAY [1..9] OF CHAR;
                END;
 VAR
   AnotherThing : ThisThing;
   Index : INTEGER;

BEGIN
   WITH AnotherThing DO
    BEGIN
      ThatThing := '?';
        FOR Index := 1 TO 9 DO
         BEGIN
           MoreThings[Index] := '$';
         END;
      WRITELN(MoreThings[5]);
    END;
END.
```

7 Program 8–37 is an example of using the record variant. Look the program over, predict what you think will happen, then try it.

Program 8–37

```
PROGRAM IE_7;
 TYPE
    Condition = BOOLEAN;

    ThisThing = RECORD
                   ThatThing : CHAR;
                   CASE Selection : Condition OF
                      TRUE : (ThisOne : INTEGER);
                      FALSE : (ThatOne : CHAR);
                END;
 VAR
    AnotherThing : ThisThing;
    Choice : BOOLEAN;

 BEGIN
    WITH AnotherThing DO
     BEGIN
        Selection := TRUE;
         ThisOne := 5;
         ThatOne := 'X';
         WRITELN(ThisOne);
         WRITELN(ThatOne);
     END;
 END.
```

8   Program 8–38 doesn't work, but the compiler doesn't catch a thing. What kind of run-time error do you get?

Program 8–38

```
PROGRAM IE_8;
  BEGIN
    WRITELN(CON, 'Where does this go?');
  END.
```

9   Program 8–39 (p. 524) doesn't work either, although again the compiler has no problems with it. What kind of run-time error do you get here?

Program 8–39

```
PROGRAM IE_9;
  VAR
   FileName
  :TEXT;

  BEGIN
    ASSIGN(FileName, 'CON');
    WRITELN(FileName, 'Where does this go?');
  END.
```

10   Program 8–40 works. Note that Pascal requires that the file be opened.

Program 8–40

```
PROGRAM IE_10;
  VAR
   FileName
  :TEXT;

  BEGIN
    ASSIGN(FileName, 'CON');
    REWRITE(FileName);
    WRITELN(FileName, 'Where does this go?');
  END.
```

11   Program 8–41 closes the opened file before data is written to it. What do you suppose the program will do? Try it and see.

Program 8–41

```
PROGRAM IE_11;
  VAR
   FileName
  :TEXT;

  BEGIN
    ASSIGN(FileName, 'CON');
    REWRITE(FileName);
    CLOSE(FileName);
    WRITELN(FileName, 'Where does this go?');
  END.
```

12   Program 8–42 has corrected the error in program 8–41. Any prediction of what it will do? Does it do what you thought it would?

Program 8–42

```
PROGRAM IE_12;
  VAR
   FileName
  :TEXT;

  BEGIN
    ASSIGN(FileName, 'CON');
    REWRITE(FileName);
    WRITELN(FileName, 'Where does this go?');
    CLOSE(FileName);
  END.
```

13   Program 8–43 opens the console keyboard as a file and gets input from it. See what happens when you try the program.

Program 8–43

```
PROGRAM IE_13;
  VAR
   FileName
  :TEXT;
   UserInput
  :STRING[79];

  BEGIN
    ASSIGN(FileName, 'CON');
    RESET(FileName);
    READ(FileName, UserInput);
    CLOSE(FileName);
  WRITELN('This is what you put into the computer: ');
  WRITELN;
  WRITELN(UserInput);

  END.
```

14   Program 8–44 (p. 526) attempts to use the console as a random-access file. It compiles nicely, but does it work? See what you get. Why do you think this happens?

Program 8–44

```
PROGRAM IE_14;
   TYPE
         RecordData = RECORD
                         SomeData : STRING[79];
                      END;

   VAR
         RecordFile : FILE OF RecordData;
         RecordInfo : RecordData;

 BEGIN
       RecordInfo.SomeData := 'This represents some data.';

       ASSIGN(RecordFile, 'CON');
       RESET(RecordFile);

       WRITE(RecordFile, RecordInfo);
       CLOSE(RecordFile);
   END.
```

15   Program 8–45 tries to use the keyboard as a random-access file. Note that the compiler lets it pass, but what happens when you try to execute it?

Program 8–45

```
PROGRAM IE_15;
   TYPE
         RecordData = RECORD
                         SomeData : STRING[79];
                      END;

   VAR
         RecordFile : FILE OF RecordData;
         RecordInfo : RecordData;

 BEGIN
       RecordInfo.SomeData := 'This represents some data.';

       ASSIGN(RecordFile, 'CON');
       RESET(RecordFile);
       READ(RecordFile, RecordInfo);
       CLOSE(RecordFile);
   END.
```

Pascal Commands

The syntax for a variable of record is

```
Identifier : RECORD
               FieldName₁:TYPE;
               FieldName₂:TYPE;
                     .
                     .
                     .
               FieldNameₙ:TYPE;
            END;
```

An example is

```
Patient_Info : RECORD
                 Name : STRING[79];
                 AmountOwed : REAL;
                 DaysHere : INTEGER;
                 ReleaseReady : BOOLEAN;
               END;
```

Each field may be referenced individually as

```
Identifierᵣ.Identifier_F
```

where Identifier$_R$ = the record identifier
        Identifier$_F$ = the field identifier

The syntax of the **WITH** statement is

WITH RecordIdentifier **DO** Statement

The syntax for a **WRITE** or **WRITELN** is

WRITE([Filevariable], variable);

and

WRITELN([Filevariable], variable);

where [Filevariable] = the file to which the data will be output.

The syntax for **READ** and **READLN** is

READ([Filevariable], Variable);

and

READLN([FileVariable], Variable);

where FileVariable = the peripheral device (file) from which the data is to be input

The following built-in Turbo Pascal procedure assigns the disk file name.

ASSIGN([FileVariable], 'FileName');

where FileVariable = a variable of type TEXT
        FileName = a legal DOS file name

Table 8–3   Disk File Commands and Conditions

Condition	Meaning	Command
1	The disk file does not exist and you want to create it on the disk and add some information.	**REWRITE**
2	The disk file already exists and you want to get information from it.	**RESET**
3	The disk file already exists and you want to add more information to it while preserving the old information already there.	**APPEND**
4	The disk file already exists and you want to get rid of all of the old information and add new information.	**REWRITE**

The following opens a new disk file for input

REWRITE(FileName**);**

where FileName = the TEXT variable that has already had the legal DOS file name assigned to it

The Pascal procedure for opening an existing disk file so that data may be read from the file is

RESET(FileName**);**

where FileName = the TEXT variable that has already had the legal DOS file name assigned to it

The Turbo Pascal procedure for opening an existing disk file so that more data may be added without destroying existing data is

APPEND([FileName]**);**

where FileName = the TEXT variable that has already had the legal DOS name assigned to it

Table 8–3 is a summary of Pascal file commands.

The Pascal command required for terminating programs with an open file or before another file is opened is

CLOSE(FileName**);**

The Pascal command for the end-of-file marker is **EOF.** An example of its use is

WHILE NOT EOF(FileName**) DO**

The built-in Turbo Pascal procedure for moving the file pointer in random-access files is

```
SEEK(FileName, RecordNumber);
```

where FileName = the variable that has the name of the disk file previously assigned
to it

RecordNumber = an integer that represents the number of the record starting
with 0

The syntax for a Turbo Pascal compiler directive is

```
{$D-}
```

An example of the use of a Turbo Pascal compiler directive is

```
{$I-} {Turn I/O error checking off.}
RESET(AnyFile)
{$I+}; {Turn I/O error checking back on.}
```

When automatic I/O error checking is off, the built-in Turbo Pascal **IORESULT**
will return a value of 0 if there are no errors and a non-zero value if there are. An
example of its use is

```
FileExists := (IORESULT = 0);
```

Self-Test

Directions

The program for this self-test contains modifications to most of the procedures for the
case study presented in section 8–8. The program will compile; however, only the
procedures that required a major modification are presented. Answer the questions
for this self-test by referring to program 8–46.

Program 8–46 (pp. 529–534)

```
PROGRAM Stock_Market;
  USES CRT;

  CONST
    MaxStocks = 10;

  TYPE

    Words = STRING[79];

    StockRecord = RECORD
                    StockName : Words;
                    Quantity : INTEGER;
                        Cost : REAL;
                    BlueChip : BOOLEAN;
                       Value : REAL;
                    GainLoss : REAL;
                  END;

    StockArray = ARRAY[1..MaxStocks] OF StockRecord;
```

Program 8–46   *continued*

```
PROCEDURE Explain_Program;
FORWARD;

PROCEDURE Display_Main_Menu(VAR UserChoice : CHAR);
FORWARD;

PROCEDURE Input_New_Stock(VAR StockInfo : StockArray;
                          VAR IndexMax : INTEGER);
FORWARD;

PROCEDURE Update_Stock_Price(VAR StockInfo : StockArray);
FORWARD;

PROCEDURE Sell_Stock(VAR StockInfo : StockArray);
FORWARD;

PROCEDURE List_Stock(VAR StockInfo : StockArray;
                     IndexMax : INTEGER);
FORWARD;

PROCEDURE Program_Repeat(VAR UserSelection : CHAR);
FORWARD;

PROCEDURE Alphabetize_Stock(VAR StockInfo : StockArray;
                            IndexMax : INTEGER);
FORWARD;

PROCEDURE Continue;
FORWARD;

PROCEDURE SaveRecords(StockInfo : StockArray;
                      IndexMax : INTEGER);
FORWARD;

PROCEDURE GetRecords(VAR StockInfo : StockArray;
                     VAR IndexMax : INTEGER);
FORWARD;

FUNCTION FileExists(FileName : Words) : BOOLEAN;
FORWARD;

  PROCEDURE Main_Programming_Sequence;
   VAR
    Decision
   :CHAR;
    Stocks
   :Stockarray;
    Index,
    IndexMax
   :INTEGER;
    StockFile
   :FILE OF StockArray;
   StockFiles : FILE OF StockRecord;
   MoreStuff : StockRecord;
       BEGIN

          IndexMax := 0;
          Explain_Program;

          IF NOT(FileExists('STOCKS.DTA'))
           THEN   {File does not exist so create it}
            BEGIN
```

Program 8–46   *continued*

```
            ASSIGN (StockFile, 'STOCKS.DTA');
            REWRITE(StockFile);
            CLOSE(StockFile);
          END
        ELSE {Count the number of records in the file.}
          BEGIN

            ASSIGN(StockFiles, 'STOCKS.DTA');
            RESET(StockFiles);  {Opens the file}
            IndexMax := FILESIZE(StockFiles);
            SEEK(StockFiles, 0);

              FOR Index := 1 TO IndexMax DO
                BEGIN
                  READ(StockFiles, MoreStuff);
                  Stocks[Index] := MoreStuff;
                END;
              CLOSE(StockFiles);
          END;


        REPEAT

            Display_Main_Menu(Decision);

            CASE UPCASE(Decision) OF

            'A' : Input_New_Stock(Stocks, IndexMax);
            'B' : Update_Stock_Price(Stocks);
            'C' : Sell_Stock(Stocks);
            'D' : List_Stock(Stocks, IndexMax);
            'E' : Program_Repeat(Decision);

              END;  {of case}

                IF UPCASE(Decision) = 'A' THEN
                    Alphabetize_Stock(Stocks, IndexMax);

        UNTIL UPCASE(Decision) = 'Y';

            SaveRecords(Stocks, IndexMax);

    END;  {of procedure Main_Programming_Sequence}

PROCEDURE Display_Main_Menu{(VAR UserChoice : CHAR)};

    BEGIN
      ClrScr;

    REPEAT
      WRITELN;
      WRITELN('            Select by letter:');
      WRITELN('A] Enter new stock.    B] Update stock price.');
      WRITELN('C] Sell stock.         D] List stock.');
      WRITELN('E] End this program.');
      WRITELN;
      WRITE('Your selection => ');
      READLN(UserChoice);
    UNTIL UPCASE(UserChoice) IN ['A', 'B', 'C', 'D', 'E'];

    END;  {of procedure Diplay_Main_Menu}
```

Program 8–46 *continued*

```
PROCEDURE Explain_Program;
  BEGIN
   ClrScr;
   WRITELN('Program explanation goes here.');
   WRITELN;
   WRITE('Press RETURN/ENTER to continue...');
   READLN;
  END;  {of procedure Explain_Program}

PROCEDURE Input_New_Stock{(VAR StockInfo : StockArray; IndexMax : INTEGER)};
 VAR
  Index
 :INTEGER;
  MoreInput
 :CHAR;

 BEGIN

    Index := IndexMax;
    ClrScr;

  REPEAT
    Index := Index + 1;
    WRITELN;

    WITH StockInfo[Index] DO
     BEGIN
      WRITE('Name of stock => ');
      READLN(StockName);
      WRITE('Quantity of stock => ');
      READLN(Quantity);
      WRITE('Cost of stock => ');
      READLN(Cost);
      WRITE('Is stock Blue Chip (Y/N) => ');
      READLN(MoreInput);
        IF UPCASE(MoreInput) = 'Y' THEN
          BlueChip := TRUE
        ELSE
          BlueChip := FALSE;
     END;

    WRITELN;
    WRITE('Do you have another entry (Y/N => ');
    READLN(MoreInput);

  UNTIL (UPCASE(MoreInput) = 'N') OR (Index = MaxStocks);

   IF Index >= MaxStocks THEN
    BEGIN
      WRITELN('You cannot enter more than ',MaxStocks);
      WRITELN('stocks in this program.');
    END;

  IndexMax := Index;

 END;

PROCEDURE Update_Stock_Price{(VAR StockInfo : StockArray)};
  BEGIN
   WRITELN('Update prices.');
   READLN;
  END;
```

Program 8–46 *continued*

```
PROCEDURE Sell_Stock{(VAR StockInfo : StockArray)};
 BEGIN
   WRITELN('Sell stocks.');
   READLN;
 END;

PROCEDURE List_Stock{(VAR StockInfo : StockArray; IndexMax : INTEGER)};
 VAR
  Index
 :INTEGER;

  BEGIN

    ClrScr;
    WINDOW(1,3,79,24);

    FOR Index := 1 TO IndexMax DO
     BEGIN
       WITH StockInfo[Index] DO
        BEGIN
          GOTOXY(3, 1 + Index);
          WRITE(StockName);
          GOTOXY(20, 1 + Index);
          WRITE(Quantity);
          GOTOXY(30, 1 + Index);
          WRITE(Cost:3:2);
          GOTOXY(40, 1 + Index);
          WRITE(BlueChip);
        END;
     END;
     Continue;
  END;

PROCEDURE Program_Repeat{(VAR UserSelection : CHAR)};
 BEGIN
   GOTOXY(5,25);
   WRITE('Are you sure you wish to exit this program (Y/N) => ');
   READLN(UserSelection);
 END;

PROCEDURE Alphabetize_Stock{(VAR StockInfo : StockArray
                                  IndexMax : INTEGER)};
 VAR
  TempRecord      {Temporary storage for record}
 :StockArray;
  Index
 :INTEGER;
  Switch
 :BOOLEAN;

  BEGIN

  REPEAT
    Switch := FALSE;

  FOR Index := 1 TO IndexMax DO
   BEGIN
    IF (StockInfo[Index].StockName > StockInfo[Index+1].StockName)
     AND (Index < IndexMax)
      THEN    {Switch the records}
       BEGIN
        TempRecord[Index] := StockInfo[Index];
        StockInfo[Index] := StockInfo[Index+1];
        StockInfo[Index+1] := TempRecord[Index];
        Switch := TRUE;
       END;
```

Program 8–46 *continued*

```
      END;
    UNTIL Switch = FALSE;

  END;

  PROCEDURE Continue;
   BEGIN
     GOTOXY(5, 24);
     WRITE('Press RETURN/ENTER to continue...');
     READLN;
   END;
PROCEDURE SaveRecords{(StockInfo : StockArray; IndexMax : INTEGER)};
 VAR
  StockFiles : FILE OF StockRecord;
  MoreStuff : StockRecord;
  Index
 :INTEGER;

   BEGIN

      ASSIGN(StockFiles, 'STOCKS.DTA');

      RESET(StockFiles);
      SEEK(StockFiles, 0);

      FOR Index := 1 TO IndexMax DO
       BEGIN
          MoreStuff := StockInfo[Index];
          WRITE(StockFiles, MoreStuff);
       END;
   END;

  PROCEDURE GetRecords{(VAR StockInfo : StockArray;
                        VAR IndexMax : INTEGER)};
   BEGIN
   END;

  FUNCTION FileExists{(FileName : Words)} : BOOLEAN;
   VAR
    AnyFile
  :FILE;

      BEGIN

        ASSIGN(AnyFile, FileName);

        {$I-}   {Turn I/O checking off.}
        RESET(AnyFile);
        CLOSE(AnyFile);
        {$I+}   {Turn I/O checking on.}

        FileExists := (IORESULT = 0);

      END;   {function FileExists}

  BEGIN
   Main_Programming_Sequence;
  END.
```

Questions
1   State the name of the disk file used by the program.
2   Explain how the program checks to see if a data file already exists.
3   What method is used for the program to know the number of records already contained in the disk file?
4   State when the program will save new information to the disk file. How did you determine this?
5   How many fields are placed in each record file? How did you determine this?
6   What are the maximum number of stocks that may be entered in this program? How did you determine this?
7   When parameters are passed between procedures is all of the information about the individual stock (its name, price, etc.) also passed? If so, explain how this is done.
8   State how individual records for each stock are saved to the disk file. How did you determine this?
9   Is there any provision in the program for allowing the program user to select the name of the disk data file?
10   When are the stocks in the program alphabetized? How did you determine this?

Problems

General Concepts

Section 8–1
1   What is a Pascal record?
2   What are the separate elements of a record called?
3   Describe the two different ways the elements of a record may be accessed.
4   Give an advantage of using Pascal records.

Section 8–2
5   State the difference between records and arrays.
6   Can an array contain records and can records contain arrays?
7   Explain the meaning of a variant record.
8   State what is meant by mutually exclusive fields.

Section 8–3
9   State the meaning of a Pascal file.
10   Explain the meaning of the term I/O.
11   What are some Pascal input files? Output files?
12   State the Pascal commands that actually output data to a file. Which commands actually input data from a file?

Section 8–4
13   What is meant by a text file?
14   Explain the significance of a disk file name.
15   What constitutes a legal DOS file name?
16   What are the four possible conditions you may encounter when working with a disk file?
17   Describe the use of each of the following Turbo Pascal procedures: a] **REWRITE (FileName);** B] **RESET(FileName);** C] **APPEND(FileName);**

Section 8–5

18 What is the difference between using **WRITE** and **WRITELN** for text files?
19 What is the difference between using **READ** and **READLN** for text files?
20 Describe the major characteristics of a text file.
21 Explain what is meant by a sequential file. Why is it important to make this distinction?
22 State the types of data that may be put into a text file.

Section 8–6

23 Describe the major characteristics of a Pascal random-access file.
24 State the purpose of the Turbo Pascal **SEEK(FileName, Pointer)** command.
25 Explain any differences between file commands for sequential and random-access files.
26 What is the built-in Turbo Pascal procedure for moving the file pointer to the end of the file?
27 State the built-in Turbo Pascal procedure that will determine the number of files in a random-access file.

Section 8–7

28 What is an instruction to the Pascal compiler called?
29 Give the general format for the Turbo Pascal compiler directive.
30 State the Turbo Pascal compiler directive that turns off automatic I/O checking. What directive turns it back on?
31 What is the built-in Turbo Pascal statement used to check and see if a disk file already exists when automatic I/O error checking is disabled? What are the values returned?

Program Analysis

32 Will program 8–47 compile? If not, why not?

Program 8–47

```
PROGRAM Analysis_1;
  VAR
    A_Record : RECORD
               Information : INTEGER;
  BEGIN
    A_Record.Information := 12;
    WRITELN(A_Record.Information);
  END.
```

33 Will program 8–48 compile and run successfully? If not, state what you would do to correct it.

Program 8–48

```
PROGRAM Analysis_2;
  VAR
    A_Record : RECORD
                  Information : INTEGER;
              END;
  BEGIN
    A_Record.Information := 12;
    WITH A_Record DO
    WRITELN(A_Record.Information);
  END.
```

34 Program 8–49 does not compile successfully. Can you spot why?

Program 8–49

```
PROGRAM Analysis_3;
  TYPE
    A_Record = RECORD
                  Information : INTEGER;
              END;
  BEGIN
    A_Record.Information := 12;
    WITH A_Record DO
    WRITELN(A_Record.Information);
  END.
```

35 Do you think program 8–50 will compile? If not, why not?

Program 8–50

```
PROGRAM Analysis_4;
  TYPE
    A_Record = RECORD
                  Information : INTEGER;
              END;
  VAR
    This_Record : A_Record;

  BEGIN
    A_Record.Information := 12;
    WITH A_Record DO
    WRITELN(A_Record.Information);
  END.
```

36   Will program 8–51 compile? If not, what is the problem?

Program 8–51

```
PROGRAM Analysis__5;
  TYPE
    A__Record = RECORD
                  Information : INTEGER;
                END;
  VAR
    This__Record : A__Record;

  BEGIN
    WITH This__Record DO
    Information := 12;
    WRITELN(Information);
  END.
```


37   Does program 8–52 compile? If not, why doesn't it?

Program 8–52

```
PROGRAM Analysis__6;
  TYPE
    A__Record = RECORD
                  Information : INTEGER;
                  More__Information : BOOLEAN;
                END;
  VAR
    This__Record : A__Record;
    Another__Record : ARRAY[1..4] OF A__Record;
    Index : INTEGER;

  BEGIN
    WITH This__Record DO
      BEGIN
       FOR Index := 1 TO 4 DO
        BEGIN
          More__Information[Index] := FALSE;
        END;
      END;
  END.
```


38   Will program 8–53 compile? If not, what would you do to correct it?

Program 8–53

```pascal
PROGRAM Analysis_7;
  TYPE
    More_Information = INTEGER;
    A_Record = RECORD
                  Information : INTEGER;
                  CASE Choice : More_Information OF
                    1 : (FirstVariable : BOOLEAN);
                    2 : (SecondVariable : CHAR);
               END;
  VAR
    This_Record : A_Record;
    Another_Record : ARRAY[1..4] OF A_Record;
    Index : INTEGER;
  BEGIN
      BEGIN
        FOR Index := 1 TO 4 DO
          BEGIN
            Another_Record[Index].Information := 12;
          END;
      END;
  END.
```

Program Design

For the following programs use the structure that is assigned to you by your instructor. Otherwise, use a structure that you prefer. Remember, it should be easy for anyone to read and understand, especially for you if it ever needs modification.

Electronics Technology

39 Develop a Pascal program using a record structure that will allow the program user to enter the values of five different resistors.

40 Create a Pascal program with a record structure that allows the program user to enter the value, voltage rating, and manufacturer's name of capacitors.

41 Design a Pascal program that uses an array of records. Each record will represent a parts package. Each parts package will have a different amount of resistors, capacitors, inductors, transistors, and diodes.

42 Modify the program in problem 39 so that the information may be saved to a disk file.

43 Add to the program of problem 40 so that the program user may save the information to a disk file.

44 Change the program for problem 41 so the program user may save the information to a disk file.

Business Applications

45 Develop a Pascal program that uses a record structure that will allow the program user to input the following data concerning a business client: name, address, phone, and credit rating (good or bad).

Computer Science

46 Create a Pascal program that demonstrates an arrayed record that contains an arrayed record element.

Drafting Technology

47   Design a Pascal program that uses a record structure to keep the following information on the status of a design project: ID Number, Project Name, Client Name, Due Date, Project Completed (Yes or No).

Agriculture Technology

48   Create a Pascal program using a record structure that keeps information on five different herds of cows, each herd with a maximum of 30 cows. The information entered by the program user is: sex, age, weight, location, and if female, milk production in quarts/day (if male, estimated market value).

Health Technology

49   Develop a Pascal program that uses a record structure for the following information about different patients of a private practice: name, address, date of birth, sex, dates of visitation, amount owed, and medical problem.

Manufacturing Technology

50   Create a Pascal program using a record structure that will keep track of 3 different production schedules, each containing the following information: production ID number, manufactured item, and customer name. The program will also have in the production schedule record a record on up to 5 employees containing the following information: employee ID Number, name, title, and hourly wages.

Business Applications

51   Modify the program in problem 45 so that the information entered by the program user may be saved to a disk file.

Computer Science

52   Develop a Pascal program that will allow the program user to use the computer as a simple word processor. The program must allow the user to save and retrieve text files, name each file, and delete old files.

Drafting Technology

53   Expand the program in problem 47 so that the information entered by the program user is saved into a disk file.

Agriculture Technology

54   Modify the program in problem 48 in order to save information entered by the program user in a disk file.

Health Technology

55   Expand the program in problem 49 so that the patient information entered by the program user is saved to a disk file.

Manufacturing Technology

56   Modify the program of problem 50 so the information entered by the program user will be saved to a disk file.

9 Color and Technical Graphics

Objectives

This chapter will give you a chance to learn:

1   How to produce text color.
2   How computer screens are used to represent graphic displays.
3   What you need in your computer system and how to find out what it has.
4   The concept of a pixel and how to use it in graphics.
5   How color is generated on the IBM and most compatibles.
6   How to draw lines on the graphic screen.
7   Some of the special Turbo Pascal built-in graphics.
8   Ways of creating bar graphs for technical analysis.
9   Methods of generating different graphic fonts.
10 Methods of graphing mathematical functions.
11 Techniques of styling your program to make it easier to read, debug, and modify.

Key Terms

Monochrome
Default
Color Constants
Text Screen
Graphics Screen
Graphic Mode
Graphic Driver
Autodetection
Graphics Adapter
Pixel

Mode
Color Palette
Area Fill
Fill-Style
Font
Bit Mapped
Stroked Font
Scaling
Coordinate Transformation

Outline

This chapter introduces the exciting world of color; Turbo Pascal can give text screens a new dimension of information with the use of color.

This chapter also introduces the power of the computer graph. One of the fastest growing areas in computers is in the world of computer graphics! Understanding computer graphics will help you present information as drawings and graphs.

Here you will be taught the secrets of adding color to graphics with the use of different text fonts. This chapter will require patience and practice—but the personal and professional rewards of this new skill will far exceed your time investment. Technical programmers who understand graphic programming skills are on the cutting edge of new technology—this chapter is your exciting first step.

9–1   Pascal Text and Color

Discussion

This section demonstrates how to employ color with your Pascal programs; color can add a new dimension of useful information to your programs. For example, think of a program to check the progress of an automated assembly line or to monitor a power plant. Green text could be used to indicate all systems are normal, yellow text that there is something requiring attention, red text could call for immediate attention, while blinking red text could signal possible danger.

What Your System Needs

In order to produce color using Turbo Pascal 4.0, your computer system must meet the following three requirements.

1. Have a color monitor.
2. Have a color-graphics card installed.
3. Be an IBM PC, AT, PS/2, or true compatible.

These color commands are built into the CRT UNIT contained in the TURBO.TPU program which must be present in your system.

The Color Monitor

There are basically two different types of monitors used with a personal computer. One is called a **monochrome** (meaning one color); the other is called a color monitor which is capable of producing a rich variety of different colors all at the same time.

System monitors use a coating of phosphor on the inside of the display screen. Electrical currents controlled by the computer strike this phosphor and cause it to glow. This action allows you to see things on your monitor screen. This glow emits light, the color of which depends upon the type of phosphor used to coat the inside of the display screen.

In a monochrome monitor, only one type of phosphor is used. Depending on the type of phosphor, the screen of a monochrome monitor may be green, amber, or any other single color. It is not possible to change the single color of a monochrome monitor.

A color monitor has three different color phosphors on the inside of the glass face in the form of thousands of tiny triads. These triads contain red, green, and blue light-emitting phosphors. These are the primary colors used for additive color mixing. With a color monitor of this type, many different colors may be displayed at the same time as text.

Display Capabilities

If your system meets requirements for producing color, it is also capable of producing two different sizes of text. One size is the 80-column by 25-row standard that you are used to. This is the **default text** size, the size automatically used by your system when it is turned on. The second size produces larger text; it is 40 columns by 25 rows. These variations and the color capabilities for text are illustrated in figure 9–1.

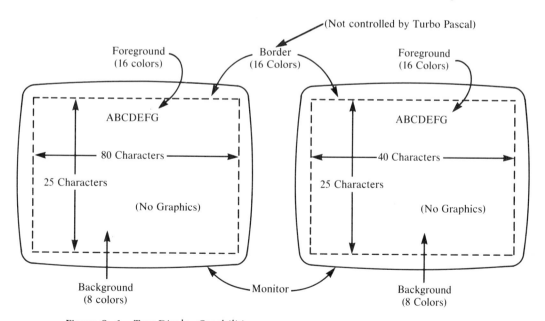

Figure 9–1   Text Display Capabilities

The Mode Procedure

Turbo Pascal has a built-in procedure for selecting either 40- or 80-column text. Since your system normally comes up in 80-column, the following program will cause it to switch to 40-column (and will only work if you have a color graphics card installed in your computer system).

Program 9–1

```
PROGRAM Wide_Text;
 USES CRT;
  BEGIN

   TEXTMODE(0);    {Changes to 40-column wide text.}
    WRITELN('This is 40 columns wide.');
    WRITELN('It only works on systems with');
    WRITELN('a color graphics card installed.');
  END.
```

The built-in Turbo procedure is

TEXTMODE(MODE);

where **MODE** = an INTEGER

Table 9–1 gives the results of particular built-in INTEGER values.

Note from table 9–1 that you cannot get 40-column text without a color graphics card installed in your system. Program 9–2 gives an example of the use of all possible variations of the built-in **TEXTMODE** procedure.

Whenever the **TEXTMODE** procedure is used, the entire screen is cleared. Thus, in program 9–2, the screen is cleared every time the mode is changed by **TEXTMODE**. Note that **TEXTMODE(7)** has been placed in the form of a comment; if you do have a color graphics card installed and try to evoke this command, your system may "hang up."

Table 9–1 Results of Integer Values for TEXTMODE

Mode Value	Results
0	Produces 40 × 25 monochrome text. An installed color graphics card is required.
1	Produces 40 × 25 color text. An installed color graphics card is required.
2	Produces 80 × 25 monochrome text. An installed color graphics card is required.
3	Produces 80 × 25 color text. An installed color graphics card is required.
7	Produces 80 × 25 monochrome text. No color graphics card installed in the system.

Program 9–2

```
PROGRAM Text_Type1;
 USES CRT;
  BEGIN

     WRITELN('This text is in the default 80 columns wide');
     WRITELN('by 25 rows down.');
     READLN;

    TEXTMODE(0);    {Changes to 40 X 25 Monochrome.}
     WRITELN('This text is now 40 columns wide');
     WRITELN('but still 25 rows down.');
     READLN;

    TEXTMODE(1);    {Changes to 40 X 25 Color.}
     WRITELN('Still a 40 column text, but on a ');
     WRITELN('color monitor display, it may reduce');
     WRITELN('color streaking.');
     READLN;

    TEXTMODE(2);    {Changes to 80 × 25 Monochrome.}
     WRITELN('This is the 80 column standard you are used to.');
     WRITELN('It is 25 rows down.');
     READLN;

    TEXTMODE(3);    {Changes to 80 × 25 Color.}
     WRITELN('Still 80 columns but on a color monitor display, it');
     WRITELN('may reduce color streaking.');
     READLN;

   { TEXTMODE(7);     80 X 25 Monochrome—no graphics card required.
   WRITELN('Still 80 columns, but for those systems that do not have');
   WRITELN('a color graphics card installed.');
   READLN;}

 END.
```

For convenience, Turbo Pascal has defined constants that represent the numerical values of the **TEXTMODE** procedure. These are defined as follows.

Mode Number	Mode Identifier
0	BW40
1	C40
2	BW80
3	CO80
7	MONO

Thus, program 9–2 could have been written as shown in program 9–3 (p. 546).

Text Color

The built-in Turbo procedure called **TEXTCOLOR** will change the color of your text as shown in program 9–4 on page 546.

Program 9–3

```pascal
PROGRAM Text_Type1;
 USES CRT;
  BEGIN

    WRITELN('This text is in the default 80 columns wide');
    WRITELN('by 25 rows down.');
    READLN;

   TEXTMODE(BW40);   {Changes to 40 x 25 Monochrome.}
    WRITELN('This text is now 40 columns wide');
    WRITELN('but still 25 rows down.');
    READLN;

   TEXTMODE(C40);   {Changes to 40 x 25 Color.}
    WRITELN('Still a 40 column text, but on a ');
    WRITELN('color monitor display, it may reduce');
    WRITELN('color streaking.');
    READLN;

   TEXTMODE(BW80);   {Changes to 80 x 25 Monochrome.}
    WRITELN('This is the 80 column standard you are used to.');
    WRITELN('It is 25 rows down.');
    READLN;

   TEXTMODE(C80);    {Changes to 80 x 25 Color.}
    WRITELN('Still 80 columns but on a color monitor display, it');
    WRITELN('may reduce color streaking.');
    READLN;

  { TEXTMODE(MONO);   80 x 25 Monochrome—no graphics card required.
   WRITELN('Still 80 columns, but for those systems that do not have');
   WRITELN('a color graphics card installed.');
   READLN;}

END.
```

Program 9–4

```pascal
PROGRAM Text_Colors;
 USES CRT;
  VAR
   ColorNumber
  :BYTE;

   ColorName
  :STRING[79];

  BEGIN
    FOR ColorNumber := 0 TO 15 DO

     BEGIN

       CASE ColorNumber OF
          0 : ColorName := 'black';
          1 : ColorName := 'blue';
```

Program 9–4   *continued*

```
        2 : ColorName := 'green';
        3 : ColorName := 'cyan';
        4 : ColorName := 'red';
        5 : ColorName := 'magenta';
        6 : ColorName := 'brown';
        7 : ColorName := 'light gray';
        8 : ColorName := 'dark gray';
        9 : ColorName := 'light blue';
       10 : ColorName := 'light green';
       11 : ColorName := 'light cyan';
       12 : ColorName := 'light red';
       13 : ColorName := 'light magenta';
       14 : ColorName := 'yellow';
       15 : ColorName := 'white';
      END; {of case}

    TEXTCOLOR(ColorNumber); {This changes the color of the text.}
      WRITELN('This text is in ',ColorName);
    END;
END.
```

In program 9–4, you will not see the first **WRITELN** statement, `This text is in black` because the background of your monitor is black. However, all the other colors will be displayed against this background.

Again, for convenience, Turbo Pascal has predefined constants for the **TEXTCOLOR** procedure, as shown in table 9–2.

Table 9–2   Turbo Pascal Built-In Color Constants

Color Number	Color Constant
0	BLACK
1	BLUE
2	GREEN
3	CYAN
4	RED
5	MAGENTA
6	BROWN
7	LIGHTGRAY
8	DARKGRAY
9	LIGHTBLUE
10	LIGHTGREEN
11	LIGHTCYAN
12	LIGHTRED
13	LIGHTMAGENTA
14	YELLOW
15	WHITE
+128	BLINK

Note that program 9–4 could have been written as shown in program 9–5.

Program 9–5

```
PROGRAM Text_Colors;
 USES CRT;
  VAR
   ColorNumber
  :BYTE;

   ColorName
  :STRING[79];

 BEGIN
   FOR ColorNumber := BLACK TO WHITE DO

    BEGIN

      CASE ColorNumber OF
            BLACK : ColorName := 'black';
             BLUE : ColorName := 'blue';
            GREEN : ColorName := 'green';
             CYAN : ColorName := 'cyan';
              RED : ColorName := 'red';
          MAGENTA : ColorName := 'magenta';
            BROWN : ColorName := 'brown';
        LIGHTGRAY : ColorName := 'light gray';
         DARKGRAY : ColorName := 'dark gray';
        LIGHTBLUE : ColorName := 'light blue';
       LIGHTGREEN : ColorName := 'light green';
        LIGHTCYAN : ColorName := 'light cyan';
         LIGHTRED : ColorName := 'light red';
     LIGHTMAGENTA : ColorName := 'light magenta';
           YELLOW : ColorName := 'yellow';
            WHITE : ColorName := 'white';
      END;   {of case}

      TEXTCOLOR(ColorNumber); {This changes the color of the text.}
       WRITELN('This text is in ',ColorName);
     END;
 END.
```

Note also the addition of **BLINK** in table 9–2. You can cause any text to blink by adding the value of 128 to the color value. Program 9–6 illustrates.

Program 9–6

```
PROGRAM Blinking_Colors;
 USES CRT;
  BEGIN
```

Program 9–6 *continued*

```
    TEXTCOLOR(RED);
      WRITELN('This text is now in red.');

    TEXTCOLOR(4);
      WRITELN('This text is also in red.');

    TEXTCOLOR(RED + BLINK);
      WRITELN('This is blinking red text.');

    TEXTCOLOR(4 + 128);
      WRITELN('This is also blinking red text.');

    TEXTCOLOR(RED);
      WRITELN('This stopped the red text from blinking.');

    TEXTCOLOR(LIGHTMAGENTA);
      WRITELN('This is light magenta and not blinking.');

    TEXTCOLOR(LIGHTGREEN + 128);
      WRITELN('This is in light green and blinking.');
  END.
```

Changing the Background

To change the background color in Turbo Pascal, use the built-in Turbo procedure **TEXTBACKGROUND**, as illustrated in program 9–7.

Program 9–7

```
PROGRAM Background_Colors;
 USES CRT;
  VAR
   ColorNumber
  :BYTE;

   ColorName
  :STRING[79];

  BEGIN
    FOR ColorNumber := 0 TO 7 DO

     BEGIN
```

Program 9–7   *continued*

```
        CASE ColorNumber OF
          0 : ColorName := 'black';
          1 : ColorName := 'blue';
          2 : ColorName := 'green';
          3 : ColorName := 'cyan';
          4 : ColorName := 'red';
          5 : ColorName := 'magenta';
          6 : ColorName := 'brown';
          7 : ColorName := 'light gray';
        END;  {of case}

      TEXTBACKGROUND(ColorNumber);
        CLRSCR;
        IF ColorNumber = 7 THEN TEXTCOLOR(BLACK);
        GOTOXY(20,12);
        WRITELN('This background is in ',ColorName);
        READLN;
      END;
  END.
```

Note from program 9–7 that there are only eight background colors possible. The **TEXTBACKGROUND** procedure is

TEXTBACKGROUND(COLOR);

where **COLOR** = type BYTE

Notice that the **TEXTCOLOR** in program 9–7 was changed when the background was light gray. Also observe that **CLRSCR** caused the background to be cleared to that color.

As with the **TEXTCOLOR** procedure, Turbo Pascal has identifier constants that represent the names of the colors. Thus **TEXTBACKGROUND(BLACK)** has the same effect as **TEXTBACKGROUND(0).**

Conclusion

This section presented important aspects of using color with text. You saw how to change the size of the text as well as to invoke 16 different colors of text and 8 different background colors. Check your understanding of this section by trying the following section review.

9–1 Section Review

1   State what is needed to obtain text color.
2   What is needed to change the text size from 80-column to 40-column?
3   How many different colors can text have? Can this happen all on the same screen?
4   How many different background colors are available? What command is used to bring the screen to the desired color?
5   What choices are available to indicate the desired color? Give an example.

9–2   Starting Turbo Pascal Graphics

Discussion

This section introduces you to the fascinating and challenging world of Turbo Pascal graphics. In this section you will see what your system, software, and programming needs are to get started. You will also be introduced to the most fundamental graphics command.

Basic Idea

If you have an installed color-graphics adapter in your computer, your monitor is capable of displaying two different kinds of screens. One screen is the **text screen** (sometimes referred to as **text mode**). The text screen is the one that you have been using up to this point. It does nothing more than display text—essentially, characters of a preassigned shape such as the letters of the alphabet, numbers, or punctuation marks. The other screen is called the **graphics screen** (sometimes referred to as the **graphics mode**). The graphics screen allows you to define your own shapes. Thus, when in the graphics mode, you can create lines, graphs, charts, diagrams, pictures, and animation—essentially, almost anything you can put on paper. Figure 9–2 illustrates the difference between the two modes of operation.

You can only be in one mode at a time—either text mode or graphics mode. Your system comes up in text mode. In order to get into graphics mode, you must use special built-in Turbo procedures to get you there.

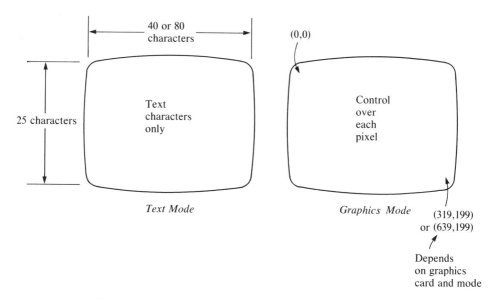

Figure 9–2   Difference Between Text Mode and Graphics Mode

What You Need to Know

Before you can get into graphics mode with Turbo Pascal, you must know what kind of color-graphics adapter your system is using; the Pascal program needs this information in order to use the correct built-in code.

Turbo Pascal 4.0 has programs made especially for different graphics adapters. You need to know which kind of graphics adapter your system has so Turbo will know which of the programs (called **graphic drivers**) it is to use.

Fortunately, Turbo Pascal has a built-in procedure that will allow you to find out which kind of installed graphics adapter is in your system, as demonstrated by program 9–8.

Program 9–8

```
PROGRAM What__Kind;
   USES GRAPH;

 PROCEDURE Find__Adapter;
  VAR
   GraphDriver,
   GraphMode
 :INTEGER;

   System
 :STRING[79];

   BEGIN

     GraphDriver := 0;   {Requests autodetection.}

     DETECTGRAPH(GraphDriver, GraphMode);

     CASE GraphDriver OF
      1 :  System := 'CGA';
      2 :  System := 'MCGA';
      3 :  System := 'EGA';
      4 :  System := 'EGA64';
      5 :  System := 'EGAMono';
      6 :  System := 'RESERVED';
      7 :  System := 'HercMono';
      8 :  System := 'ATT400';
      9 :  System := 'VGA';
     10 :  System := 'PC3270';
     END;   {of case}

     IF System = -2 THEN WRITELN('No hardware detected.')
     ELSE
     WRITELN('Your installed graphics system is ',System);
```

Program 9–8   *continued*

```
END;  {of procedure Find_Adapter}

BEGIN
  Find_Adapter;
END.
```

The Turbo built-in procedure used in program 9–8 is:

DETECTGRAPH(GraphDriver, GraphMode);

where **GraphDriver** = a variable INTEGER that returns a number that represents the type of installed graphics hardware

GraphMode = a variable INTEGER that returns the mode of operation for the detected installed graphics hardware

Note that the two parameters (**GraphDriver, GraphMode**) are both value parameters. This means that they must be variables, because both of them will be passing values back to the calling procedure. In program 9–8, the parameter **GraphDriver** was set to 0. This is required for Turbo to do an autodetection of the type of graphics adapter installed in your system.

Note that program 9–8 also required a **USES GRAPH**. This means that the program **GRAPH.TPU** must be present on your active disk.

Selecting Your Graphics Driver

Once you know which installed graphics adapter you have in your system, you must copy the appropriate graphics driver program from the original Turbo Pascal 4.0 disks to your working disk. Turbo Pascal 4.0 system disks have the following graphics drivers.

ATT.BGI = Graphics driver for AT&T 6300 graphics.
CGA.BGI = Graphics driver for CGA and MCGA graphics.
EGAVGA.BGI = Graphics driver for EGA and VGA.
HERC.BGI = Graphics driver for Hercules monographics.
PC3270.BGI = Graphics driver for 3270PC graphics.

Remember that you will also need **GRAPH.TPU** on your working disk.

The Graphics Screen

Your graphics screen is divided into **pixels**. You can think of a pixel as the smallest point possible that can be displayed on your graphics screen. Figure 9–3 (p. 554) presents the concept of pixels.

How many pixels your graphic screen has depends upon the graphics adapter installed in your system. For example, CGA, MCGAC, and ATT400 graphics adapters can have 320 pixels horizontally and 200 vertically. Each pixel is identified by a

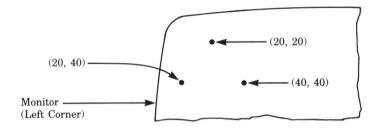

Figure 9–3   Concept of Pixels

coordinate system (X,Y) where X is the horizontal value of the pixel and Y is the vertical value of the pixel. The graphics screen for this hardware is shown in figure 9–4.

Note that the upper left pixel is identified as (0,0), the top right as (319,0), the bottom left as (0,199), and the bottom right as (319,199). Each pixel on the graphics screen has a unique location that can be identified by the (X,Y) system.

Lighting up a Pixel

The simplest graphics command is to light up a pixel on the graphics screen. To do this, there are three things your Pascal program must do: 1. Get your system into the

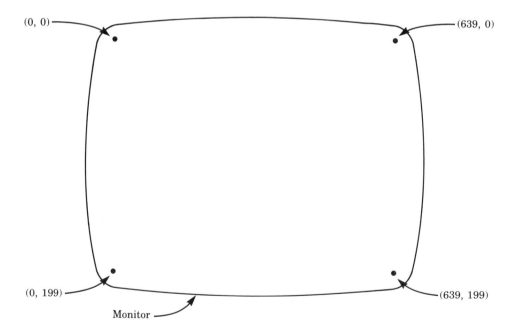

Figure 9–4   Graphics Screen for Specific Hardware

graphics mode with the correct graphics driver; 2. Activate the selected pixel; 3. Leave the graphics mode.

As you may suspect, Turbo Pascal 4.0 has three built-in procedures for doing these three operations. They are

1. To get into the graphics mode

INITGRAPH(GraphDriver, GraphMode, DriverPath);

where `GraphDriver` = a variable INTEGER that designates the type of graphics driver to be used

`GraphMode` = a variable INTEGER that designates the mode of operation for the graphics driver

`DriverPath` = a STRING variable that indicates the directory path where the graph driver can be found (if **' '** are used, the active disk is used)

2. To light up a pixel on the graphics screen

PUTPIXEL(X,Y, Pixel);

where `X` = an INTEGER representing the horizontal value of the pixel

`Y` = an INTEGER representing the vertical value of the pixel

`Pixel` = type WORD, a number indicating the color of the pixel

3. To leave the graphics mode

CLOSEGRAPH;

This procedure shuts down the graphics system and restores the original screen mode.

Program 9–9 is for a system that has a CGA color graphics adapter installed. Note that since `GraphDriver` and `GraphMode` (used with **INITGRAPH**) are both value parameters, they must be variables.

Program 9–9

```
PROGRAM Get_Into_Graphics;
  USES GRAPH;

 PROCEDURE Initialize;

  VAR

   GraphDriver,
   GraphMode
 :INTEGER;

  BEGIN

  GraphDriver := CGA; {For a system with an IBM Color Graphics Adapter.}
  GraphMode := 2;     {Select a 320 × 200 screen with green, red, brown.}

   INITGRAPH(GraphDriver, GraphMode, ''); {Puts you into graphics.}
```

Program 9–9   *continued*

```
     PUTPIXEL(160,100,0);   {Puts a black dot near center of screen.}
     PUTPIXEL(162,100,1);   {Puts a green dot near center of screen.}
     PUTPIXEL(164,100,2);   {Puts a red dot near center of screen.}
     PUTPIXEL(166,100,3);   {Puts a brown dot near center of screen.}

  READLN;

  CLOSEGRAPH; {Restores the original screen mode and frees the
            graphics drivers from memory.}

 END;  {of procedure initialize}

BEGIN
 Initialize;
END.
```

Note that a **READLN;** statement is used so that you may view the graphic. After you press the RETURN/ENTER key, the **CLOSEGRAPH** procedure is initialized.

Program 9–9 used `GraphMode := 2;`. The reason for this will be explained shortly.

Using Color

The kind and amount of color available in graphics mode is much less than the rich color combinations you can get in text mode. As a matter of fact, the kind and amount of color (if any) depends upon the installed graphics adapter in your system. Some graphics adapters have different modes of operation. For example, the CGA graphics adapter used in program 9–9 has five different graphics modes numbered 0 through 4. Table 9–3 lists the modes available for different graphics hardware.

As you can see from table 9–3, the CGA driver that will be used in the following example programs can display a maximum of only three colors at a time, from 4 different modes.

It should be pointed out before using the built-in Turbo Pascal 4.0 procedure **INITGRAPH(**GraphDriver, GraphMode, DrivePath**);** that Turbo has defined graphic driver constants corresponding to the number of the driver. These are listed for you.

```
DETECT = 0
CGA = 1
MCGA = 2
EGA = 3
EGA64 = 4
EGAMONO = 5
RESERVED = 6
HERCMONO = 7
ATT400 = 8
```

```
VGA = 9
PC3270 = 10
```

Thus, **INITGRAPH(CGA,0, '');** means initialize graphics mode using the **CGA.BGI** graphics driver in mode 0 (320 × 200 screen, colors light green, light red and yellow), with **CGA.BGI** driver program on the active drive.

INITGRAPH(1, 0, '');

means exactly the same thing, and so does

INITGRAPH(CGA, CGAC0, '');

Table 9–3   Graphics Modes for Given Installed Hardware

Hardware (Driver) #	Mode # Const	Screen Size (Pixels)	Colors
CGA 1	0 CGA0	320 × 200	LightGreen, LightRed, Yellow
	1 CGA1	320 × 200	LightCyan, LightMagenta, White
	2 CGA2	320 × 200	Green, Red, Brown
	3 CGA3	320 × 200	Cyan, Magenta, LightGray
	4 CGAHI	640 × 200	(One color on black)
MCGA 2	0 MCGAC0	320 × 200	LightGreen, LightRed, Yellow
	1 MCGAC1	320 × 200	LightCyan, LightMagenta, White
	2 MCGAC2	320 × 200	Green, Red, Brown
	3 MCGAC3	320 × 200	Cyan, Magenta, LightGray
	4 MCGAMED	640 × 200	(One color on black)
	5 MCGAHI	640 × 480	(One color on black)
EGA 3	0 EGALO	640 × 200	16 colors
	1 EGAHI	640 × 350	16 colors
EGA64 4	0 EGA64LO	640 × 200	16 colors
	1 EGA64HI	640 × 350	4 colors
	3 EGAMONO	640 × 350	(One color on black)
HERCMONO 7	0 HERCMONOHI	720 × 348	(White on black)
ATT400 8	0 ATT400C0	320 × 200	LightGreen, LightRed, Yellow
	1 ATT400C1	320 × 200	LightCyan, LightMagenta, White
	2 ATT400C2	320 × 200	Green, Red, Brown
	3 ATT400C3	320 × 200	Cyan, Magenta, LightGray
	4 ATT400MED	640 × 200	(One color on black)
	5 ATT400HI	640 × 400	(One color on black)
VGA 9	0 VGALO	640 × 200	16 colors
	1 VGAMED	640 × 350	16 colors
	2 VGAHI	640 × 480	16 colors
PC3270 10	0 PC3270HI	720 × 350	(White on black)

Making an Autodetect Procedure

Program 9–10a illustrates a procedure that will automatically detect the type of graphics hardware in your system and then automatically select the proper graphics driver. For this to work on your system, you must have all of the graphic drivers on the active disk, or indicate with the **INITGRAPH** procedure where they are located.

Program 9–10a

```
PROGRAM Use_This;
  USES GRAPH;

  PROCEDURE Auto_Initialization;
   {Note:  You must have Turbo GRAPH.TPU and the graphics
           driver for your system on the active disk.}
   VAR
    GraphDriver,        {Gets graphics driver number.}
    GraphMode           {Gets the mode for the driver.}
   :INTEGER;

   BEGIN
      GraphDriver := DETECT;   {Initiates autodetect.}
      INITGRAPH(GraphDriver, GraphMode, '');
   END;   {of procedure Auto_Initialization}

BEGIN
  Auto_Initialization;
END.
```

The procedure **Auto_Initialization;** will be used with all of the graphics programs in the remainder of this chapter.

Conclusion

This section presented what you need to get into graphics mode using Turbo Pascal 4.0 and IBM systems or true compatibles. Once you get past this point, you will be able to enjoy the world of technical graphics available to you. Test your understanding of this section by trying the following section review.

9–2 Section Review

1   Name the two different kinds of screens that can be available to you on your monitor.
2   State the difference between the graphics mode and text mode.
3   Before you can get your system into graphics operation with Turbo Pascal 4.0, what is the one piece of hardware and two kinds of software you will need?
4   State how many different colors are available with color graphics.
5   Define the term pixel.

9–3   Knowing Your Graphics System

Discussion

This section will help you familiarize yourself with your graphics system. Here you will learn about the different modes of operation for your graphics system and how to use them to your advantage in the creation of technical graphics. This section will also introduce some of the most fundamental graphics concepts and commands.

Your System's Screen Size and Colors

It's important that you know about your system's screen size and colors. If you don't, you may mistakenly try drawing lines on your graphics screen in black and never see anything against the black background. Or you may try drawing images that do not fit on the size of your graphics screen. The size of the graphics screen and colors (if any) available depend largely on the kind of graphics hardware installed in your system. Table 9–3 listed the kinds of hardware and their various modes of operation. For example, you can see in table 9–4 that a CGA (Color Graphics Adapter) has five modes of operation.

Table 9–4 is an excerpt from table 9–3. It shows that for the first 4 modes of operation (mode 0 through 3), the screen size is 320 pixels horizontally and 200 pixels vertically. All modes have only three colors available at any one time. These different color modes are referred to as the **color palette** or simply **palette**. Thus, the palette for mode 0 of the CGA hardware is light green, light red, and yellow, while the palette for mode 1 is light cyan, light magenta, and white. You cannot have all palettes at the same time (you must be in one of the 4 color modes). Black is also available in both modes.

Now note that the fifth mode (mode 4) has more pixels horizontally (640) and the same amount vertically (200), but only one color (the background will be black and visible graphics will be in the chosen color). In this mode you can get more detail (for a CGA adapter) but lose some ability to produce color.

Table 9–4   Graphics Modes for Given Installed Hardware

Hardware (Driver) #	Mode # Const	Screen Size (Pixels)	Colors
CGA 1	0 CGA0	320 × 200	LightGreen, LightRed, Yellow
	1 CGA1	320 × 200	LightCyan, LightMagenta, White
	2 CGA2	320 × 200	Green, Red, Brown
	3 CGA3	320 × 200	Cyan, Magenta, LightGray
	4 CGAHI	640 × 200	(One color on black)

From all this information, you can see how important it is to be able to

1. Determine the modes of your system.
2. Change the modes of your system.

First, let us see how to determine the modes of your system.

Your System's Modes

Program 9–10 will determine what mode your graphics system is in as well as give you other useful graphics information.

Program 9–10

```
PROGRAM Get_Information;
  USES GRAPH;

PROCEDURE Auto_Initialization;
  VAR
    GraphDriver,
    GraphMode
  :INTEGER;
    BEGIN
        GraphDriver := DETECT;
        INITGRAPH(GraphDriver, GraphMode, '');
    END;


PROCEDURE Mode_Range;
  VAR
   LoMode,
   HiMode
  :INTEGER;

  BEGIN
      GETMODERANGE(CGA, LoMode, HiMode);
      WRITELN('Lowest mode is ',LoMode);
      WRITELN('Highest mode is ',HiMode);
  END;
PROCEDURE Current_Mode;
  VAR
   ModeNumber
  :INTEGER;

    BEGIN
      ModeNumber := GETGRAPHMODE;
      WRITELN('You are currently in mode ',ModeNumber);
    END;

PROCEDURE Maximum_Colors;
  VAR
   MaxColorNumber
   :WORD;

   BEGIN
     MaxColorNumber := GETMAXCOLOR;
  WRITELN('The maximum color number is ',MaxColorNumber,' in this mode.');
   END;
```

Program 9–10 *continued*

```
  PROCEDURE Graphic_Screen_Size;
    VAR
     XMax,
     YMax
    :INTEGER;

       BEGIN
         XMax := GETMAXX + 1;
         YMax := GETMAXY + 1;
         WRITELN('Your current screen size is ',XMax,' X ',YMax,' pixels.');
       END;
  BEGIN
   Auto_Initialization;
   Mode_Range;
   Current_Mode;
   Maximum_Colors;
   Graphic_Screen_Size;
  END.
```

The results you get with program 9–10 will depend upon the installed graphics hardware in your system. If your system has an IBM CGA, the resulting output will be

```
Lowest mode is 0
Highest mode is 4
You are currently in mode 4
The maximum color number is 1 in this mode.
Your current screen size is 640 × 200 pixels.
```

Program 9–10 lets you know exactly where you are. An explanation of each of the procedures used in program 9–10 now follows.

Auto Initialization

This procedure (explained in the last section) automatically detects the type of graphics hardware and gets you into graphics mode. It will be used with every graphics program in this text.

Mode Range

This procedure gives you the range of the modes available with your installed graphics hardware. The built-in Turbo Pascal 4.0 procedure used here is

GETMODERANGE(GraphDriver, LoMode, HiMode);

where **GraphDriver** = an INTEGER that indicates the graph driver used for your system (such as 1 or CGA)

LoMode = a variable parameter of type INTEGER that will return the numerical value of the lowest mode available for the installed graphics adapter

HiMode = a variable parameter of type INTEGER that will return the numerical value of the largest mode available for the installed graphics adapter

Current Mode

This procedure identifies the active mode which your system currently is in. The built-in Turbo Pascal 4.0 function that returns the numerical value of the current mode is

GETGRAPHMODE;

where the value returned is of type INTEGER and is the numerical value of the active graph mode of your system. (Refer to table 9–3 for the significance of each graph mode.)

Maximum Colors

This procedure identifies the maximum color number available in the current mode. The built-in Turbo Pascal 4.0 function that returns this value is

GETMAXCOLOR

where the value returned is of type WORD and is the numerical value of the largest value color for the current mode. (In those modes where only white graphs will appear on a black background, 0 represents black and 1 represents white. In any mode, black is always 0, while the other numbers indicate colors that depend upon the installed graphics hardware and active palette.)

Graphic Screen Size

This function gives the maximum size of the current graphics screen. There are two built-in Turbo Pascal functions that allow you to get these values.

GETMAXX;
GETMAXY;

where the value returned is of type INTEGER and is the numerical value *X* of the location of the rightmost pixel or the numerical value *Y* of the bottommost pixel.

Creating Lines

In Turbo Pascal there is a built-in procedure that allows you to create a line on the graphics screen. To use this procedure your system must be in graphics mode.

LINE(X1, Y1, X2, Y2);

where **X1, Y1** = INTEGER values of the starting point of the line
   X2, Y2 = INTEGER values of the ending point of the line
   Program 9–11 causes a horizontal line to be created on the graphics screen.
   How far the line extends across the screen will depend upon your installed graphics hardware. If you want to insure that the line goes all the way across the screen, modify the **Make_Line** procedure as follows.

```
PROCEDURE Make_Line;
  BEGIN
    LINE(0,50,GETMAXX,50);
  END;
```

Program 9–11

```
PROGRAM Draw_a_Line;
  USES GRAPH;
  PROCEDURE Auto_Initialization;
   VAR
    GraphDriver,         {Gets graphics driver number.}
    GraphMode            {Gets the mode for the driver.}
   :INTEGER;

    BEGIN
       GraphDriver := DETECT;  {Initiates autodetect.}
       INITGRAPH(GraphDriver, GraphMode, '');
    END;  {of procedure Auto_Initialization}

  PROCEDURE Make_Line;

     BEGIN

        LINE(0,50,299,50);

     END;

 BEGIN
  Auto_Initialization;
  Make_Line;
 END.
```

Now the line will always extend across the whole graphics screen regardless of the current mode. Program 9–12 makes a box around the graphics screen, using the built-in Turbo Pascal functions **GETMAXX** and **GETMAXY**.

Program 9–12

```
PROGRAM Draw_a_Box;
  USES GRAPH;

  PROCEDURE Auto_Initialization;
   VAR
    GraphDriver,       {Gets graphics driver number.}
    GraphMode          {Gets the mode for the driver.}
   :INTEGER;

    BEGIN
       GraphDriver := DETECT;  {Initiates autodetect.}
       INITGRAPH(GraphDriver, GraphMode, '');
    END;  {of procedure Auto_Initialization}

  PROCEDURE Make_Line;
   VAR
    MaximumX,
    MaximumY
   :INTEGER;
```

Program 9–12   *continued*

```
      BEGIN
         MaximumX := GETMAXX;
         MaximumY := GETMAXY;
         LINE(0,0,MaximumX,0);    {Horizontal line across top of screen.}
         LINE(0,MaximumY,MaximumX,MaximumY);   {Horizontal line across bottom
                                                of screen.}
         LINE(0,0,0,MaximumY);   {Vertical line along left side of screen.}
         LINE(MaximumX,0,MaximumX, MaximumY);   {Vertical line along right
                                                 side of screen.}

      END;

BEGIN
 Auto_Initialization;
 Make_Line;
END.
```

Conclusion

This section presented some very important built-in Turbo Pascal procedures and functions for understanding your graphics system. You were also introduced to some fundamental graphics commands. In the next section, you will learn some of the powerful graphics commands of Turbo Pascal that will allow you to create different shapes with a variety of colors and line styles. For now, check your understanding of this section by trying the following section review.

9–3 Section Review

1   State why it's important to know about your graphics system's modes and colors.
2   Explain what is meant by a color palette.
3   How many different color palettes are available for the IBM CGA color graphics card? How many colors per palette?
4   Are there any modes in which color is not available? Explain how you would find this information.
5   How can you find out how many modes your system has?

9–4   Built-In Shapes

Discussion

In this section you will learn about Turbo Pascal's built-in shapes. Here you will discover how to create different line styles, rectangles, and circles with ease. You will also see how to fill in areas of the graphic screen with different patterns and colors. This is an exciting section. Using the information learned here, you can begin to construct programs that will produce powerful technical graphics.

Changing Graphics Modes

Recall from table 9–3 that the graphics hardware in your system may offer several different modes of operation. In the last section, you were introduced to built-in Turbo Pascal 4.0 procedures that allow you to find how many different modes your particular installed graphics system has.

For the example programs in this section, it will be assumed that you have IBM CGA graphics hardware in your system. The commands are the same no matter what acceptable system you have; if your system is different, only the modes and pixel sizes may be different.

Program 9–13 demonstrates changing graphics modes.

Program 9–13

```
PROGRAM Mode_Change;
  USES GRAPH;

  PROCEDURE Auto_Initialization;
  VAR
    GraphDriver,
    GraphMode
  :INTEGER;

    BEGIN
      GraphDriver := DETECT;  {Initiates autodetect.}
      INITGRAPH(GraphDriver, GraphMode, '');
    END;  {of procedure Auto_Initialization}

  PROCEDURE Change_Modes;

    {IMPORTANT!  The following comments apply only to an IBM CGA color
            graphics adapter.  The results with your system's installed
            graphics hardware may be different.  Refer to table 9-3,
            or your Turbo Pascal 4.0 Owner's Manual.}

    BEGIN
      SETGRAPHMODE(CGAC0);  {320 X 200 pixel screen, palette 0}
        LINE(0,0,GETMAXX,GETMAXY);  {Produces a yellow line}
        READLN;

      SETGRAPHMODE(CGAC1);  {320 X 200 pixel screen, palette 1}
        LINE(0,0,GETMAXX,GETMAXY);  {Produces a white line}
        READLN;

      SETGRAPHMODE(CGAC2);  {320 X 200 pixel screen, palette 2}
        LINE(0,0,GETMAXX,GETMAXY);  {Produces a brown line.}
        READLN;

      SETGRAPHMODE(CGAC3);  {320 X 200 pixel screen, palette 3}
        LINE(0,0,GETMAXX,GETMAXY);  {Produces a light gray line.}
        READLN;

      SETGRAPHMODE(CGAHI);  {640 X 200 pixel screen, white on black}
        LINE(0,0,GETMAXX,GETMAXY);  {Produces a white line.}
```

Program 9–13 *continued*

```
    END;
BEGIN
 Auto_Initialization;
 Change_Modes;
END.
```

The comments in program 9–13 apply to a system with an IBM CGA color graphics adapter. If you refer to table 9–3, you see that this adapter has 5 modes of operation. Each of these modes may be referred to by a number (0 through 4) or by built-in predefined Turbo graphic constants (CGAC0 through CGAHI). The built-in Turbo Pascal 4.0 procedure that sets the graphics mode is

SETGRAPHMODE(Mode**)**

where **Mode** = an INTEGER that must be a valid mode for the installed graphics
hardware; you may use a number or one of the predefined graphics
constants

Each time **SETGRAPHMODE** is activated, the graphics screen is cleared and the graph is redrawn in the default graphics color. In program 9–13, **SETGRAPH-MODE(CGAC0)** produces the same result as **SETGRAPHMODE(0);**, and **SETGRAPHMODE(CGAHI);** the same results as **SETGRAPHMODE(4);.**

Line Styles

Program 9–13 produced a diagonal line across the screen. Turbo Pascal 4.0 allows you to select from several different line styles. For example, you can choose to have your lines drawn as solid lines, dotted lines, dashed lines, or center lines. You can also select one of two line thicknesses. This is demonstrated by program 9–14.

The results of program 9–14 are illustrated in figure 9–5.

Program 9–14

```
PROGRAM Line_Style;
  USES GRAPH;

  PROCEDURE Auto_Initialization;
   VAR
    GraphDriver,
    GraphMode
  :INTEGER;

   BEGIN
      GraphDriver := DETECT;   {Initiates autodetect.}
      INITGRAPH(GraphDriver, GraphMode, '');
   END;   {of procedure Auto_Initialization}

PROCEDURE Draw_Lines;
```

Program 9–14   *continued*

```
    BEGIN

        SETGRAPHMODE(CGAC0);   {320 X 200 pixel screen, palette 0}

            SETLINESTYLE(SOLIDLN,0,NORMWIDTH);
            LINE(0,10,150,10);   {Draws a solid line of normal width}

            SETLINESTYLE(DOTTEDLN,0,NORMWIDTH);
            LINE(0,20,150,20);   {Draws a dotted line of normal width}

            SETLINESTYLE(CENTERLN,0,NORMWIDTH);
            LINE(0,30,150,30);   {Draws a center line style of normal width}

            SETLINESTYLE(DASHEDLN,0,NORMWIDTH);
            LINE(0,40,150,40);   {Draws a dashed line of normal width}

            SETLINESTYLE(SOLIDLN,0,THICKWIDTH);
            LINE(0,50,150,50);   {Draws a solid line of thick width}

            SETLINESTYLE(DOTTEDLN,0,THICKWIDTH);
            LINE(0,60,150,60);   {Draws a dotted line of thick width}

            SETLINESTYLE(CENTERLN,0,THICKWIDTH);
            LINE(0,70,150,70);   {Draws a center line style of thick width}

            SETLINESTYLE(DASHEDLN,0,THICKWIDTH);
            LINE(0,80,150,80);   {Draws a dashed line of thick width}

    END;

BEGIN
 Auto_Initialization;
 Draw_Lines;
END.
```

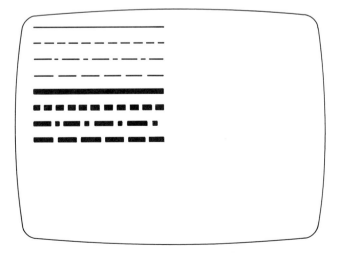

Figure 9–5   Different Graphic Line Styles

Table 9–5   Line Style Constants

Number	Constant	Result
0	SOLIDLN	Graphic lines will be solid.
1	DOTTEDLN	Graphic lines will be dotted.
2	CENTERLN	Graphic lines will be center lines (long line followed by a short one).
3	DASHEDLN	Graphic lines will be dashes.
4	USERBITLN	A user-defined line style.

The built-in Turbo Pascal 4.0 procedure for setting the line style is

SETLINSTYLE(LineStyle, Pattern, Thickness**);**

where LineStyle = a type WORD that is one of those shown in table 9–5.

Pattern = the bit pattern for the line; for the purpose of this text, set to 0 (type WORD)

Thickness = a type WORD that is used to determine one of two available line thicknesses: 1 = **NORMWIDTH** (a line of "normal" width); 2 = **THICKWIDTH** (a line thicker than the "normal" width)

For example, **SETLINESTYLE(SOLIDLN,0,NORMWIDTH);** is identical to the command **SETLINESTYLE(0,0,1);**.

Making Rectangles

Turbo Pascal has built-in graphics procedures for making rectangles. This is a great timesaving feature. As you may suspect, Turbo allows you to select the line styles for each of these rectangles, as illustrated in program 9–15.

Program 9–15

```
PROGRAM Make_Rectangles;
  USES GRAPH;

  PROCEDURE Auto_Initialization;
   VAR
    GraphDriver,
    GraphMode
   :INTEGER;

   BEGIN
     GraphDriver := DETECT;   {Initiates autodetect.}
     INITGRAPH(GraphDriver, GraphMode, '');
   END;   {of procedure Auto_Initialization}
```

Program 9–15   *continued*

```
PROCEDURE Draw_Rectangles;

   BEGIN

      SETGRAPHMODE(CGAC0);   {320 X 200 pixel screen, palette 0}

         SETLINESTYLE(SOLIDLN,0,NORMWIDTH);
         RECTANGLE(0,10,100,30);
      {Draws a solid line rectangle of normal width}

         SETLINESTYLE(DOTTEDLN,0,NORMWIDTH);
         RECTANGLE(0,40,100,60);
      {Draws a dotted line rectangle of normal width}

         SETLINESTYLE(CENTERLN,0,NORMWIDTH);
         RECTANGLE(0,70,100,90);
      {Draws a center line rectangle style of normal width}

         SETLINESTYLE(DASHEDLN,0,NORMWIDTH);
         RECTANGLE(0,100,100,120);
      {Draws a dashed line rectangle of normal width}

         SETLINESTYLE(SOLIDLN,0,THICKWIDTH);
         RECTANGLE(110,10,210,30);
       {Draws a solid line rectangle of thick width}

         SETLINESTYLE(DOTTEDLN,0,THICKWIDTH);
         RECTANGLE(110,40,210,60);
      {Draws a dotted line rectangle of thick width}

         SETLINESTYLE(CENTERLN,0,THICKWIDTH);
         RECTANGLE(110,70,210,90);
      {Draws a center line rectangle style of thick width}

         SETLINESTYLE(DASHEDLN,0,THICKWIDTH);
         RECTANGLE(110,100,210,120);
      {Draws a dashed line rectangle of thick width}

   END;
BEGIN
 Auto_Initialization;
 Draw_Rectangles;
END.
```

Figure 9–6 on page 570 illustrates the results of program 9–15.

The built-in Turbo Pascal procedure for drawing a rectangle on the graphics screen is

RECTANGLE(X1, Y1, X2, Y2);

where **X1, Y1** = INTEGER type coordinates of the top left corner of the rectangle

      X2, Y2 = INTEGER type coordinates of the bottom corner of the rect-
angle

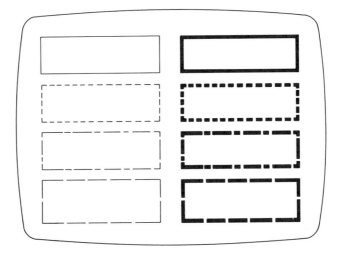

Figure 9–6   Built-In Rectangle Shapes with Different Line Styles

As you can see from program 9–15, the line style of the rectangle is determined by the last **SETLINESTYLE** procedure used in the program.

Creating Circles

As you may suspect, Turbo Pascal has a built-in procedure for creating circles. However, unlike the **RECTANGLE** procedure the **CIRCLE** procedure is affected by the **SETLINESTYLE** procedure in only one of two ways: the circle will always be drawn with a solid line that is one of two thicknesses. This is illustrated in program 9–16, where the resulting circles ignore the types of lines issued by the **SETLINESTYLE** procedures.

Program 9–16

```
PROGRAM Make_Circles;
  USES GRAPH;

  PROCEDURE Auto_Initialization;
   VAR
    GraphDriver,
    GraphMode
  :INTEGER;

   BEGIN
     GraphDriver := DETECT;   {Initiates autodetect.}
     INITGRAPH(GraphDriver, GraphMode, '');
   END;   {of procedure Auto_Initialization}
```

Program 9–16   *continued*

```
PROCEDURE Draw_Circles;
    BEGIN

        SETGRAPHMODE(CGAC0);   {320 x 200 pixel screen, palette 0}

            SETLINESTYLE(SOLIDLN,0,NORMWIDTH);
            CIRCLE(30,30,30);
        {Draws a solid line circle of normal width}

            SETLINESTYLE(DOTTEDLN,0,NORMWIDTH);
            CIRCLE(30,60,30);
        {Draws a solid line circle of normal width}

            SETLINESTYLE(CENTERLN,0,NORMWIDTH);
            CIRCLE(30,90,30);
        {Draws a solid line circle style of normal width}

            SETLINESTYLE(DASHEDLN,0,NORMWIDTH);
            CIRCLE(30,120,30);
        {Draws a solid line circle of normal width}

            SETLINESTYLE(SOLIDLN,0,THICKWIDTH);
            CIRCLE(110,30,30);
        {Draws a solid line circle of thick width}

            SETLINESTYLE(DOTTEDLN,0,THICKWIDTH);
            CIRCLE(110,60,30);
        {Draws a solid line circle of thick width}

            SETLINESTYLE(CENTERLN,0,THICKWIDTH);
            CIRCLE(110,90,30);
        {Draws a solid line circle style of thick width}

            SETLINESTYLE(DASHEDLN,0,THICKWIDTH);
            CIRCLE(110,120,30);
        {Draws a solid line circle of thick width}

    END;

BEGIN
 Auto_Initialization;
 Draw_Circles;
END.
```

Figure 9–7 (p. 572) shows the result of program 9–16. Again, note that there are only two line styles for the circle.

The built-in Turbo Pascal procedure for creating circles on the graphic screen is

```
CIRCLE(X, Y, Radius);
```

where X, Y = the INTEGER coordinates of the circle
Radius = the circle radius of type WORD

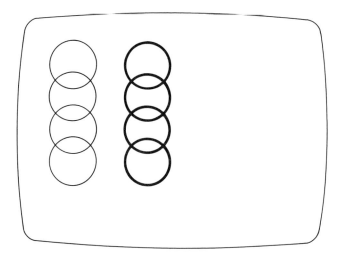

Figure 9–7   Circles Created by the Circle Procedure

Area Fills

There may be times when you want the area enclosed by a shape such as a circle, rectangle, or any shape you may create to be of a different color from the background. Turbo Pascal has a built-in graphics procedure for filling in enclosed areas; you can, for instance, make the area inside a circle different from its background. This is demonstrated by program 9–17.

The result of program 9–17 is shown in figure 9–8.

Program 9–17

```
PROGRAM Fill_Areas;
  USES GRAPH;

  PROCEDURE Auto_Initialization;
   VAR
    GraphDriver,
    GraphMode
  :INTEGER;

    BEGIN
      GraphDriver := DETECT;   {Initiates autodetect.}
      INITGRAPH(GraphDriver, GraphMode, '');
    END;   {of procedure Auto_Initialization}

PROCEDURE Fill_Circles;

    BEGIN

      SETGRAPHMODE(CGAC0);   {320 X 200 pixel screen, palette 0}
```

Program 9–17   *continued*

```
                SETLINESTYLE(SOLIDLN,0,NORMWIDTH);
                CIRCLE(30,30,30);    {Draws a solid line circle of normal width}
                FLOODFILL(30,30,3);    {Must be in the color of the object}

        END;

BEGIN
 Auto_Initialization;
 Fill_Circles;
END.
```

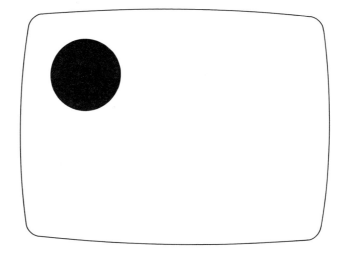

Figure 9–8   Results of Filling an Area of an Object

The built-in Turbo procedure for filling in an enclosed area is

FLOODFILL(X, Y, Border);

where **X, Y** = any coordinates inside the area of the object to be filled (INTEGER)
       Border = color of the fill (must be the color of the object border)

You can use the **FLOODFILL** procedure to fill areas around enclosed objects
as demonstrated by program 9–18.

The results of program 9–18 are shown in figure 9–9 on page 574.

Program 9–18

```
PROGRAM Fill_Areas;
  USES GRAPH;

  PROCEDURE Auto_Initialization;
   VAR
```

Program 9–18   *continued*

```
    GraphDriver,
    GraphMode
 :INTEGER;

   BEGIN
       GraphDriver := DETECT;  {Initiates autodetect.}
       INITGRAPH(GraphDriver, GraphMode, '');
   END;  {of procedure Auto_Initialization}

PROCEDURE Fill_Shapes;

   BEGIN

       SETGRAPHMODE(CGAC0);  {320 X 200 pixel screen, palette 0}

          SETLINESTYLE(SOLIDLN,0,NORMWIDTH);
          CIRCLE(30,30,30);
   {Draws a solid line circle of normal line width}

          RECTANGLE(100,20,200,50);
   {Draws a solid line rectangle of normal line width}

          FLOODFILL(50,50,3);
   {Must be in the color of the object}
   {Fills the area around the circle and the rectangle.}

     END;

BEGIN
 Auto_Initialization;
 Fill_Shapes;
END.
```

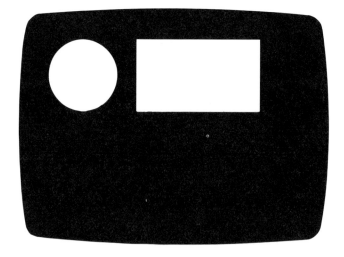

Figure 9–9   Another Application of Filling Areas

Conclusion

This section presented the basic graphics building blocks offered by Turbo Pascal. Here you saw how to create different line styles, rectangles that used the line styles, and circles that only used part of the line styles. You also saw how to fill areas (or around them) while in the graphics mode.

   In the next section, you will see how easy it is to create technical bar graphs using the built-in features of Turbo Pascal. For now, test your understanding of this section by trying the following section review.

9–4 Section Review

1   State what effect changing a graphics mode can have on the graphics screen.
2   How many different line styles are available in Turbo Pascal? State what they are.
3   Give the built-in Turbo Pascal shapes introduced in this section. Which one(s), if any, have limited line styles?
4   What does a flood fill do? How do you determine what area will be filled?

9–5   Bars and Text in Graphics

Discussion

This section shows you how to create bar charts and put text into your graphics. Here you will see how to create the bars used in bar charts. Being able to manipulate text, by changing its size, style (font), as well as being able to write vertically will greatly enhance your technical graphics. You will also learn how to change color and use built-in bar-fill patterns. All of these features are presented here.

Creating Bars

Bar charts are a common method of presenting information. An example of a bar chart is shown in figure 9–10 on page 576.

   As you can see from figure 9–10, the basic graphic for a bar chart is a rectangle. You can construct a rectangle in Turbo Graphics by simply constructing 4 lines. However, there is an easier way, as shown in program 9–19.

Program 9–19

```
PROGRAM Simple_Bars;
  USES GRAPH;
  PROCEDURE Auto_Initialization;
  VAR
    GraphDriver,
    GraphMode
  :INTEGER;
```

Program 9–19   *continued*

```
   BEGIN
      GraphDriver := DETECT;
      GraphMode := 0;
      INITGRAPH(GraphDriver, GraphMode, '');
   END;  {of procedure Auto_Initialization}

PROCEDURE Make_Bars;
  BEGIN
    SETGRAPHMODE(0);
    BAR(20,20,70,100);
    BAR3D(90,20,140,100,10,FALSE);
    BAR3D(160,20,220,100,10,TRUE);
  END;

BEGIN
 Auto_Initialization;
 Make_Bars;
END.
```

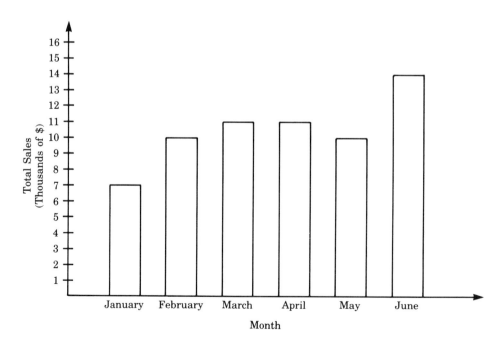

Figure 9–10   Typical Bar Chart

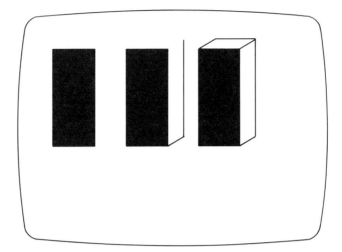

Figure 9–11   Resulting Graph for Program 9–19

The graph generated by program 9–19 is shown in figure 9–11.

Note from figure 9–11 that there were three bars generated; one "flat" and then two with some depth, although the second has no top, while the third does have a top. The built-in Turbo procedure that produces a flat bar is

```
BAR(X1, Y1, X2, Y2);
```

where $X1$, $Y1$ = the top left point of the bar (INTEGER)
$\quad\quad$ $X2$, $Y2$ = the bottom right point of the bar (INTEGER)

The procedure that produces a bar with depth is

```
BAR3D(X1, Y1, X2, Y2, Depth, Top);
```

where $X1$, $Y1$ = the top left point of the bar (INTEGER)
$\quad\quad$ $X2$, $Y2$ = the bottom right point of the bar (INTEGER)
$\quad\quad\quad$ $Depth$ = the number of pixels deep for the 3–D outline (WORD)
$\quad\quad\quad\quad$ Top = BOOLEAN variable—if TRUE, a top is put on the bar

These bars were drawn in the current color and fill style. The use of fill styles follows.

Filling in the Bars

There may be times when using a bar-graph display that bars must be used to represent many different types of information on the same graph. Since the use of color is limited on some graphics adapters, Turbo has a built-in procedure for automatically filling the area within a bar with one of 12 different **fill-styles**. This is demonstrated by program 9–20 on page 578.

The graph generated by program 9–20 is shown in figure 9–12 on page 578.

Program 9–20

```
PROGRAM Simple_Bars2;
  USES GRAPH;
  PROCEDURE Auto_Initialization;
   VAR
    GraphDriver,
    GraphMode
  :INTEGER;

    BEGIN
       GraphDriver := DETECT;
       GraphMode := 0;
       INITGRAPH(GraphDriver, GraphMode, '');
    END;    {of procedure Auto_Initialization}

  PROCEDURE Make_Bars;
    BEGIN
      SETGRAPHMODE(0);
      SETFILLSTYLE(LINEFILL,2);
      BAR(20,20,70,100);
      BAR3D(90,20,140,100,10,FALSE);
      BAR3D(160,20,220,100,10,TRUE);
    END;

  BEGIN
   Auto_Initialization;
   Make_Bars;
  END.
```

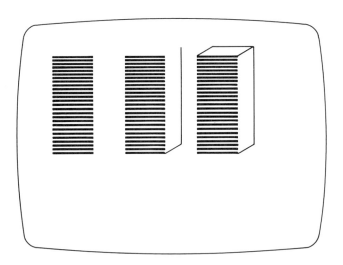

Figure 9–12 Bar Graph Generated by Program 9–20

Table 9–6   Fill-Pattern Constants

Number	Constant	Bar is Filled
0	EmptyFill	in the background color.
1	SolidFill	with a solid color.
2	LineFill	with horizontal lines.
3	LtSlashFill	with light slashes.
4	SlashFill	with thick slashes.
5	BkSlashFill	with thick back slashes.
6	LtBkSlashFill	with light back slashes.
7	HatchFill	with light hatch marks.
8	XHatchFill	with heavy cross hatches.
9	InterLeaveFill	with an interleaving line.
10	WideDotFill	with widely-spaced dots.
11	CloseDotFill	with closely-spaced dots.

You can see from figure 9–12 that each bar has a distinctive pattern. The commands to create the bars are exactly the same as in program 9–19, but the pattern inside the bars created by program 9–20 is different. This was done by the built-in Turbo Pascal procedure **SETFILLSTYLE**.

SETFILLSTYLE(Pattern, Color);

where **Pattern** = a type WORD (from 0 to 11) that sets the style of the fill pattern
      Color = a type WORD that sets the color of the filled pattern

Turbo Pascal has defined a set of constants in the **GRAPHS UNIT** as shown in table 9–6.

Program 9–21 illustrates the use of several of these constants and the reason for having tops on or off a 3–D bar.

The resulting graph from program 9–21 is shown in figure 9–13 (p. 580). Note that the two bottom stacked bars had the top FALSE and the top one had this item TRUE.

Program 9–21

```
PROGRAM Simple_Bars3;
  USES GRAPH;
  PROCEDURE Auto_Initialization;
   VAR
     GraphDriver,
     GraphMode
   :INTEGER;
```

Program 9–21   *continued*

```
   BEGIN
      GraphDriver := DETECT;
      GraphMode := 0;
      INITGRAPH(GraphDriver, GraphMode, '');
   END;  {of procedure Auto_Initialization}

PROCEDURE Make_Bars;
  BEGIN
    SETGRAPHMODE(0);

    SETFILLSTYLE(SLASHFILL,3);
    BAR3D(10,100,50,150,10,FALSE);

    SETFILLSTYLE(LINEFILL,1);
    BAR3D(10,50,50,100,10,FALSE);

    SETFILLSTYLE(HATCHFILL,2);
    BAR3D(10,10,50,50,10,TRUE);

    SETFILLSTYLE(WIDEDOTFILL,2);
    BAR3D(50,85,90,150,10,TRUE);

  END;

BEGIN
 Auto_Initialization;
 Make_Bars;
END.
```

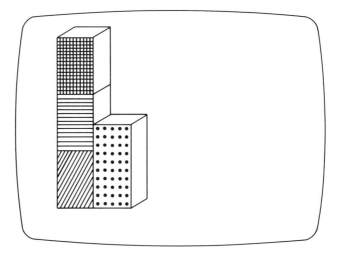

Figure 9–13   Results of Program 9–21

Text with Graphics

The use of text with graphics is an important feature to have when developing technical graphics. Turbo Pascal provides many advanced methods for presenting text on your graphics screens. The basic text graphics command is illustrated in program 9–22.

Program 9–22

```
PROGRAM Text_Output;
  USES GRAPH;
  PROCEDURE Auto_Initialization;
   VAR
     GraphDriver,
     GraphMode
   :INTEGER;

     BEGIN
         GraphDriver := DETECT;
         INITGRAPH(GraphDriver, GraphMode, '');
     END;   {of procedure Auto_Initialization}

  PROCEDURE Fonts;
   BEGIN
     OUTTEXT('This is the default font...');
     OUTTEXT('See where this is...');
   END;

  BEGIN
   Auto_Initialization;
   Fonts;
  END.
```

Execution of program 9–22 results in the following text appearing, starting at the top left portion of the monitor:

This is the default font . . . See where this is . . .

Note that there is no carriage return (as you would find in a **WRITELN** statement). Instead, the built-in Turbo procedure **OUTTEXT** sends a string of characters to the current position of the **graphics pointer**. The built-in procedure is

OUTTEXT(TextString);

where **TextString** = type STRING

The graphics pointer is the point on the graphics screen from which some action will take place. It is similar to the text cursor only it isn't visible. The **OUTTEXT** procedure will produce a string of text from the position of the graphics pointer and leave the pointer at the end of the text string. One method of moving the text string is to use the built-in Turbo Pascal procedure **OUTTEXTXY**.

OUTTEXTXTY(X1, Y1, TextString);

  where X1, Y1 = the position on the graphics screen from which the string is to start
    TextString = the STRING type

Changing Text Fonts

Turbo Pascal allows you to select different text styles (called **fonts**). Turbo Pascal actually has two major types of graphic text styles. One is **bit-mapped**; this is the default font. The other type is a **stroked font**. Characters in bit-mapped fonts are composed of rectangular pixel arrays. Characters in stroked fonts are composed of line segments whose sizes and directions are defined as relative to some starting point.

Turbo Pascal allows you to change the size of graphic text. When you enlarge bit-mapped fonts, the pixels are simply enlarged. Doing this makes the text look "blocky" with large stairstep effects. However, with a stroked font, text enlargement is made by lengthening each component line segment, making these fonts look more natural. The difference is illustrated in figure 9–14.

The main advantage of bit-mapped fonts is speed. Stroked fonts take longer because of their many line segments.

The built-in Turbo Pascal procedure that selects the font is

SETTEXTSTYLE(Font, Direction, Size);

  where Font = the style of the font (WORD)

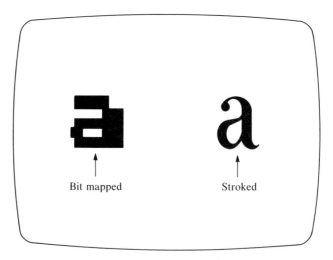

Figure 9–14 Difference Between Bit-Mapped and Stroked Fonts

`Direction` = the direction of the text to be displayed—(0 left to right, 1 bottom to top—WORD)

`Size` = size of the text (1 to 10)—a size of 1 produces a character of 8 × 8 pixels, a size of 2 a character of 16 × 16 pixels and so on)

The different text styles supported by Turbo Pascal are

`0` = `DefaultFont` (Bit-mapped)
`1` = `TriplexFont` (All the rest are stroked)
`2` = `SmallFont`
`3` = `SanSeriFont`
`4` = `GothicFont`

The active disk must contain these Turbo Pascal fonts, which come on your Turbo Pascal 4.0 distribution disks.

`TRIP.CHR    LITT.CHR    SANS.CHR    GOTH.CHR`

Other built-in constants for the **SETTEXTSTYLE** procedure are

`0` = `HorizDir`
`1` = `VertDir`
`1` = `NormSize`

Thus, **SETTEXTSTYLE(**`TriplexFont, HorizDir, Norm-`**Size);** means exactly the same as **SETTEXTSTYLE(**`1, 0, 1`**);**.

Program 9–23 illustrates the use of graphics text with a bar graph.

Program 9–23

```
PROGRAM Simple_Bars4;
  USES GRAPH;
  PROCEDURE Auto_Initialization;
   VAR
    GraphDriver,
    GraphMode
  :INTEGER;

    BEGIN
      GraphDriver := DETECT;
      GraphMode := 0;
      INITGRAPH(GraphDriver, GraphMode, '');
    END;  {of procedure Auto_Initialization}

  PROCEDURE Make_Bars;
    BEGIN
      SETGRAPHMODE(0);

      SETFILLSTYLE(SLASHFILL,3);
      BAR3D(30,100,70,150,10,FALSE);
```

Program 9–23   *continued*

```
      SETFILLSTYLE(LINEFILL,1);
      BAR3D(30,50,70,100,10,FALSE);

      SETFILLSTYLE(HATCHFILL,2);
      BAR3D(30,10,70,50,10,TRUE);

      SETFILLSTYLE(WIDEDOTFILL,2);
      BAR3D(80,85,120,150,10,TRUE);
    END;

PROCEDURE Label_Bars;
  BEGIN
    SETTEXTSTYLE(DefaultFont, VertDir, 1);
    SETCOLOR(1);
    OUTTEXTXY(20,10,'TOM  DICK  HARRY');
    SETCOLOR(2);
    OUTTEXTXY(145,85,'Others');
  END;

BEGIN
 Auto_Initialization;
 Make_Bars;
 Label_Bars;
END.
```

Figure 9–15 shows the result of program 9–23. Note that the vertical text displays actually start at the bottom and go up.

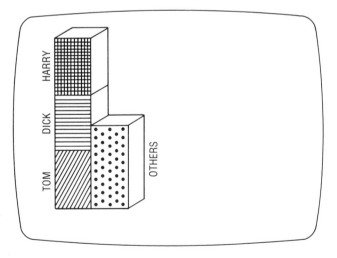

Figure 9–15   Use of Vertical Text in Graphics Mode

The built-in Turbo Pascal procedure **SETCOLOR** determines the graphics drawing color.

SETCOLOR(Color**);**

where Color = the selected color (WORD)

Conclusion

This section presented important information concerning the creation of bars used in bar graphs. Here you saw how to create 3–D bars and fill in their areas with different built-in fill patterns. You also saw how to create text in graphics. The different fonts were presented as well as how to change the size of the text and make it appear either vertically or horizontally. Test your knowledge of this section by trying the following section review.

9–5 Section Review

1 Why does Turbo Pascal have a built-in procedure for creating bars?
2 State what kind of bars can be presented using built-in Turbo Pascal procedures.
3 What options are available for presenting a 3–D bar using Turbo Pascal built-in procedures? Why is this available?
4 Describe how you could distinguish one bar from another using built-in Turbo Pascal procedures.
5 State some of the options available to you when using text in Turbo Pascal graphics.

9–6   Graphing Functions

Discussion

Turbo Pascal is a powerful language system for graphing mathematical functions. This section presents the fundamentals that are necessary to consider when programming for this type of technical graph. Recall that a mathematical function shows the relationship between two or more quantities. Knowing how to do this gives you a programming power that can enhance your technical programming skills.

Fundamental Concepts

Scaling is using numerical methods to insure that all of the required data appear on the graphics screen and utilize the full pixel capability. Remember that for any computer system, the number of pixels is limited. For example, in the high-resolution mode of the CGA adapter, there are 640 horizontal pixels and 200 vertical pixels.

To understand how to apply scaling to produce practical graphics, consider the graph of the function: $Y = X + 2$. The plot of this function is shown in figure 9–16.

The graph shown in figure 9–16 (p. 586) uses three of the four quadrants of the Cartesian coordinate system. Suppose you needed to develop a computer program

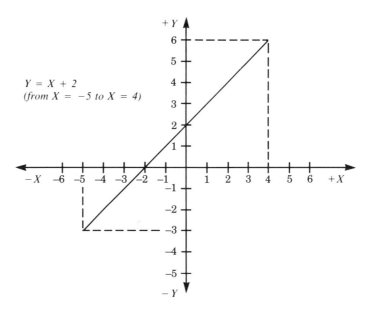

Figure 9–16   Graph for Scaling Example

that would display such a graph, and that you wanted the full graphics screen to be used in this display. This means that the actual values used to plot the graph would not be the values used in the equation. Figure 9–17 illustrates this important point.

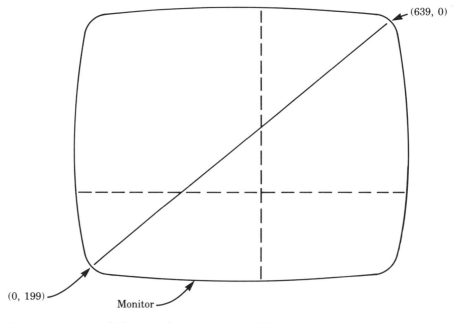

Figure 9–17   Actual Plotting Values for Graphical Display

Observe from figure 9–17 that the extreme left of the graph (the point that represents $X = -5$ and $Y = -3$) must actually have the plotted values of $X_P = 0$ and $Y_P = 200$ (the P subscript is used to denote the actual plotted values). The process used to achieve this transformation is called scaling. Also note from figure 9–17 that the origin of the coordinate system is not in the exact center of the screen. This was done to make practical use of the full size of the monitor. Not only has scaling been used, but so has **coordinate transformation**. Both processes are explained below.

Scaling

Observe figure 9–18.

The **horizontal scale factor** can be expressed mathematically as

$$HS = P_H/|X_2 - X_1|$$

where HS = the horizontal scale factor
       P_H = number of pixels in the horizontal direction
       X_1 = minimum value of X
       X_2 = maximum value of X

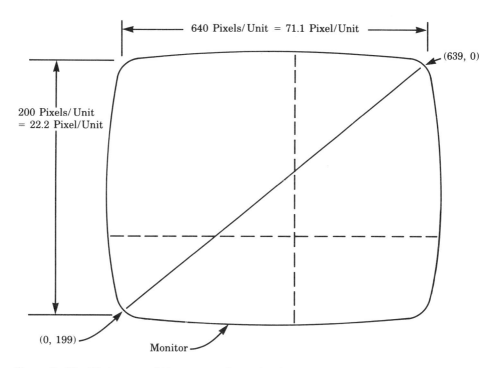

Figure 9–18   Minimum and Maximum Values of Scaling Example

The **vertical scale factor** can be expressed mathematically as

$$VS = P_V/|Y_2 - Y_1|$$

where VS = the vertical scale factor
  P_V = number of pixels in the vertical direction
  Y_1 = minimum value of Y
  Y_2 = maximum value of Y

Calculate the horizontal scale factor for the graph of figure 9–18.

$$\begin{aligned} HS &= P_H/|X_2 - X_1| \\ &= 640/|(4)-(-5)| \\ &= 640/|4 + 5| = 640/9 = 71.1 \end{aligned}$$

Calculate the vertical scale factor next.

$$\begin{aligned} VS &= P_V/|Y_2 - Y_1| \\ &= 200/|(6)-(-3)| \\ &= 200/|6 + 3| = 100/9 = 22.22 \text{ (See figure 9–18.)} \end{aligned}$$

These calculations mean that every major division along the X-axis will. be 71 pixels and every major division along the Y-axis will be 22 pixels. This will use the full capabilities of the graphics screen, as shown in figure 9–19.

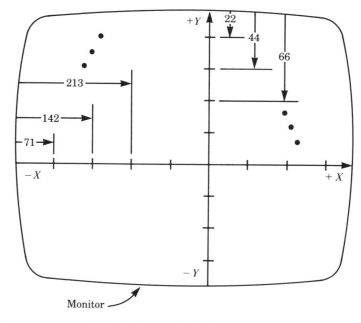

Figure 9–19   Meaning of Scaling for Example Graph

Coordinate Transformation

Note from figure 9–19 that the origin of the coordinate system used to display the example graph is not in the exact center of the graphics screen. This was intentionally done in order to display the full range of required data utilizing the full graphics screen. To accomplish this required the use of coordinate transformation, the process of using numerical methods to cause the origin of the coordinate system to appear at any desired place on the graphics screen.

The process of transforming coordinates for the horizontal transformation can be expressed mathematically as

$$XT = |(HS)\,(X_1)|$$

where XT = horizontal transformation
  HS = horizontal scaling factor
  X_1 = minimum value of X.

The mathematical expression for the vertical transformation is

$$YT = |(VS)\,(Y_2)|$$

where YT = vertical transformation
  VS = vertical scaling factor
  Y_2 = maximum value of Y

The horizontal transformation for the example in figure 9–19 is then

$$XT = |(HS)\,(X_2)|$$
$$= |(71.1)(-5)| = |-355|$$
$$= 355$$

The vertical transformation is

$$YT = |(VS)(Y_2)|$$
$$= |(22.2)(6)| = |133|$$
$$= 133$$

These calculations mean that the origin of the coordinate system will be at the location of X_o = 355 and Y_o = 133, as illustrated in figure 9–20 on page 590.

The only thing left to do now is to develop equations that can be used to display the data on the graphics screen.

Putting it Together

To display the graph of any continuous function related by two variables when scaling and coordinate transformations are used, the following equations are necessary.

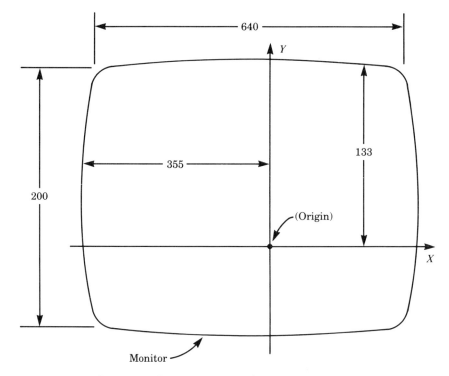

Figure 9–20   Actual Location of Origin for Example Graph

Horizontal Value of Graph

$$XG = XT + (X)(HS)$$

where XG = graphical X value
XT = horizontal transformation
HS = horizontal scaling factor

Vertical Value of Graph

$$YG = YT - (Y)(VS)$$

where YG = graphical Y value
YT = vertical transformation
VS = vertical scaling factor

Program 9–24 plots the graph of the function $Y = X + 2$. The program asks the user for the maximum and minimum values of X, then draws the transformed coordinate system and scales the plot of the graph. Note that it makes no difference what pixel size the graphics screen has as long as Turbo Pascal recognizes it.

The graph resulting from program 9–24 is shown in figure 9–21 on page 593.

Analyzing the Program

The first procedure in program 9–24 is `Auto_Initialization`, done so the Turbo system knows what kind of graphics system the computer has (if any at all).

Program 9–24

```pascal
PROGRAM Make_Graph;

USES GRAPH;

PROCEDURE Auto_Initialization;
 VAR
  GraphDriver,
  GraphMode
 :INTEGER;

  BEGIN
    GraphDriver := DETECT;
    INITGRAPH(GraphDriver, GraphMode, '');
  END;

FUNCTION Y(X:REAL):REAL;
  BEGIN
    Y := X + 2;
  END;

PROCEDURE Scaling_Values(MaximumX, MaximumY,
                         MinimumX, MinimumY
                         :REAL;
                         VAR
                         HorScale, VerScale,
                         HorTrans, VerTrans
                         :REAL);

  BEGIN

   IF MaximumX = MinimumX THEN HorScale := GETMAXX
   ELSE
    HorScale := GETMAXX/(MaximumX-MinimumX);

   IF MaximumY = MinimumY THEN HorScale := GETMAXY
   ELSE
    VerScale := GETMAXY/(MaximumY-MinimumY);

   HorTrans := ABS(HorScale * MinimumX);
   VerTrans := ABS(VerScale * MaximumY);

  END;

PROCEDURE Get_Values;
 VAR
  X,
  XMax,
  YMax,
  XMin,
  YMin,
  HorScale,
  VerScale,
  HorTrans,
  VerTrans
 :REAL;
```

Program 9–24   *continued*

```
HorPixel,
VerPixel
:INTEGER;

 BEGIN
     WRITELN;
     WRITE('Maximum X = ');
     READLN(XMax);
     YMax := Y(XMax);
     WRITE('Minimum X = ');
     READLN(XMin);
     YMin := Y(XMin);

        Scaling_Values(XMax, YMax, XMin, YMin, HorScale, VerScale,
                   HorTrans, VerTrans);

     WRITELN('Horizontal scale = ',HorScale:3:2);
     WRITELN('Vertical scale = ',VerScale:3:2);
     WRITELN('Horizontal transformation = ',HorTrans:3:2);
     WRITELN('Vertical transformation = ',VerTrans:3:2);

     {Draw coordinate system...}
     SETLINESTYLE(SOLIDLN,0,THICKWIDTH);
     LINE(0,ROUND(VerTrans),GETMAXX,ROUND(VerTrans));
     LINE(ROUND(HorTrans),0,ROUND(HorTrans),GETMAXY);

     {Draw the graph...}

        X := XMin;

      REPEAT

        HorPixel := ROUND(HorTrans + X * HorScale);
        VerPixel := ROUND(VerTrans - Y(X) * VerScale);
         PUTPIXEL(HorPixel, VerPixel, 1);
         X := X + 0.1;

      UNTIL HorPixel >= GETMAXX;
  END;

BEGIN
 Auto_Initialization;
 Get_Values;
END.
```

The equation to be graphed is put in the form of a function so that it is easy to change the relationship between *X* and *Y* if you wish to plot a different formula.

The procedure `Scaling_Values` computes the horizontal and vertical scale as well as the horizontal and vertical transformations necessary for the graphics screen. Note that if the maximum and minimum values of *X* or *Y* are equal (this could happen when plotting a quadratic) the scales are changed to the maximum pixel values of the given graphics driver.

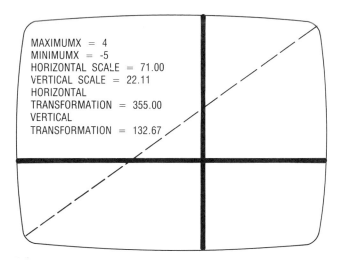

MAXIMUMX = 4
MINIMUMX = -5
HORIZONTAL SCALE = 71.00
VERTICAL SCALE = 22.11
HORIZONTAL
TRANSFORMATION = 355.00
VERTICAL
TRANSFORMATION = 132.67

Figure 9–21   Resulting Computer-Generated Graph

The procedure `Get_Values` simply gets the maximum and minimum values of the X variable from the program user. In order to see the chosen values, **WRITELN** statements are used to display the transformation and scaling information on the upper left portion of the screen. The graph is then constructed using the built-in Turbo Pascal procedure **PUTPIXEL**. However, the **PUTPIXEL** procedure requires a type INTEGER and all of the graph values up to this point have been of type REAL. The built-in Turbo Pascal procedure **ROUND** is used to convert the REAL values to INTEGER. The value of X is incremented by 0.1 each time. This can be changed depending upon the resolution you want (and time it takes) for the actual graph construction.

Conclusion

This section introduced you to the practical requirements for displaying the graph of a formula. You were introduced to the concepts of scaling and coordinate transformation. These techniques will prepare you for plotting technical data that relates one variable to another. Test your understanding of this section by trying the following section review.

9–6 Section Review

1   Define scaling as it applies to computer graphics.
2   Define coordinate transformation as it applies to computer graphics.
3   State the factors that determine the values of the scaling factors.
4   State the factors that determine the coordinate transformation values.

9-7   Program Debugging and Implementation— Programming Style

Discussion

As you gain experience in writing your own Pascal programs, you will begin to develop a unique style. You will begin to standardize the format of your program commands, the length of your declared identifiers, your capitalization, and when you use the underscore (__) to break up your identifiers.

There are no hard and fast rules for these individualistic elements of programming—just be sure your programs are easy to understand and modify. It's important to see many different programming styles; read programming magazines to see other programmers' styles. This section presents some ideas on programming styles you may want to consider using in your own work.

Pascal Reserved Words

Throughout this text, all Pascal reserved words were capitalized and **BOLDFACED**. It's not necessary to boldface your Pascal commands, but it is suggested that you distinguish them from the body of the program. Much published material does not do this and can be more difficult to understand.

Procedure and Function Identifiers

Though not consistently done in this text, you may consider always using the underscore (__) in user-defined procedure or function identifiers. This would guarantee that you could easily spot any calls to them. You would know, for instance, that `Do__Calculations;` was an identifier for one of your procedures or functions. For your declared identifiers, don't use the __ to separate words. This means that using `Do__Calculations(Information);`, rather than `Do__Calculations(User__Information);`, may make it clearer to see that `Information` is just a program variable or constant.

Naming Things

Pascal will let you use very long identifiers, but this can be a mixed blessing. Consider the following two program lines.

```
E := I * R; {Compute voltage}
Voltage := Current * Resistance;
```

The first of the two lines uses familiar single-letter symbols to represent an electronic formula (familiar at least to electronics students and professionals). The second formula spells out the relationship. If you are used to using certain formulas with specific single-letter variables (and doing so is a generally-accepted standard), single-letter variables may make your source code more readable. For example,

```
Xt := Xl - Xc; {Total reactance}
```

may be more recognizable to an electronics technician than

```
TotalReactance := InductiveReactance - CapacitiveReactance;
```

Your Structure

One of the main advantages of the Pascal language is the rich variety of program structures available. Usually, beginning programmers tend to shy away from user-defined scalar types, **CASE** statements, and records. However, if you just keep these in mind, they can be very useful in the construction of a Pascal program.

Consider, for example, program 9–24, used to generate the graph of a mathematical relationship between two variables. Its programming style could have been improved by using a record as shown in program 9–25.

Program 9–25

```
PROGRAM Make_Graph;

 USES GRAPH;

  TYPE
      CoordinateRecord = RECORD
                           MaximumX,
                           MaximumY,
                           MinimumX,
                           MinimumY : REAL;
                         END;

      TranslationRecord = RECORD
                           HorScale,
                           VerScale,
                           HorTrans,
                           VerTrans : REAL;
                         END;

 PROCEDURE Auto_Initialization;
  VAR
   GraphDriver,
   GraphMode
 :INTEGER;

  BEGIN
     GraphDriver := DETECT;
     INITGRAPH(GraphDriver, GraphMode, '');
  END;

FUNCTION Y(X:REAL):REAL;
  BEGIN
    Y := X + 2;
  END;
```

Program 9–25   *continued*

```
PROCEDURE Scaling_Values(CoordinateData : CoordinateRecord;
                    VAR TranslationData : TranslationRecord);

   BEGIN

   WITH CoordinateData DO
      BEGIN
        WITH TranslationData DO
         BEGIN
         IF MaximumX = MinimumX THEN HorScale := GETMAXX
         ELSE
           HorScale := GETMAXX/(MaximumX-MinimumX);

         IF MaximumY = MinimumY THEN HorScale := GETMAXY
         ELSE
           VerScale := GETMAXY/(MaximumY-MinimumY);

         HorTrans := ABS(HorScale * MinimumX);
         VerTrans := ABS(VerScale * MaximumY);
        END;
      END;
   END;

PROCEDURE Get_Values;
 VAR
  X
 :REAL;

  CoordinateInput
 :CoordinateRecord;

  TranslationOutput
 :TranslationRecord;

 HorPixel,
 VerPixel

 :INTEGER;

   BEGIN

     WITH CoordinateInput DO
       BEGIN
         WITH TranslationOutput DO
          BEGIN

            WRITELN;
            WRITE('Maximum X = ');
            READLN(MaximumX);
              MaximumY := Y(MaximumX);
             WRITE('Minimum X = ');
             READLN(MinimumX);
              MinimumY := Y(MinimumX);

             Scaling_Values(CoordinateInput, TranslationOutput);
```

Program 9–25   *continued*

```
                WRITELN('Horizontal scale = ',HorScale:3:2);
                WRITELN('Vertical scale = ',VerScale:3:2);
                WRITELN('Horizontal transformation = ',HorTrans:3:2);
                WRITELN('Vertical transformation = ',VerTrans:3:2);

                {Draw coordinate system...}
                SETLINESTYLE(SOLIDLN,0,THICKWIDTH);
                LINE(0,ROUND(VerTrans),GETMAXX,ROUND(VerTrans));
                LINE(ROUND(HorTrans),0,ROUND(HorTrans),GETMAXY);

        {Draw the graph...}

          X := MinimumX;

        REPEAT

            HorPixel := ROUND(HorTrans + X * HorScale);
            VerPixel := ROUND(VerTrans - Y(X) * VerScale);
            PUTPIXEL(HorPixel, VerPixel, 1);
            X := X + 0.1;

        UNTIL HorPixel >= GETMAXX;
        END;
      END;
    END;
BEGIN
  Auto_Initialization;
  Get_Values;
END.
```

Note the saving of program code in program 9–25 by using type RECORD compared to program 9–24 in the last section. Observe also that there are no global variables in the program; all variables are local, thus protecting them.

Conclusion

This section presented some suggestions you may want to consider to help improve your programming style, suggestions for using identifiers as well as for taking advantage of other types of program structure. Check your understanding of this section by trying the following section review.

9–7 Section Review

1   State the guiding principle in the style you choose for developing Pascal programs.
2   What is one method of writing user-defined identifiers for procedures and functions?
3   Should variables always be more than one or two letters long? Explain.
4   State what kind of structure you should use in a Pascal program.

Summary

1   To produce color, your system needs internal color capabilities and a color monitor.
2   The Turbo Pascal color commands are contained in the CRT UNIT.
3   A monochrome monitor is capable of producing only one color while a color monitor can produce a rich variety of color.
4   The IBM PC and true compatibles are capable of producing two different-size text screens: 80 × 25 and 40 × 25.
5   The IBM PC and true compatibles are capable of producing 16 different text colors and 8 different background colors.
6   There are basically two different kinds of screens, the text screen and the graphics screen.
7   Turbo Pascal comes with programs, called graphic drivers (.BGI), made especially for the operation of different graphic adapters installed in your IBM or true compatible system.
8   A pixel is the smallest unit on the graphics screen.
9   The number of pixels available on your graphics screen depends upon the type of installed graphics adapter in your system.
10   The amount of color available for your graphics, if any, depends upon the kind of installed graphics adapter in your system and the selected mode of operation.
11   When using a Turbo Pascal program for graphics, you must develop a procedure that either detects the graphics adapter you're using or instructs the system which adapter to use.
12   When using a Turbo Pascal program for graphics, the appropriate Turbo graphics driver (.BGI) must be available on the disk.
13   You can use built-in Turbo Pascal procedures to detect the range of modes and colors available for a given system.
14   There are built-in Turbo Pascal procedures for creating lines, circles, and other standard shapes.
15   Turbo Pascal has built-in constants for changing line styles and drawing colors.
16   There are built-in Turbo Pascal procedures for creating a rich variety of bar graphs.
17   Turbo Pascal comes equipped with 5 different text fonts. One is the default and the other 4 are code (.CHR) that must be on the active disk to be used in the graphics program.
18   With Turbo Pascal built-in procedures, graphic text may be displayed vertically as well as horizontally and may have its size changed.
19   To take full advantage of the size of the graphics screen when graphing mathematical functions, a process of scaling and coordinate transformation is used.
20   The type of programming style you select should always lead to source code that is easy to understand and modify.

Interactive Exercises

Directions

Because of the nature of this chapter and the differences in computer systems that display color text and graphics, this section differs from the other interactive exercise sections.

This section asks questions about the computer system(s) to which you have access. They are designed to help you discover information useful for developing the programs in the end-of-chapter problems.

Exercises

1   Is your system capable of displaying text in color? How did you find this out?
2   How many different colors of text can your system display? Can all these different colors be displayed at the same time?
3   If your system can display text in color, can it also display different background colors? How many?
4   What kind of graphics adapter does your system have? How did you find this out?
5   How many different modes does the graphics adapter in your system have? How did you find this out?
6   What .BGI file do you need on your disk in order to operate your graphics system?
7   What is the maximum number of horizontal and vertical pixels available for your graphics system? How did you determine this?
8   What is the maximum number of drawing colors available with your system? How was this determined?
9   How many palettes does your system have? How did you determine this?
10 State the maximum number of colors that can be displayed on your graphics screen at the same time. How was this determined?

Pascal Commands

To change the mode of the text, use

TEXTMODE(MODE**);**

where MODE = an INTEGER
   To change the text color, use

TEXTCOLOR(COLOR**);**
TEXTBACKGROUND(COLOR**);**

where COLOR = type BYTE
   To determine the type of graph driver and default mode, use

DETECTGRAPH(GraphDriver, GraphMode**);**

where GraphDriver = a variable INTEGER that returns a number that represents the type of installed graphics hardware
    GraphMode = a variable INTEGER that returns the mode of operation for the detected installed graphics hardware
   To get into the graphics mode, use

INITGRAPH(GraphDriver, GraphMode, DriverPath**);**

where GraphDriver = A variable INTEGER that designates the type of graphics driver to be used
    GraphMode = a variable INTEGER that designates the mode of operation for the graphics driver
    DrivePath = a STRING variable that indicates the directory path where the graph driver can be found; if " are used, the active disk is used

To light up a pixel on the graphics screen, use

```
PUTPIXEL(X,Y, Pixel);
```

where X = an INTEGER representing the horizontal value of the pixel
Y = an INTEGER representing the vertical value of the pixel.
Pixel = type WORD, a number indicating the color of the pixel

To leave the graphics mode, use

```
CLOSEGRAPH;
```

This procedure shuts down the graphics system and restores the original screen mode before graphics was initialized.

Table 9–7   Graphics Modes for Given Installed Hardware

Hardware Driver #	Mode # Const	Screen Size (Pixels)	Colors
CGA 1	0 CGA0	320 × 200	LightGreen, LightRed, Yellow
	1 CGA1	320 × 200	LightCyan, LightMagenta, White
	2 CGA2	320 × 200	Green, Red, Brown
	3 CGA3	320 × 200	Cyan, Magenta, LightGray
	4 CGAHI	640 × 200	(One color on black)
MCGA 2	0 MCGAC0	320 × 200	LightGreen, LightRed, Yellow
	1 MCGAC1	320 × 200	LightCyan, LightMagenta, White
	2 MCGAC2	320 × 200	Green, Red, Brown
	3 MCGAC3	320 × 200	Cyan, Magenta, LightGray
	4 MCGAMED	640 × 200	(One color on black)
	5 MCGAHI	640 × 480	(One color on black)
EGA 3	0 EGALO	640 × 200	16 colors
	1 EGAHI	640 × 350	16 colors
EGA64 4	0 EGA64LO	640 × 200	16 colors
	1 EGA64HI	640 × 350	4 colors
	3 EGAMONO	640 × 350	(One color on black)
HERCMONO 7	0 HERCMONOHI	720 × 348	(White on black)
ATT400 8	0 ATT400C0	320 × 200	LightGreen, LightRed, Yellow
	1 ATT400C1	320 × 200	LightCyan, LightMagenta, White
	2 ATT400C2	320 × 200	Green, Red, Brown
	3 ATT400C3	320 × 200	Cyan, Magenta, LightGray
	4 ATT400MED	640 × 200	(One color on black)
	5 ATT400HI	640 × 400	(One color on black)
VGA 9	0 VGALO	640 × 200	16 colors
	1 VGAMED	640 × 350	16 colors
	2 VGAHI	640 × 480	16 colors
PC3270 10	0 PC3270HI	720 × 350	(White on black)

Turbo has defined graphic driver constants corresponding to the number of the driver, as listed.

```
DETECT = 0
CGA = 1
MCGA = 2
EGA = 3
EGA64 = 4
EGAMONO = 5
RESERVED = 6
HERCMONO = 7
ATT400 = 8
VGA = 9
PC3270 = 10
```

The mode range procedure gives you the range of modes available with your installed graphics hardware. The built-in Turbo Pascal 4.0 procedure used here is

GETMODERANGE(GraphDriver, LoMode, HiMode);

where GraphDriver = an INTEGER that indicates the graph driver used for your system (such as 1 or CGA)

LoMode = a variable parameter of type INTEGER that will return the numerical value of the lowest mode available for the installed graphics adapter

HiMode = a variable parameter of type INTEGER that will return the numerical value of the largest mode available for the installed graphics adapter

The current mode procedure identifies the active mode in which your system is current. The built-in Turbo Pascal 4.0 function that returns the numerical value of the current mode is

GETGRAPHMODE;

where the value returned is of type INTEGER and is the numerical value of the active graph mode of your system; refer to table 9–3 for the significance of each graph mode.

The maximum colors procedure identifies the maximum color number available in the current mode. The built-in Turbo Pascal 4.0 function that returns this value is

GETMAXCOLOR

where the value returned is of type WORD and is the numerical value of the largest value color for the current mode. (In those modes where only white graphs will appear on a black background, 0 represents black and 1 represents white. In any mode, black is always 0, while the other numbers indicate colors that depend upon the installed graphics hardware and active palette.)

The graphics screen-size procedure gives the maximum size of the current graphics screen. There are two built-in Turbo Pascal functions that allow you to get these values.

GETMAXX;
GETMAXY;

where the value returned is of type INTEGER and is the numerical value X of the location of the rightmost pixel, or the numerical value Y of the bottommost pixel.

To create a line, use

LINE(X1, Y1, X2, Y2);

where X1, Y1 = INTEGER values of the starting point of the line
       X2, Y2 = INTEGER values of the ending point of the line

To create bars, use

BAR(X1, Y1, X2, Y2);

where X1, Y1 = the top left point of the bar (INTEGER)
       X2, Y2 = the bottom right point of the bar (INTEGER)

BAR3D(X1, Y1, X2, Y2, Depth, Top);

where X1, Y1 = the top left point of the bar (INTEGER)
       X2, Y2 = the bottom right point of the bar (INTEGER)
       Depth = the number of pixels deep for the 3–D outline (WORD)
         Top = BOOLEAN variable—if TRUE, a top is put on the bar

To fill the bars, use

SETFILLSTYLE(Pattern, Color);

where Pattern = a type WORD (from 0 to 11) that sets the style of the fill pattern
       Color = a type WORD that sets the color of the filled pattern

Turbo Pascal has defined a set of constants in the **GRAPHS UNIT** as shown in table 9–8.

To output text in the current font, use

OUTTEXT(TextString);

where TextString = type STRING

Table 9–8   Fill-Pattern Constants

Number	Constant	Bar is filled:
0	EmptyFill	in the background color.
1	SolidFill	with a solid color.
2	LineFill	with horizontal lines.
3	LtSlashFill	with light slashes.
4	SlashFill	with thick slashes.
5	BkSlashFill	with thick back slashes.
6	LtBkSlashFill	with light back slashes.
7	HatchFill	with light hatch marks.
8	XHatchFill	with heavy cross hatches.
9	InterLeaveFill	with an interleaving line.
10	WideDotFill	with widely spaced dots.
11	CloseDotFill	with closely spaced dots.

To output text in the current font to a place on the graphics screen, use

OUTTEXTXTY(X1, Y1, TextString);

where X1, Y1 = the position on the graphics screen from which the string is to start
TextString = the STRING type

The built-in Turbo Pascal procedure that selects the font is

SETTEXTSTYLE(Font, Direction, Size);

where Font = the style of the font (WORD)
Direction = the direction of the text to be displayed − (0 left to right, 1 bottom to top− WORD)
Size = size of the text (1 to 10)—a size of 1 produces a character of 8 × 8 pixels, a size of 2 a character of 16 × 16 pixels and so on

The different text styles supported by Turbo Pascal are

```
0 = DefaultFont          (Bit-mapped)
1 = TriplexFont          (All the rest are stroked)
2 = SmallFont
3 = SanSeriFont
4 = GothicFont
```

The active disk must contain the following Turbo Pascal fonts.

TRIP.CHR   LITT.CHR   SANS.CHR   GOTH.CHR

Other built-in constants for the **SETTEXTSTYLE** procedure are

```
0 = HorizDir
1 = VertDir
1 = NormSize
```

The built-in Turbo Pascal procedure **SETCOLOR** determines the graphic drawing color.

SETCOLOR(Color);

where Color = the selected color (WORD)

Self-Test

Directions

Answer the questions for this self-test by referring to program 9–26 on page 604. The questions will be found here and at the end of the program, on page 605.

Questions

1 Describe what program 9–26 does.
2 Where in program 9–26 is the graphics screen initialized?
3 What is the purpose of the **ROUND** procedure in program 9–26? Why is this necessary?
4 Would program 9–26 work for every graphics adapter? Explain.

Program 9–26

```
PROGRAM Use_This;
 USES GRAPH, CRT;
 PROCEDURE Auto_Initialization;
  VAR
   GraphDriver,        {Gets graphics driver number.}
   GraphMode           {Gets the mode for the driver.}
  :INTEGER;

    BEGIN
      GraphDriver := DETECT;  {Initiates autodetect.}
      INITGRAPH(GraphDriver, GraphMode, '');
    END;

 PROCEDURE Sine_Wave(Amplitude, Phase, Cycles : INTEGER);
  VAR
   Degrees,
   Radians
  :REAL;
   XValue,
   YValue
  :INTEGER;

    BEGIN
      Auto_Initialization;
      SETGRAPHMODE(3);

      SETBKCOLOR(1);
      FOR XValue := 0 TO 360 DO
        BEGIN
          Degrees := XValue;
          Radians := XValue * PI/180;
          YValue := ROUND(Amplitude*SIN(Cycles * Radians + Phase));
          PutPixel(XValue, 75-YValue, 2);
        END;

    END;

 PROCEDURE User_Input;
  VAR
   Amplitude,
   Cycles,
   Phase
  :INTEGER;

      BEGIN
        ClrScr;
        WRITE('Amplitude => ');
        READLN(Amplitude);
        WRITE('Cycles => ');
        READLN(Cycles);
        WRITE('Phase => ');
        READLN(Phase);

        Sine_Wave(Amplitude, Phase, Cycles);

      END;

BEGIN
 User_Input;
END.
```

5 A sine function has + and − values, but the graphics screen has only + values. What statement in program 9–26 adjusts for this fact?

6 What is the maximum amplitude allowed for the generated sine wave in program 9–26?

7 Does program 9–26 automatically adjust itself for different size graphics screens? Explain.

8 Where on the graphics screen does the sine wave start? How did you determine this?

9 How many different colors will be displayed by program 9–26? Explain.

Problems

General Concepts

Section 9–1

1 What must your computer system contain to produce text in color using Turbo 4.0 Pascal?

2 How many columns of text can be displayed with an installed graphics adapter?

3 What is the maximum number of colors text may have? Can all of these text colors be displayed at the same time?

4 Explain how you would cause colored text to blink.

Section 9–2

5 Explain the difference between a text screen and a graphics screen.

6 State what your systems needs, in terms of hardware and software, to get a Turbo Pascal program into the graphics mode of operation.

7 Define the term pixel.

8 Which color-graphics adapters supported by Turbo Pascal can produce 16 colors in graphics?

Section 9–3

9 What built-in Turbo Pascal procedure would you use to determine the modes available on your system?

10 State what Turbo Pascal command you would use to determine the number of available colors for any given graphics mode of your system.

11 Explain the purpose of the built-in Turbo Pascal functions **GETMAXX** and **GETMAXY**.

12 State the meaning of the term palette as used in Turbo Pascal.

Section 9–4

13 Explain how you would change the palette on the graphics screen using Turbo Pascal.

14 What built-in Turbo Pascal procedure would you use to create dotted lines?

15 Name two ways of creating a rectangle using Turbo Pascal.

16 When creating a rectangle in Turbo Pascal, can you change the style of the line used to create the rectangle? Explain.

17 What are the line styles for creating a circle in Turbo Pascal?

Section 9–5

18 There are three varieties of bars that are built into Turbo Pascal; state what they are.

19 Explain what is meant by a Turbo Pascal fill style.

20 When wouldn't you want the top of a 3–D bar to appear?

21 How many different fill styles for bars are available in Turbo Pascal?

22 What built-in Turbo Pascal procedure determines the fill style of a bar?

Section 9–6

23 State what is meant by scaling when graphing a mathematical function.
24 Explain what is meant by coordinate transformation when graphing a mathematical function.
25 Why is coordinate transformation used?
26 How are the values of the scaling factors determined?
27 State how the coordinate transformation values are determined.

Section 9–7

28 What is meant by programming style?
29 State what is concerned to be good programming style.
30 Are there any general rules to follow when developing your own variable identifiers? Explain.
31 What Pascal structures should you keep in mind when developing programs?

Program Analysis

32 Will program 9–27 compile? If not, why not?

Program 9–27

```
PROGRAM First1;
 BEGIN
    TEXTMODE(BW40);
    WRITELN('Here is some text.');
 END.
```

33 What happens on your system when executing program 9–28?

Program 9–28

```
PROGRAM Second1;
USES CRT;
 BEGIN
    TEXTMODE(?);
    WRITELN('This is text.');
 END.
```

34 Will program 9–29 compile? If it does, what does it do when executed?

Program 9–29

```
PROGRAM Third1;
USES CRT;
 VAR
  Mode
:INTEGER;
```

Program 9–29   *continued*

```
BEGIN
   FOR Mode := BW40 TO C80 DO
      BEGIN
         TEXTMODE(Mode);
         WRITELN('This is text.');
         DELAY(1000);
      END;
END.
```

35 Will program 9–30 compile? If not, why not?

Program 9–30

```
PROGRAM ThisAgain;
  BEGIN
    LINE(0,0,50,50);
  END.
```

36 Will program 9–31 compile? If not, why not?

Program 9–31

```
PROGRAM ThisHere;
 USES GRAPH;
  BEGIN
    LINE(0,0,50,50);
  END.
```

37 Will program 9–32 compile? If not, why not?

Program 9–32

```
PROGRAM ThisAndThat;
 USES GRAPH;
  PROCEDURE Initialize;
    BEGIN
      INITGRAPH(CGA, CGAC0, '');
    END;
```

Program 9–32   *continued*

```
BEGIN
  Initialize;
  LINE(0,0,50,50);
END.
```

38   Will program 9–33 compile? If not, why not?

Program 9–33

```
PROGRAM TheLastOne;
 USES GRAPH;
  PROCEDURE Initialize;
   VAR
    GraphCard,
    CardMode
   :INTEGER;
   BEGIN
     GraphCard := CGA;
     CardMode := CGAC0;
     INITGRAPH(GraphCard, CardMode, '');
   END;

  BEGIN
    Initialize;
    LINE(0,0,50,50);
  END.
```

Program Design

When designing these programs use a structure assigned by your instructor. Otherwise, use whatever structure is preferred by you; it should be easy for anyone to read and understand (especially for you if it ever needs modification). Note that for these programs, you will need a system with color graphics capabilities.

Electronics Technology

39   Develop a Pascal program that makes use of text color to display the wattage value of a resistor. The user input is the resistance of the resistor and the voltage across the resistor. The color of the answer is to be influenced by the resulting wattage values as given below.

microwatts = blue
milliwatts = purple
1 to 10 watts = green
10 to 100 watts = yellow
100 to 1000 watts = red
over 1000 watts = flashing red

The relationship for power is

$$P = I^2R$$

40 Create a Pascal program that will display a graph of the relationship of the voltage across a resistor versus the current in the resistor and the value of the resistor. User input is the value of the resistor and the range of currents to be graphed. The X-axis represents the current, the Y-axis the voltage. This relationship is known as Ohm's Law.

$$E = IR$$

41 Make a Pascal program that will display the reactance of an inductor for a given range of frequencies. The user input is the value of the inductor and the range of frequencies to be graphed. The X-axis represents the frequency, the Y-axis the inductive reactance. The relationship is

$$X_L = 2\pi FL$$

42 Develop a Pascal program that will display the impedance of a series RLC circuit. User inputs are the values of the resistor, inductor, and capacitor. The X-axis represents frequency, the Y-axis impedance. The relationship is

$$X_T = \sqrt{R^2 + (X_L - X_C)^2}$$

43 Create a Pascal program that will display the resultant of two sine waves. The user input is the amplitude, frequency, and relative phase of each sine wave.

Business Applications
44 Make a Pascal program that will display a bar graph of the amount of sales, measured in dollars, for five salespersons in a given year. The X-axis is to represent each salesperson and the Y-axis the amount of money. The user input is the amount of sales for each salesperson.

Computer Science
45 Develop a Pascal program that will place a coordinate-axis system anywhere on the graphics screen. User input is the line style and location of the coordinates.

Drafting Technology
46 Create a Pascal program that will display a rectangle with dimension lines. User input is the size of the rectangle in pixels. The dimension lines are to contain the size of the rectangle in pixels.

Agriculture Technology
47 Develop a Pascal program that displays the land defined by four posts. The program user may enter the coordinates of each post and the program returns with a display of lines connecting each of the posts.

Health Technology
48 The department head needs a Pascal program that will display the temperature readings of three different patients taken 5 times during the day. The X-axis represents the time a

temperature reading was made, the *Y*-axis represents the temperature reading in degrees F. Use a different line style for each patient.

Manufacturing Technology

49   A machine shop requires a Pascal program that will display a stacked bar graph of the number of different parts used in an assembly process each day (during the assembly process, some of the parts are lost or damaged). There are 5 different parts used in the process and the graph is to display the results for five days. The *X*-axis represents the day and the *Y*-axis the number of parts. User input is the day and number of parts used that day.

Business Applications

50   Expand the program in problem 44 so that the name of each salesperson may be displayed vertically along the bar representing his/her sales for the time period.

Computer Science

51   The built-in Turbo Pascal procedure for developing an arc is

$$\texttt{ARC(X,Y, StartAngle, EndAngle, Radius);}$$

where $\texttt{X, Y}$ = INTEGER
$\texttt{StartAngle, EndAngle}$ = WORD
$\texttt{Radius}$ = WORD

The procedure draws a circular arc from the coordinates *X, Y* with the radius $\texttt{Radius}$. The arc travels from $\texttt{StartAngle}$ to $\texttt{EndAngle}$ and is drawn in the current drawing color.

   Create a Pascal program that will allow the program user to select the variables for the above arc and then have the program display the arc along with the values of the variables selected by the program user.

Drafting Technology

52   The built-in Turbo Pascal procedure for creating an ellipse is

$$\texttt{ELLIPSE(X,Y,StartAngle,EndAngle, XRadius,YRadius);}$$

where $\texttt{X,Y}$ = INTEGER
$\texttt{StartAngle, EndAngle}$ = WORD
$\texttt{XRadius, YRadius}$ = WORD.

   The procedure draws an elliptical arc with *X* and *Y* as the center point. $\texttt{XRadius}$ and $\texttt{YRadius}$ are the horizontal and vertical axes. The ellipse travels from $\texttt{StartAngle}$ to $\texttt{EndAngle}$ and is drawn in the current color.

   Create a Pascal program that will allow the program user to select the variables for the above ellipse and then have the program display the ellipse along with the values of the variables selected by the program user.

Agriculture Technology

53   Expand the program in problem 47 so that the program user may enter any number of posts. The program will then display the land enclosed by the posts.

Health Technology

54   Expand the program in problem 48 so that each patient's blood pressure (taken the same number of times each day) is displayed on the same graph as the temperature. Use a different line color to represent the patient's blood pressure.

Manufacturing Technology

55   Create a Pascal program that will display a bar graph of the production output of ten factory workers for a given month. The Y-axis is to represent each individual factory worker, and the X-axis the amount of production output (assume 100 units maximum). The program user is to enter the production output for each of the ten workers along with the name of each worker. Make sure that each bar in the graph is easily distinguishable from the others and that the name of the worker appears vertically along the corresponding bar.

Appendices

Appendix A—Introduction to IBM and MS-DOS

Introduction

The word **DOS** actually stands for Disk Operating System. It is a program or list of instructions contained on a disk that tells your computer how to operate the disk. A disk operating system enables you to do all the things shown in figure 1–3 of Chapter 1. IBM DOS is made by IBM and MS-DOS is Microsoft's version.

The information here is just enough to get you started. When you complete this section, you will know how to boot your system, initialize a disk, and copy a program from one disk to another—just scratching the surface of everything DOS can do. After you understand how to do what is shown here, you should refer to the DOS manual that comes with the computer you are using.

What to Do

Your computer system has one of the following setups.

    1. A single disk drive.
    2. Two disk drives.
    3. A single disk drive and a hard disk drive.

The disk drives (the colon : is part of the drive name) are called:
              A : Drive on the left (or top).
              B : Drive on the right (or bottom).
              C : Hard disk drive (if present).

DOS is contained on a floppy disk sold with the computer. It will be referred to here as the **System Disk**. In most school situations there will be a copy of this disk.

Care of the Disk

    1. When not using the disk, keep it inside its protective jacket.

2. Never touch an exposed surface of the disk (or allow anything else to come in contact with it).
3. Do not bend the disk or place it in a position where it could be bent.
4. Do not use sharp objects (such as pencils or ball point pens) on the disk or on any of its labels. If you must write on it, use a felt-tip pen.
5. Do not expose the disk to direct sunlight.
6. Do not expose the disk to any magnetic fields (watch out for all metal objects such as scissors, staple removers, or screw drivers—they could be magnetized).
7. Do not expose the disk to extreme heat (such as in the glove compartment of your car) or extreme cold.

Starting the System

1. Be sure the computer is OFF.
2. Insert the System Disk in drive **A:** as shown in figure A–1.
3. Gently close the load lever. If it does not close, it means the disk is not correctly inserted.
4. Locate the power switch on your system. Refer to figure A–2.
5. Your next step depends upon the system you are using. Refer to table A–1 on page 616.
6. Turn the system on. Nothing will happen for what may seem like a long time, perhaps up to a minute, depending on how much memory the computer has. Once things begin to happen, the red light located on drive **A:** (where you put the System Disk) will come on. This means the disk is being used by the computer. (It is now in the process of reading the DOS from the disk and loading it into its memory.)

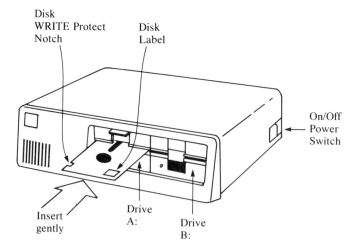

Figure A–1   Inserting the System Disk

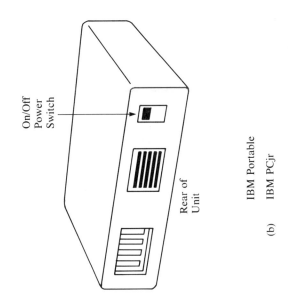

On/Off
Power
Switch

Rear of
Unit

IBM Portable

(b)   IBM PCjr

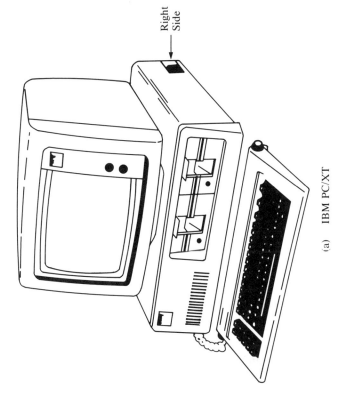

Right
Side

(a)   IBM PC/XT

Figure A–2   Location of Power Switch

615

Table A–1   What Needs Turning On

System	What to Do
IBM PC/XT with a monochrome monitor	Turn on the single power switch that provides power to both the computer and the monitor.
IBM Portable PC	Turn on the single power switch that powers everything.
IBM PCjr	Turn on two switches—one for the computer, one for the monitor.
If you are using an IBM color monitor.	Turn on two switches—one for the computer, one for the color monitor.

7. Look at figure A–3 and make sure you know where the major control keys are on your keyboard.

8. The process of your computer reading DOS into its memory is called **booting**. It comes from "picking yourself up by your bootstraps"—essentially, this is what the computer is doing. It is getting instructions on how to operate a floppy disk from a floppy disk.

9. Once your computer is booted, the red light on the disk drive will turn off and you will be asked to enter the date on the screen. Do this as shown on the screen (`Month-Day-Year`), for example, `9-1-90`, then press the RETURN/ENTER key.

10. Next the screen will ask for an input for the time. Enter this information as shown on the screen (`Hour:Minutes`), for example, `13:15` (means 1:15 PM)

11. You will then see the famous **DOS prompt**.

    A>

    This means the computer is waiting for a DOS instruction from you; it also means that drive **A:** is the active (default) drive.

12. Your computer has now been booted (it contains DOS). When you remove the System Disk, DOS is still inside your computer memory. It will stay there until you turn the computer off. Once the computer is turned off (on purpose or accidentally) the DOS is lost and you must start this process over again.

Preparing a New Disk

1. The first step in using a new disk is to format it—to get the disk ready to put programs on it. If your disk has been used before, formatting it will erase everything on it. Make sure there are no programs you'll need later on the used disk, because they won't be there after formatting.

2. Make sure your system is booted, leave the System Disk in drive A:, enter

    FORMAT B:

    and press RETURN/ENTER.

    (Note that it makes no difference if you use uppercase or lowercase letters. You could have entered **format b:**)

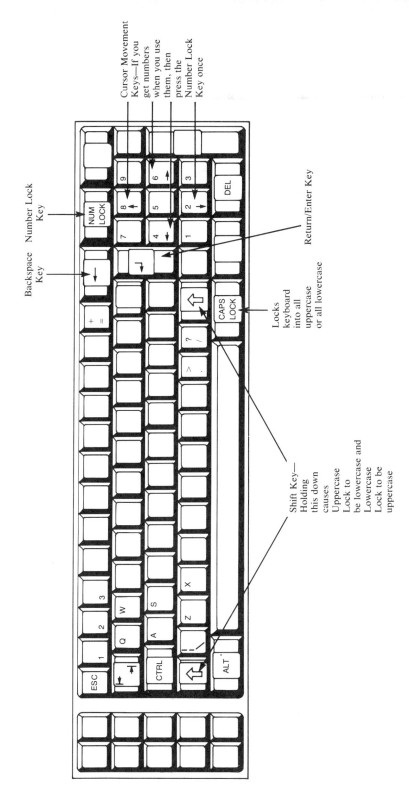

Figure A–3 IBM PC Keyboard

3. Follow the instructions on the screen:

```
     For a one-drive system, remove the System Disk
from drive A: and place your disk in drive A: and then
press a key to continue...
     For a two-drive system, leave the System Disk in
drive A: and put your disk into drive B: and then press a
key to continue...
```

4. When the formatting process begins, the red light of the disk drive with your disk in it will light up, indicating that it is in use. Your disk is now being prepared to be used by the computer system; this process takes almost a minute.

5. When the light turns off, the computer will ask if you want to format another disk. In any case, your disk is now ready to have programs saved onto it.

Copying Programs to Your Disk

1. Make sure your system is booted and your disk is formatted.
2. Put the disk you want to copy programs from in drive **A**:. This will be called the **source disk.**
3. Put the formatted disk you want to copy programs to in drive **B**:. This will be called the **destination disk.**
4. To see the programs that are on the source disk, enter

DIR/W

and press the RETURN/ENTER key.

5. The names of the programs on the source disk will appear on the screen. Note that there may be two parts to the name of each program.

NAME.EXT

The **NAME** of the program will usually have letters and/or digits; the **EXT** (after the period) is optional. The **NAME** can have a maximum of eight places, the **EXT** a maximum of three places.

6. To copy a selected program from the source disk to the destination disk, enter

COPY SOURCE.EXT B: SOURCE.EXT

where **SOURCE.EXT** is the name of the program on the source disk you want to copy. For example, if the name of the program you want to copy is **PASCAL.40**, you would enter

COPY PASCAL.40 B: PASCAL.40

and press the RETURN/ENTER key.

7. The red light on drive **A**: will come on, indicating that the selected program is being placed into the computer's memory. The red light on drive **B**: will then come on to indicate that the selected program is being copied to your disk. The screen will then tell you if the copy was successful. (Note that some programs are copy-protected; the manufacturer does not want copies made of the program and you will not be able to copy it to your disk.)

Appendix B—Standard Pascal

Turbo Pascal contains many enhancements of standard Pascal. The reserved words of standard Pascal (sometimes called "textbook Pascal") are:

```
AND ARRAY BEGIN CASE CONST DIF DN DOWNTO ELSE END
FILE FOR FUNCTION GOTO IF IN LABEL MOD NIL NOT OF OR
PACKED PROCEDURE PROGRAM RECORD REPEAT SET THAN TO TYPE
UNTIL VAR WHILE WITH
```

Appendix C—ASCII Character Set

American Standard Code for Information Interchange

Character	Code	Character	Code	Character	Code
blank	32	@	64	`	96
!	33	A	65	a	97
"	34	B	66	b	98
#	35	C	67	c	99
$	36	D	68	d	100
%	37	E	69	e	101
&	38	F	70	f	102
'	39	G	71	g	103
(	40	H	72	h	104
)	41	I	73	i	105
*	42	J	74	j	106
+	43	K	75	k	107
,	44	L	76	l	108
—	45	M	77	m	109
.	46	N	78	n	110
/	47	O	79	o	111
0	48	P	80	p	112
1	49	Q	81	q	113
2	50	R	82	r	114
3	51	S	83	s	115
4	52	T	84	t	116
5	53	U	85	u	117
6	54	V	86	v	118
7	55	W	87	w	119
8	56	X	88	x	120
9	57	Y	89	y	121
:	58	Z	90	z	122
;	59	[	91	{	123
<	60	\	92	|	124
=	61	]	93	}	125
>	62	↑	94	-	126
?	63	—	95	DEL	127

Appendix D—The Seven Phases of APM

Table D–1   Seven Phases of APM

Phase	Action	Discussion	Guide
1	Understand the problem.	This is the most important step in the solution of any problem.	Problem Statement
2	Solve the problem by hand.	This confirms that you understand the problem, and helps you identify all of the steps necessary for the solution of the problem. You may of course use a pocket calculator or even the computer in the solution.	Problem Solution
3	Record all of the necessary steps needed for solving the problem by hand.	Here you are creating a recipe called an algorithm of what to do. It lists, step-by-step, each part of the process to go through to solve the problem. This is really the first phase of developing the computer program.	Algorithm Development
4	Decide which of the three control methods will be needed.	Recall the Boehm and Jacopini Theorem of action blocks, branch blocks, and loop blocks. In this step you are documenting which of these blocks you will need in order to replicate the algorithm from step 3.	Control Blocks
5	List and define all program constants and variables.	You must have a clear and well-documented record of all formulas, as well as what each part of each formula means. This is very important for all computer languages, but especially so for Pascal.	Constants and Variables
6	Write the Pascal program just using procedures and enough code to compile.	This is the all-important documentation that all good programs require. In top-down design, the program documentation is done first and the coding is done last.	Program Design Steps
7	Enter the Pascal program code and test the program.	If the first six steps have been done well, this step will require very little time.	

Appendix E—APM Guides

PROBLEM STATEMENT GUIDE

1. What will the program do?
2. What is needed for input?
3. What will be given as output?

PROBLEM SOLUTION GUIDE

Formula:
Values Used:

Step	Computation	What did you actually do?
1		

Formula:
Values Used:

Step	Computation	What did you actually do?

ALGORITHM DEVELOPMENT GUIDE

  I. Explain program to user.
    a. What does the program do?
      [The process]
    b. What does the program user need to do?
      [The input]
    c. What will be displayed?
      [The output]
 II. Get values from user.
III. Do computations.
 IV. Display answer(s).
  V. Ask if program is to be repeated.

CONTROL BLOCKS GUIDE (PROCEDURES)

Action blocks

Number	Action taken	
1		[Procedure Name;]

Counting
Loop Blocks

No: What is repeated:		
Starting Value:		
Increment:		
Ending Value:		[Procedure Name;]

No: What is repeated:		
Starting Value:		
Increment:		
Ending Value:		[Procedure Name;]

CONTROL BLOCKS GUIDE (PROCEDURE) *continued*

Sentinel	
Loop Blocks	

No: What is repeated:	
Under what condition:	[Procedure Name;]

No: What is repeated:	
Under what condition:	[Procedure Name;]

Branch	
Blocks	

No: Condition for branch;	
If condition is met:	
If condition is not met:	[Procedure Name;]

No: Condition for branch:	
If condition is met:	
If condition is not met:	[Procedure Name;]

CONSTANTS AND VARIABLES GUIDE

Formula or [Other]	Variables and Constants	Meaning	Type					Block(s) Used
			R	I	C	B	S	

Appendix F—Pascal Design Guide

PASCAL DESIGN GUIDE

Step 1. Give the program a descriptive name, ending with a ;
   Example: **PROGRAM** `Resistor_Power;`
         (Note: You are safe if you use at least two
            words separated by an underscore.)
Step 2. Give your name and the development date using comments.
   Example: `{*****************************************`
            `Developed by:`
            `Date:`
         `*****************************************}`
Step 3. State what the program is to do and what is needed.
   Example: Refer to your Problem Statement Guide.
Step 4. Define all constants and variables.
   Example: Refer to your Constants and Variables Guide

Step 5. Write a program block for each of your procedures. The generalized format of a program block (procedure) is

```
PROCEDURE Name_of_Procedure;
{----------------------------------------
 Purpose of procedure:
 Constants used:
 Variables used:                                    }

 BEGIN

 {List procedure steps you will want the procedure
   to do.
       .
       .
       .                                            }
 {----------------------------------------}
   END; { End of Name_of_Procedure }
```

Observe that the **END** of a procedure finishes with a **;**
not a period!

Step 6. Write the main programming sequence using the names of each of your procedures. Put this between the Pascal **BEGIN** and **END.** statements. Note there must be a period following the reserved word **END.**

```
Example: { || Main Programming Sequence ||
         =========================================}
           BEGIN
             Name_of_a_Procedure;
             Name_of_Anotherone;
             Name_of_the_Last_One;
           {=========================================}
           END. {of main programming sequence }
```

Step 7. Compile the program, correct any errors, save it on your disk and make a printed copy. You are now ready to enter program code. The design phase of your program is completed.

Appendix G—General Pascal Structure

Program G–1

```
PROGRAM Example_One;

  {*************************************************************
            Developed by:  Name of program developer.
                    Date:  Date of development.
```

Program G–1   *continued*

```
       ************************************************************
          Program description:
          What the program will do, what is needed for input
          and what will be given for output.
       ************************************************************
                          Constants used:
       -------------------------------------------------------------
              Constant1 = Value              [Type]
              Constant2 = Value              [Type]
                    .           .                .
                    .           .                .
                    .           .                .
              ConstantN = Value              [Type]
       ************************************************************
                          Variables used:
       -------------------------------------------------------------
          Variable1 = Explain what variable is.   [Type]
          Variable2 = Explain what variable is.   [Type]
              .                       .                .
              .                       .                .
              .                       .                .
          VariableN = Explain what variable is.   [Type]
       ***********************************************************}

       PROCEDURE First_One;
     {------------------------------------------------------------
        Explain what this procedure does.
        [Action, looping and/or branching]
        ------------------------------------------------------------
        Constants used:
        Variables used:                                            }

       BEGIN

          { Step one.....;}
          { Step two.....;}
          {     .        ;}
          {     .        ;}
          {     .        ;}
          { Step N.......;}

     {  ---------------------------------------------------------}
       END;   {of Procedure First_One;}

       PROCEDURE Second_One;
     {  ------------------------------------------------------------
        Explain what this procedure does.
        [Action, looping and/or branching]
        ------------------------------------------------------------
        Constants used:
        Variables used:                                            }

       BEGIN
```

Program G–1   *continued*

```
      { Step one.....;}
      { Step two.....;}
      {      .       ;}
      {      .       ;}
      {      .       ;}
      { Step N.......;}
  { ----------------------------------------------------------}
    END;    {of Procedure Second_One;}
                        .
                        .
                        .

    PROCEDURE Last_One;
  { -----------------------------------------------------------
    Explain what this procedure does.
    [Action, looping and/or branching]
    -----------------------------------------------------------
    Constants used:
    Variables used:                                           }

  BEGIN

      { Step one.....;}
      { Step two.....;}
      {      .       ;}
      {      .       ;}
      {      .       ;}
      { Step N.......;}
  { ----------------------------------------------------------}
    END;    {of Procedure Last_One;}

 { || Main Programming Sequence ||
    ==========================================================}
  BEGIN
      { These may be in any order you need. }
          First_One;
          Second_One;
                  .
                  .
                  .
          Last_One;
  { ==========================================================}
    END.  {of main programming sequence;}
```

Glossary

Action Block   A program block that consists of a linear execution of programming statements.

Algorithm   A step-by-step procedure in words of a computer program.

Algorithm Development Guide   A structured format for the development of an algorithm.

AND   A logical operator that will be TRUE if and only if all of its elements are TRUE; otherwise it is FALSE.

Applications Software Programs that cause the computer to perform a particular function.

Applied Programming Method A structured procedure for converting a problem into a computer program.

Area Fill A programming method of filling defined areas on the graphic screen with a particular pattern or color.

Array Identical data elements that are grouped sequentially and are accessible by an index value.

Assignment Operator The := symbol; it assigns a value to a variable or function of the same type.

Auto Debug A method of programming so that you can activate or deactivate debugging features in your program.

Autodetection In Turbo Pascal, a procedure within the program that will automatically determine the type of graphics hardware available in the computer system.

Bit-Mapped Graphics developed by specifying the location of each pixel.

Block Source code grouped together in such a manner that it is easy for people to see the purpose of different parts of a program.

Block Structure In programming, putting the source code in a paragraph-like structure (called a block) in which each block does a specific program task; this makes the entire program easier for people to read, understand, and modify.

Boolean A type that has only two conditions: TRUE or FALSE. Used in branch blocks.

Boolean Algebra A system of algebra that utilizes only two conditions: TRUE and FALSE. Named after its inventor, George Boole, who introduced it in 1847.

Branch In programming, signifying a condition where there are two or more choices.

Branch Block A program block that has the capability of altering the forward sequencing of program statements.

Bubble Sort A method of sorting where sorted elements are bubbled to the top much the same as bubbles in water.

Byte A type in Pascal that conventionally consists of eight bits.

Calling a Procedure In Pascal, activating a specific pre-defined PROCEDURE from within the program.

Carriage Return The ENTER key on the computer keyboard.

Char A type in Pascal that consists of a single character.

Character A literal that occupies only one space.

Character Set An agreed-to group of symbols (usually those available from the keyboard) used to enter a program into the computer.

Character Strings A sequence of keyboard entries that are treated as a single unit.

Chip Accepted technical slang for integrated circuit.

Closed Branch A programming branch that will take one of two options and then continue forward in the program.

Color Constants Predefined data types that indicate a specific color on the graphics screen.

Color Palette A graphics mode of operation that defines a specific range of available colors.

Compiler   A program that converts a program in a high-level language into another program in a lower-level language that is easily executed. This conversion need be done only once, and then the lower-level language program can be used for day-to-day operation.

Compiler Directive   In Pascal, an instruction to the compiler that is imbedded in a program.

Compile-Time Error   A programming error which will stop the compiling of a program. In most systems, the user will be informed what is the most likely cause of the error.

Completeness Theorem   A theorem that states that any program logic, no matter how complex, can be resolved into action blocks, branch blocks, and loop blocks.

Compound Statement   In Pascal, a series of statements bordered by a matching set of the reserved words BEGIN and END.

Computer   A device capable of accepting information, applying a prescribed process to the information, and supplying the results of this process.

Computer Language   A set of characters that can form symbols and the rules for using these symbols so that the programmer and the computer can communicate.

Computer Science   The entire range of theoretical and applied disciplines concerned with the development and applications of computers. Includes mathematics, logic, language analysis, computer design, systems engineering, and information systems as well as computer programming.

Concatenation   The joining of two or more character strings.

Congruent Arrays   Arrays that have elements of the same type, dimensionality, and number of elements in each dimension.

Constant   In a program, represents a fixed value.

Constants and Variables Guide   An aid for defining all constants and variables to be used in the program.

Control Blocks Guide   An aid for the development of procedures in Pascal.

Coordinate Transformation   The process of relocating the graphic screen coordinates to accommodate the requirements of a specific display.

Counting Loop   A programming loop with a defined starting value, incrementing value, and ending value.

Data   Fundamental elements of information that can be processed or produced by the computer.

Debug Routine   Procedure in a Pascal program used to locate a run-time error.

Declarations   Stating what identifiers are to be used in the Pascal program. These occur after the program header and before the rest of the program.

Declare   To enter into a Pascal program the definition, name, and type of a program identifier.

Default   Value assignments automatically given unless overridden by the user.

Disk Drive   The peripheral device used to store and retrieve information to and from a magnetic disk.

Disk Operating System (DOS)   A program which contains instructions so the computer and user can interact with the disk drives to store, retrieve, modify, and transfer information.

Editor   A program that allows you to enter programming code and other information from the keyboard. It usually contains features that will help you correct errors and make modifications to existing programs.

End-of-File Marker   A character that denotes the end of a complete file.

End-of-Line Marker   A character that denotes the end of a line.

Exponential Notation   Expressing a number as a base raised to a specific value denoted by the use of an exponent.

Field   The identifier used within a record.

Files   Collections of data that can be stored and retrieved from a disk or other similar memory element.

Fill-Style   The type of pattern to be used in an area fill or similar graphic operation.

Firmware   Hardware that contains programs stored in Read-Only Memory (ROM).

Floppy Disk (Diskette)   A thin, flexible disk coated with magnetic material used for long-term storage of computer programs or other information.

Flowcharts   A pictorial representation of a computer program.

Font   The style used to present text characters.

Forward Declarations   Declaring procedures or functions and their parameters ahead of the actual location of the body of the procedure or function.

Function   In Pascal, a subroutine that computes and returns a value.

Global   The term given to a program variable that is declared in the main program block and can be accessed from anywhere within the program.

Graphic Driver   A program specifically developed to work with a designated type of graphic hardware.

Graphic Mode   A specific operating condition from a list of available operating conditions for a given graphics adapter.

Graphics Adapter   The actual system hardware used to display graphics.

Graphics Screen   The monitor display of graphic information.

Guide   A document used by programmers as an aid in developing a computer program.

Hardware   The electronic and mechanical equipment used for processing information.

Hexadecimal Number System   A number system to the base 16 used in computer programming because of its ease of transforming code between programmer and machine.

High-Level Language   In programming, using commands that are close to everyday writing used by people.

Identifier   A name given by the programmer to a specific item in a program.

Index   Identifies a position within a list of elements.

Initialize   Giving a known value to a variable or data structure.

Instruction   A coded program step that tells the computer how to perform a single operation of the program.

Integer   A type of whole number that has a specific number range from a minimum negative value to a maximum positive value.

Integer Number   A whole number without the use of the decimal point.

Integrated Circuit   A combination of interconnected circuit elements inseparably associated on or within a continuous physical material called a substrate.

Interface   A common hardware boundary between two systems, such as a computer and a peripheral device.

Interpreter   A program that converts a high-level language program into an executable machine code every time the high-level language program is executed.

I/O   Input/Output.

Keyboard   The peripheral device used to enter typed information into the computer.

Language Level   The degree to which a programming language is easily understood by the microprocessor compared to the similarity of the language to the natural written language of people.

Library   Part of a program that is referenced by other parts of the program for frequently used procedures and functions as well as user-defined types and constants.

Local   The term given to a program variable that is declared within a specific procedure or function and whose scope is limited to that procedure or function.

Loop   That part of the program that has the capability of repeating itself.

Loop Block   A program block that has the capability of going back and repeating a previously executed part of the program.

Low-Level Language   In programming, commands that are easily implemented by the microprocessor with little-or-no further processing.

Machine Language   Information that can be used directly by the microprocessor without further processing.

Machine-Language Program   A computer program written in binary numbers (1 and 0). Sometimes programs written in hexadecimal numbers are called machine-language programs.

Main Programming Block   In Pascal, that part of the program that calls the procedures in the order of their execution.

Memory   Hardware used to store information.

Microprocessor   The controlling portion of a computer that consists of a single integrated circuit.

Mode   The selected operating condition. (See graphic mode.)

Monitor   The peripheral device used to display information so that the program user can interact with the computer.

Monochrome   A monitor capable of producing only one color.

Nested Loops   A program loop that contains other program loops.

Nesting   The placement of one unit of a program within another.

NOT A logical operator that means the opposite condition. NOT FALSE is TRUE and NOT TRUE is FALSE.

Null String A string that does not contain any characters.

Object Code In Pascal a P-code (for pseudomachine) is produced as an intermediate code. This P-code is executed interpretively on the host computer. Ideally, this allows the P-code to be used by different computer systems.

Open Branch A branch block that has the capability of executing a different programming sequence and/or continuing on with the program.

Operator Symbols such as + and − that are used to form expressions.

OR A logical operator that will produce a TRUE if one or more of its elements are TRUE; otherwise it produces a FALSE.

Ordinal A range of values that has a predetermined order.

Packed Arrays Arrays that have been stored in memory in such a manner as to minimize the amount of memory space used.

Parameter List The variables and values declared in the heading of a procedure or function.

Pascal A programming language designed to teach good programming habits emphasizing the aspects of structured programming.

Pascal Design Guide An outline that list the steps for creating the initial Pascal program using comments and procedures with only enough code to cause the program to compile.

Pascal Operating System A set of programs that allow the editing, compiling, and other processes necessary to use the Pascal language with a microcomputer.

Peripheral Devices Devices external to the main part of the computer that are used to store information, retrieve information, display information, or otherwise interact with the user or external environment.

Pixel The smallest graphic element that can be controlled by the program.

Predecessor That which comes before.

Printer The peripheral device used to print information supplied by the computer or other peripheral devices associated with the computer.

Problem Solution Guide An aid for documenting the steps required for the solution of the problem by hand.

Problem Statement Guide An aid in stating in words the problem to be programmed.

Procedure In Pascal, a program block that does an identifiable task.

Processor A physical device which receives data, manipulates it, and supplies results. It usually contains internally stored instructions accessible to the programmer.

Program A list of coded instructions or steps that tells the computer exactly what process(es) to perform.

Programmer's Block A series of comments that begin every program; contains all the information anyone would need about the entire program. Usually written before any other part of the program is started and updated when the program is completed.

Program Structure   The way in which the source code is displayed.

Random-Access Files   Data collections that may be directly accessed without sequentially searching the entire element containing the desired collection of data.

Read-Only Memory (ROM)   Computer memory that stores information on a permanent basis—usually programmed at the factory.

Read-Write Memory (RAM)   Computer memory where information can be stored or retrieved as long as there is electrical energy being supplied to it.

Real   A type of number that contains decimal places.

Real Number   A number that contains a decimal point.

Record   A collection of elements called fields that may be of different types.

Recursion   The act of a procedure or function calling itself.

Relational Operators   The symbols $<=$, $<$, $>$, $<>$, $<=$, $>=$, and IN that are used to form Boolean expressions.

Remarks   That part of the program that will have no effect on the execution of the program and is used only for the benefit of people reading the source code.

Reserved Words   Identifiers reserved by the compiler that have very specific predefined meanings.

Resident Monitor Program   Program stored in ROM that starts up the computer system so it is ready to receive instructions from a peripheral device, usually the disk drive or keyboard.

Scalar   In Pascal, a type that consists of ordered components.

Scaling   The process of mathematically modifying a formula that is to be graphically displayed on the screen, to insure that the entire graph will be displayed.

Scope   The accessibility of an identifier within a program.

Sectors   Coded areas on a disk used by the computer to identify where information such as a stored program is located.

Sentinel Loop   A program loop for which it is not known how many times the loop will be required, if at all.

Sequential Files   Files that must be accessed one after the other, as opposed to random-access files.

Software   Different programming aids supplied by manufacturers or programmer to facilitate efficient operation of computer equipment or to cause the computer system to modify itself to act as a specialized machine (such as a word processor).

Standard Pascal   The Pascal programming language as defined in the report by Jensen and Wirth.

Statement   The most basic unit in a Pascal program, separated by semicolons.

String   Sequence or group of characters treated as a single unit.

Stroked Font   The development of text on the graphics screen using vectors.

Structure   In programming, the way in which the source code is displayed. Good programming structure requires that the source code make it easy for people to read, understand, and modify.

Structured Program   A program displaying source code in such a manner that it makes it easy for people to read, understand, and modify.

Subscript   A number or letter written below and to the right of a quantity used to identify it from similar quantities.

Successor   That which comes after.

System Software   Program(s) that are used to operate a particular type of computer and may include the operation of peripheral equipment.

Text File   A file that contains string characters.

Text Screen   The monitor display of text information, usually developed by factory-defined characters. (See graphics screen.)

Top-Down Design   The process of designing computer programs whereby the most general ideas of the program are the first step, and the details of inserting program code are entered as the last step.

Type   The kind of variable or constant used in a Pascal program.

Unit   A Turbo Pascal library.

Unstructured Program   A program in which little or no attempt has been made to make the source code easy for people to read, understand and modify. This is also considered poor program structure.

Value Parameter   A variable whose value is passed between procedures or functions. Its value cannot be changed once assigned by the calling procedure.

Variable   Identifiers used in the program whose values may change during the course of program execution.

Variable Parameter   A variable whose value may be changed as it passes between procedures or functions.

Variant Record   A record in which some of the fields share the same area in memory.

XOR   A logical operator that will produce a TRUE only if one of its two terms are TRUE, but not both; otherwise, it produces a FALSE condition.

Answers

Answers to Section Reviews—Chapter 1

Section Review 1–2

1. The major parts are the monitor, keyboard, disk drive, printer, and the computer itself, consisting of the processor and memory (RAM and ROM).
2. Hardware is the computer and all physical things attached to it. Software is a list of instructions called a program that the computer processor will perform.
3. Some examples are a printer, disk drive, monitor, and keyboard. (There are also others such as game paddles and card readers.)
4. The purpose is to make the computer act as a word processor, accounting spread sheet, game, or some other specialized system.
5. A microprocessor is a small integrated circuit that can perform specified sets of instructions.
6. RAM is memory that can store instructions entered by the computer user or a peripheral device. These instructions will be lost when the computer is turned off. ROM is memory that has been installed at the factory and will be retained even when power is not applied to the computer.
7. The purpose is to have a set of instructions (a program) built into the computer ROM, so that the computer will know what to do once it is turned on by the user.
8. The purpose is to have a program that will give instructions to the computer on how to operate the computer's disk drive system.
9. 1. Formatting (initializing) a disk.
   2. Copying information from memory to the disk.
   3. Copying information from the disk to memory.
   4. Copying information from one disk to another.
   5. Displaying information from the disk on the monitor.
   6. Copying information from a disk to the printer.

10. Storage takes place on concentric tracks that are divided into sectors.

Section Review 1–3

1. A computer language is a set of characters that are used to form symbols, and the rules for combining these symbols into meaningful communications between the programmer and the computer.
2. The two things are ON and OFF or 1 and 0.
3. The language is machine language.
4. Language levels refer to how close the symbolism and structure of the computer language is to the everyday common language of the user, contrasted with how close the computer language is to the ON and OFF instructions understood by the microprocessor.
5. An interpreter is a program stored in the computer that converts (interprets) the symbolism of a higher-level language into the ON and OFF code understood by the microprocessor.
6. A BASIC interpreter is a program in the computer that converts the symbolism of BASIC into the ON and OFF symbolism understood by the microprocessor.
7. The computer must have an interpreter for that language.
8. Some of these languages are FORTRAN, used for solving mathematical formulas; BASIC, which is easy to learn; Pascal, which teaches good programming habits; PILOT, used as an aid in developing computer-aided instruction; and C, used to easily control the computer.
9. A program that converts the source code into an object code. This object code may be saved and executed over and over again.

Section Review 1–4

1. An editor is a program that lets you enter and change information from the computer keyboard.
2. A compiler is used to convert the information entered by you from the keyboard to a code that is understood by the computer.
3. The compiler converts the information entered by you into a code understood by the computer, thus creating another program. This results in two programs: your original, and the compiled version of your original.
4. Source code is the information you enter. Object code is information compiled from the source code and used by the computer to execute your instructions.

Section Review 1–5

1. Pascal is intended to teach good programming habits and to be portable between different computer systems.
2. Portability means that a program can operate on computers with different systems.
3. Pascal programs must start with **PROGRAM** `Name;`.
4. All program instructions must go between **BEGIN** and **END.**
5. `{ This is a Pascal comment. }`
6. The brackets enclose comments that make the program easier for people to understand.

Section Review 1–6

1. Structure is the format used when entering a program.
2. It isn't necessary to give any particular kind of structure.
3. A block structure has certain parts of the program located in specific areas of the program document. An example of block structure is the structure of a letter.
4. Structured programming makes the program easier to read, understand, and change.
5. The programmer's block consists of remarks that give all needed information about the program. It contains the name of the program, the developer's name, the date, and what the program will do; it also describes the purpose of all the program constants and variables.

Section Review 1–7

1. Before using the Turbo Pascal operating system, you should make a working copy of the files you will use and put the original disks away in a safe place.
2. The main menu is the top line of the Turbo Pascal screen that contains **File**, **Edit**, **Run**, **Compile**, and **Options**.
3. You enter the editor by using the left/right arrow keys to highlight the word Editor on the main menu, then press the RETURN/ENTER key.
4. The key that will get you back to the main menu is **F10**.
5. The preferred method of leaving the Turbo environment is to get to the file menu and use arrow keys to select the **Quit** option, then press RETURN/ENTER.

Answers to Self-Test—Chapter 1

1. The program will display the following on the monitor.

```
This program will compute the volume
of a room. You need to enter the value
of the room height, width, and length, and
the computer program will do the rest.
```

2. The same display as that in question 1 will appear on the monitor.
3. L = length of room; W = width of room; H = height of room. V is the volume of the room and was not defined in the programmer's block.
4. No. It omits the program name, name of the programmer, and the date, and it fails to explain one variable (V, the volume of the room).
5. Yes. The first bracket at the very beginning of the program is backwards (**}**). **PROGRAM Name;** is missing and though this is not a problem in the TURBO operating system, it should be inserted.
6. The program will ask the program user to enter the value of the room height, length and width, then compute the volume and display the answer on the monitor screen.

Answers to Problems—Chapter 1

General Concepts

1. See figure 1–1 for the major parts of a computer system.

3. The resident monitor program causes your system to start up. For the IBM PC it does some internal checking, then beeps the speaker and causes the disk drive on the left to come on. If no disk is in this drive, it automatically evokes the IBM BASIC program.

5. Most agree that BASIC is the easiest to learn. Pascal was designed to teach good programming habits, while COBOL was developed for business applications.

7. The main parts of the Pascal operating system are the editor (program that allows you to type in Pascal commands from the computer keyboard), compiler (program that develops an object code from your source code), and disk operating system (program that assists you in saving your programs to the disk and copying other programs from the disk). [Note: In Turbo Pascal, this job is accomplished by the FILER, which is contained within the Turbo Pascal operating system.]

9. A suggested program is

```
{ This Pascal program will print a message on
the screen. }
PROGRAM Answer;
 BEGIN
  WRITELN('Applied Pascal for technology.');
 END.
```

Program Analysis

11. The following will appear on the screen.

```
{ This is a Pascal program. }
```

13. The program needs to have the { and } added around the comment.

```
{ This is a comment }
```

Note that the semicolon is removed—none is used after a comment.

Answers to Section Reviews—Chapter 2

Section Review 2–1

1. Computers are good at (as listed in table 2–1) repetition, arithmetic, logical decisions, sorting information, and graphical representation.

2. Computers are not good at (as listed in table 2–2) patterns, judgement, adaptation, and working from incomplete information.

3. The outcome of a logical decision is always predictable, but the outcome of a judgement is not.

Section Review 2–2

1. The completeness theorem describes the three tools available to the programmer: action, branching, and looping.
2. An action block is one programming sequence following another, a branch block has the ability to do something different based on a pre-determined condition, and a loop block has the potential of going back and repeating some prior instructions.
3. An open branch may do something different but it will always continue on with the program; a closed branch will always do one of two unique choices and then continue on with the program.
4. When using a counting loop, you know ahead of time how many times a part of the program will be repeated; when using a sentinel loop, it is not known how many times a part of the program will be repeated.

Section Review 2–3

1. Flowcharting shows how the program is structured and is language independent.
2. A) Decision   B) Terminal   C) Input/Output   D) Subroutine
3. A process block flowchart symbol represents a computer instruction indicating an action, such as adding two numbers together.
4. A decision logic block flowchart symbol can be used to show the structure of a loop or branch block.
5. An input/output block flowchart symbol shows input or output external to the computer, such as from the keyboard or to the printer.
6. A terminal point tells you where the program starts and ends.

Section Review 2–4

1. The main idea of top-down design is to start with the most generalized concept of the problem as the first step and to enter program code as the last step.
2. The first step in top-down design is to explain the problem to be programmed in writing; this is done to insure that you and others fully understand the problem before beginning to write a program to solve it.
3. The most common programming sequence is: explain program, get values, do computations, display answer, and ask for program repeat.
4. A Pascal procedure is a programming block that does a specific program activity. It must have the reserved words **PROCEDURE** followed by a name, **BEGIN**, and **END;**. It ends with a semicolon (**;**).
5. The main programming sequence block is the last part of the Pascal program; it starts with the reserved word **BEGIN** and ends with the reserved word **END.** followed by a period (.). It uses the names of predefined procedures to execute the program.

Section Review 2–5

1. APM presents a structured procedure for converting a problem into a computer program.

2. The first six steps used in APM are: understand the problem, solve the problem by hand, record all the steps in solving the problem, decide which control methods are needed, list and define all program constants and variables, and write the Pascal program just using procedures and enough code to compile.

3. A. The Problem Statement Guide helps you state the problem in words.
   B. The Problem Solution Guide helps make sure you understand how to solve the problem.
   C. The Algorithm Development Guide helps you create a correct programming sequence for the solution of the problem.
   D. The Control Blocks Guide assists you in the development of Pascal procedures.
   E. The Constants and Variables Guide helps make sure you have carefully considered all of the constants and variables you will need in the program.
   F. The Pascal Design Guide gives you a format for the initial design of the actual Pascal program.

4. The steps used in Pascal program design are: give the program a descriptive name, give your name and the development date, create the problem statement, enter definitions of all constants and variables, enter the main programming sequence by using the names of your procedures, enter each of the procedures ahead of the main programming sequence, then compile the program.

Section Review 2–6

1. The first three phases of APM are to understand the problem, solve the problem by hand, and record all of the necessary steps needed for solving the problem by hand.

2. The guides used for the first three phases of APM are the Problem Statement Guide, Problem Solution Guide, and Algorithm Development Guide.

3. The second three phases of APM are to decide which control methods are needed, list and define all program variables, and write the Pascal program using just comments, procedures and a main programming sequence with just enough code in it to compile.

4. Guides used for phases four, five, and six are the Control Blocks (Procedures) Guide, Constants and Variables Guide, and Pascal Design Guide.

Section Review 2–7

1. It is considered good practice to keep your Pascal programs on a separate disk from your Pascal working disk to prevent the Pascal operating system files from being accidentally modified.

2. When saving source code to the disk, the Turbo system automatically gives it the extention `.PAS`.

3. The extension `.BAK` on a Pascal file means that this is a backup file of a modified `.PAS` file of the same file name.

4. To create an executable Pascal program you must save the compiled program to the disk. This is done by changing the **Compile** menu option of **Destination memory** to **Destination disk**.

Answers to Self-Test—Chapter 2

1. There are six procedures, not counting the main programming sequence.
2. The procedures are `Explain_Program_to_User`, `Get_Resistor_Value`, `Check_It_Out`, `Display_the_Answer`, `Calculate_It`, and `Ask_for_Repeat`.
3. Yes. `Calculate_It` uses two other procedures: `Display_the_Answer` and `Check_It_Out`.
4. There are four action blocks: `Explain_Program_to_User`, `Get_Resistor_Value`, `Display_the_Answer`, and `Ask_for_Repeat`.
5. Yes, there are two loop blocks: the main programming sequence, and `Calculate_It`. `Calculate_It` is a counting loop and the other is a sentinel loop.
6. There are four constants used. `Starting_Current_Value = 1;` gives the starting value of the current. `Increment_Current_Value = 1;` gives the amount the current will be incremented each time. `Ending_Current_Value = 10;` states the current value that will cause the loop to end. `Maximum_Voltage = 100;` gives the decision value to be used by the branch block.
7.

CONTROL BLOCKS GUIDE

Action blocks Number	Action taken
1	Explain the program to program user. [`Explain_Program_to_User;`]
2	Get value of resistor from program user. [`Get_Resistor_Value;`]
3	Display the value of the voltage across resistor [`Display_the_Answer;`]
4	Ask the program user if the program is to be repeated. [`Ask_for_Repeat;`]

Counting Loop Blocks

No: 1   What is repeated: `Resistor_Voltage` calculation,
                [`Display_the_Answer;`]
                [`Check_it_Out;`]
    Starting Value: `Resistor_Current = 1`
    Increment: **by** 1
    Ending Value: `Resistor_Current = 11`
                [`Calculate_It;`]

CONTROL BLOCKS GUIDE *continued*

Sentinel
Loop Blocks

No: 1   What is repeated: `[Get_Resistor_Value;]`
`[Calculate_It;]`
`[Ask_for_Repeat;]`
Under what condition: **While** `Program_Repeat` **is not N**
`[Main Programming Sequence]`

Branch
Blocks

No: 1   Condition for branch: **If** `Resistor_Value` **is greater than 100.**
If condition is met: **Display** `Resistor voltage is larger than 100!`
If condition is not met: **Display** `Resistor voltage is safe.`
`[Calculate_It]`

8.

ALGORITHM DEVELOPMENT GUIDE

I.   Explain program to user.
a. What does the program do?
[The process]
Computes the resistor voltage for ten different current values from 1 to 10. Displays the answer and warns the user if the resistor voltage is larger than 100 volts.
b. What does the program user need to do?
[The input]
The value of the resistor in ohms.
c. What will be displayed?
[The output]
The value of the voltage for each current value. A message stating that the voltage is safe if it is not more than 100 volts. A message warning the program user that the voltage is larger than 100 volts otherwise.

II.   Get values from user.
a. **Get value of the resistor.**

III.   Do computations.
a. **Repeat the computation 10 times for different values of current from 1 amp to 10 amps.**
1. **Voltage = Current × Resistance**

IV.   Display answer(s).
a. **Value of voltage for each calculation.**

V.   Ask for repeat.
a. **Ask program user to enter Y or N.**

9.

PROBLEM STATEMENT GUIDE

1. What will the program do?
   Calculate the value of a resistor voltage for ten different values of current from 1 amp to 10 amps in steps of 1 amp each.
2. What is needed for input?
   Value of the resistor in ohms.
3. What will be given as output?
   Value of the voltage for each computation. A message stating that the resistor voltage is safe if voltage is 100 volts or less or, if more, a message stating that the resistor voltage is larger than 100 volts.

10.

CONSTANTS AND VARIABLES GUIDE

Formula	Constant or Variable	Meaning	Type R I C B S	Block(s) Used
	Starting_Current_ Value [CONST]	Beginning value for current.	X	[Calculate_It;]
	Increment_Current _Value [CONST]	Value to increase current each time.	X	Calculate_It;]
	Ending_Current_ Value [CONST]	Ending value for current.	X	[Calculate_It;]
V = I × R	Resistor_Voltage [VAR]	Voltage of Resistor.	X	[Calculate_It;] [Display_the_Answer;]
	Resistor_Current [VAR]	Current in resistor.	X	[Calculate_It;]
	Resistor_Value [VAR]	Value of the resistor.	X	[Get_Resistor Value;] [Calculate_It;]
	Program_Repeat [CHAR]	To check if user wants repeat.	X	[Ask_for_Repeat;]

Answers to Problems—Chapter 2

General Concepts

1. Three activities that computers do well are activities requiring repetition, arithmetic, and logical decisions. Others include sorting information and graphical representation.

3. Your words may be somewhat different but they should include the essence of the theorem, that any program logic, no matter how complex, can be resolved into action blocks, branch blocks, and loop blocks.

5. A branch block is a decision between two alternatives. An example would be: If a given computation produces a number larger than 100 have the screen display **TOO LARGE!**.

7. There are two types of branch blocks. An open branch might state: If it is raining, get an umbrella, then go to work. A closed branch might state: Take your car or take a bus, and then go to work.

9. (A) Input/Output   (B) Terminal points.   (C) Process   (D) Decision   (E) Subroutine.

11. The difference between a loop block and a branch block is that a loop block may go back and repeat a part of the program. A branch block will always continue forward in the program.

13. The general idea of top-down design is to start with the most general idea of the program and then add the program details; putting in programming code is the very last step.

15. A Pascal procedure is a program block that starts with the reserved word **PROCEDURE** followed by a name given by you.

17.

APM Step	Programming Guide
1. Understand the problem	Problem Statement Guide
2. Solve problem by hand	Problem Solution Guide
3. Record all of the necessary steps needed for solving the problem by hand.	Algorithm Development Guide
4. Decide which of the three control methods will be needed.	Control Blocks Guide
5. List and define all program constants and variables.	Constants and Variables Guide
6. Write the Pascal program just using procedures and enough code to compile.	Program Design Steps

Program Analysis

19. `This is not from a procedure.`
    `This is from a procedure.`

21. Each of the procedures needs a **BEGIN** before the **WRITELN** statement.

23. **PROGRAM** Straighten_Me_Out;

```
  PROCEDURE Where_Are_You;
   BEGIN
    {This procedure will be used for some calculations}
   END;

  PROCEDURE This_is_Another;
   BEGIN
    {This procedure will do some screen displays.}
   END;

BEGIN
   WRITELN('This is the main programming block.');
END.
```

Answers to Section Reviews—Chapter 3

Section Review 3–1

1. A Pascal statement is an instruction that tells the computer to perform a specific operation.
2. Ten Pascal statements were listed. Three types listed were assignment statements, procedure call statements, and compound statements.
3. A semicolon is used to separate one Pascal statement from another.
4. In non-Turbo Pascal, the semicolon cannot be used to separate the last statement from the END. With Turbo Pascal, you may or may not use a semicolon to separate the last statement from the END.

Section Review 3–2

1. Data is the information that you want to do something with (such as values or names of things); statements are the instructions telling the computer what to do with the data.
2. A constant is a value that doesn't change during program execution.
3. A variable represents a memory location that will contain different data during program execution.
4. A Pascal identifier is the name you give to data (constant or variable) and to the main parts of the program (such as the names of the program and procedures).
5. An identifier must start with a letter of the alphabet or an underscore and may be followed by letters of the alphabet, numbers, and/or underscores. No spaces are allowed.
6. A reserved word is a Pascal command that the Pascal compiler understands as some kind of instruction. Five Pascal reserved words are: **PROGRAM**, **BEGIN**, **END**, **PROCEDURE**, and **REPEAT**.

Section Review 3–3

1. A data type is a way of classifying data.

2. The data types presented in this section are: **BOOLEAN**, **BYTE**, **CHAR**, **INTEGER**, **REAL**, and **STRING**.

3. A BOOLEAN type can have two different values: **TRUE** or **FALSE**.

4. In Pascal, assigning a constant means setting an identifier equal to a given amount of data. This must be done after the reserved words **CONST** and **PROGRAM** and before a **PROCEDURE** or **BEGIN**.

5. In Pascal, declaring a variable means stating what type of variable each identifier is to be. This must be done after the reserved word **VAR**, and will come after the **CONST** assignments and before a **PROCEDURE** or **BEGIN**.

Section Review 3–4

1. **WRITELN** is a built-in Turbo Pascal procedure. When programming, it is convenient to think of it as a reserved word.

2. The first **WRITELN** statement

```
WRITELN('This_One');
```

will display the string

```
This_One
```

when executed. The second **WRITELN** statement:

```
WRITELN(This_One);
```

will treat its argument `This_One` as a variable. The difference is the presence or absence of the single quotation marks (').

3. The Pascal command for allowing the user to input data from the keyboard is

```
READLN(Variable);
```

where `Variable` is a previously defined variable.

4. When executed, it would display

```
This_Value is This string This_String.
```

Section Review 3–5

1. The Pascal assignment operator replaces the equal sign (=) in an equation. It means: evaluate the expression on the right of the assignment operator (`:=`) and store the results in the memory location reserved for the identifier on the left of the assignment operator.

2. `Sum := X + Y + Z;` (The semicolon must be included because it signifies the end of a Pascal statement.)

3. The equation means: add 2 to the number already stored in the memory location signified by the identifier `This_One.`

Section Review 3–6

1. The six arithmetic operators used by Pascal are: multiplication, division, **DIV**, **MOD**, addition, and subtraction.

2. The two types of variables used in arithmetic operations are REAL and INTEGER.

3. When doing arithmetic operations containing an INTEGER and a REAL, the answer must be of type REAL.

4. To keep an INTEGER answer in division, use **DIV**.

Section Review 3–7

1. **WRITELN** forces a carriage return, while **WRITE** does not.
2. **READLN** forces a carriage return, while **READ** does not.
3. `4.0000000000E+00.`
4. `_____3.15`
5. `ClrScr;`
   `GOTOXY(40,12);`

Section Review 3–8

1. You can get help when working in the Turbo environment by pressing the F1 key.
2. The first method for getting any menu item is by highlighting the item with the arrow keys and then pressing RETURN/ENTER. The second method is by depressing the letter on the keyboard that is the same as the highlighted capital letter of the desired menu or menu item.
3. A Turbo hot key requires the Alt key to be depressed along with another. This is a method of gaining quick access to specific items from anywhere in the Turbo system.
4. You can change the bottom line to see the Alt key combination functions simply by holding down the Alt key for a few seconds.
5. A quick way of loading and saving your Pascal program is to use the F2 key to save it and the F3 key to load it.

Answers to Self-Test—Chapter 3

1. The program will solve for the total resistance of three resistors in series or in parallel. The program user is to input the value of each resistor and then select if they are in series or in parallel.
2. There are five different data types: REAL, INTEGER, CHAR, STRING, and BOOLEAN.
3. Yes, a type BYTE could have been used for the integer variable `User_Selection`, because it is a positive number less than 255.
4. Four different types of statements are used in program 3–35.
   Assignment statement: `R_T := R_1 + R_2 + R_3:`
   Procedure call statement: `User_Prompt;`
   Compound statement: Main programming sequence.

   IF-THEN-ELSE: PROCEDURE Do_Selection;

5. Yes, the procedure `Get_Selection;` calls procedure `User_Prompt`.
6. `R1 = 1.00 R2 = 2.00 and R3 = 3.00 ohms.`
   `The total resistance is = 6.000 ohms.`

7. R1 = 9.00 R2 = 9.00 and R3 = 9.00 ohms.
   The total resistance is = 3.000 ohms.
8. Pascal reserved words used in program 3–35 are: **PROGRAM, CONST, VAR, BOOLEAN, REAL, INTEGER, STRING[20], CHAR, PROCEDURE, BEGIN, END, WRITELN, GOTOXY, WRITE, READ,** and **CLRSCR**. The list includes Turbo built-in procedures as well as Pascal reserved words.
9. The purpose of the procedure User_Prompt is to cause the text 'Press RETURN/ENTER to continue . . . ' to appear at the bottom of the monitor screen and wait for the program user to press the RETURN/ENTER key.
10. The program knows which type of circuit was selected by the value of Series_Solution (either **TRUE** or **FALSE**). The selection is done with an **IF-THEN-ELSE** statement in **PROCEDURE** Do_Selection;. It is a closed branch.

Answers to Problems—Chapter 3

General Concepts

1. A Pascal statement is an instruction that tells the computer to perform a specific operation.
3. Semicolons are used to mark the end of a Pascal statement. The exception to this is the last statement just before an **END**. In this case, the **END** marks the statement ending. Turbo Pascal doesn't care if a semicolon is used in this case.
5. A constant is data that will not change during the execution of the program. A variable is a memory location that may contain different data during the execution of the program.
7. Pascal reserved words are the words that make up Pascal's instruction set. Some of them are: **BEGIN, END, PROGRAM, PROCEDURE, VAR**, and **CONST**.
9. A Pascal type indicates the kind of data that will be used by a constant or variable.
11. A) BOOLEAN   B) STRING   C) INTEGER   D) REAL
13. The maximum value a BYTE can have is 255.
15. Constants are assigned as follows.

CONST
    Identifier = Constant;

17. To make a variable called **A_Variable** appear on the screen, you must do three things.
    [1] Declare it.

VAR
    A_Variable
:INTEGER;

[2] Give it a value.

```
A_Variable := 25;
```

[3] Cause it to be displayed on the monitor. Note the absence of the single quotes.

```
WRITELN(A_Variable);
```

19. A Pascal statement that would allow a program user to enter a value during program execution would be

```
WRITELN('How old are you?');
READLN(A_Variable);
```

21. A Pascal assignment statement that decreases a memory location by three would be

```
VAR
  Memory_Location
:INTEGER;
{Body of program}
  Memory_Location := Memory_Location - 3;
```

23. When two INTEGERs are added, subtracted, or multiplied, the result is an INTEGER. When any of these processes is done with an INTEGER and a REAL, the result is a REAL.

25. When dividing using type REAL, both the whole number and fractional part of the answer are given. When dividing with type INTEGER, only the whole number part of the answer is presented.

27. Precedence of operations means the order in which calculations are performed. In Pascal, the precedence of operations is such that all work is done inside parentheses first, then all multiplication and division, followed by addition and subtraction. Other than that, all operations are taken from left to right.

29. To demonstrate the difference between **WRITE** and **WRITELN**, the following statements in a Pascal program

```
WRITE('This is WRITE');
WRITELN(' and this is WRITELN.');
WRITELN('This is also WRITELN.');
```

when executed, would produce

```
This is WRITE and this is WRITELN.
This is also WRITELN.
```

31. To make a carriage return to appear in the middle of a string constant, treat the carriage return as a separate variable.

```
WRITELN('This is how to place a ',
       ,'carriage return in here.');
```

33. A type REAL in Turbo Pascal is displayed with eleven-place accuracy using the **E** notation.

    `1.0000000000E+00`

35. A Turbo Pascal statement that would output a type REAL to the screen with five spaces and three decimal point accuracy would be

    `WRITELN(A_Real_Number:5:3);`

37. The Turbo Pascal procedure for clearing the screen is

    `USES CRT;`

    and then, in the body of the program

    `ClrScr;`

39. To output a string constant to the printer, you must use

    `USES Printer;`

    and then, in the **WRITELN**

    `WRITELN(LST,'This will now go to the printer.');`

Program Analysis

41. Program 3–37 will not compile because it has a missing ending single quote in its string constant.
43. Yes, program 3–39 will compile and execute.
45. The corrected program is

```
PROGRAM Problem_45;
  VAR
    Hello : BOOLEAN;
  BEGIN
    Hello := TRUE;
  END.
```

47. `Number = 33 4.7200000000E+01 is the Value`
49. `7`
51. `2.500`
53.   `Here I am...Over there...`
    `and now here!`

Answers to Section Reviews—Chapter 4

Section Review 4–1

1. Defining a procedure is writing an actual procedure containing instructions (statements) that tell the computer to do something and giving it an identifier.

Calling a procedure is using the given identifier to evoke the instructions of the procedure.
2. There is no limit to the number of times that a procedure may be called within a program. The only consideration is whether the resulting program will be too large to be accommodated by the computer's memory.
3. Yes, a procedure can be called from another procedure, provided the procedure being called has already been defined. However, a procedure can not be called through another procedure.
4. For the purposes of this section, a procedure may be called anytime after it has been defined.
5. Calling through a procedure is an attempt to call a nested procedure from a level outside the procedure in which it is nested. The Pascal compiler will reject this because the procedure identifier is only known to the nesting procedure.
6. Recursion is a procedure calling itself.

Section Review 4–2

1. The declaration is the part of the program that declares the constants and variables.
2. A global variable is one that is declared in the main programming block.
3. Any procedure within the program can use a global variable.
4. A local variable is declared within a procedure and can only be used by it and procedures nested within it.
5. Scope refers to the range that a constant or variable has within a Pascal program.
6. It is considered good programming practice to keep program declarations as local as possible.

Section Review 4–3

1. A function is used to define a mathematical operation.
2. They can both have parameters passed to them from the calling procedure.
3. A function will always return a value and must have a type.
4. Yes, a function can call itself; this is called recursion.
5. A built-in Turbo Pascal function is one that was programmed into the Turbo system at the factory. An example would be: **SQRT(Number :INTEGER or REAL):REAL;**

Section Review 4–4

1. There is no difference between passing values to a procedure or a function.
2. A value parameter is a value passed to a called procedure.
3. A variable parameter will cause a value to be returned back to the calling procedure (much like a function returns a value to the calling procedure).

Section Review 4–5

1. A library routine is a procedure or function that is routinely used in different programs.

2. A "good" routine would have constant usage, good documentation, good structure, a user-friendly interface, and no bad side effects.

3. Two of the most common built-in Pascal procedures are **WRITELN** and **READLN**.

4. The built-in Turbo procedure for activating the computer speaker is **SOUND(Frequency :INTEGER);**. The variable `Frequency` determines the tone of the sound emitted from the speaker.

Section Review 4–6

1. Normally, a procedure or function may be called after having been defined.

2. The purpose of a **FORWARD** declaration is to allow a procedure or function to be called before it is defined.

3. The parameter list of a **FORWARD**-declared procedure or function cannot be defined again in the procedure definition.

4. Other than the reserved words **PROCEDURE** and **FUNCTION**, there is no difference in a **FORWARD** of these two routines.

Section Review 4–7

1. The major Turbo editor features presented in this section include cursor movement, inserting and deleting, and block commands.

2. The main purpose of the cursor-movement commands used in the Turbo editor is to control the position of the cursor on the editor screen.

3. Yes, the coding of a Pascal procedure can be saved to the disk separately from the program in which it is contained. This may be accomplished by using the block command Ctrl-K W after the block has been marked.

4. If you accidentally used a Pascal reserved word as an identifier throughout your Pascal program you could automatically correct this error with the Turbo editor find and replace command. This would allow you to replace the improper identifier with a correct one supplied by you.

Answers to Self-Test—Chapter 4

1. There are ten procedures and ten functions defined in program 4–39.

2. The Turbo Pascal built-in procedures used are: **CLRSCR**, **WINDOW**, **WRITELN**, **GOTOXY**, **WRITE**, **READ**, **READLN**, **SOUND**, **DELAY**, **NOSOUND**, **LOWVIDEO**, and **NORMVIDEO**. The Turbo Pascal built-in functions used are **PI**, **SQRT**, **SQR**, **ARCTAN**.

3. There are no global variables used in the program.

4. At the beginning of the program, a description of what the program does is given as a comment.

5. The procedures that use value parameters are:
   `Do_Calculations(R, L, C, F, VP :REAL);`
   The meanings of the input parameters are: R = Circuit resistance, L = circuit

inductance, C = Circuit capacitance, F = Applied frequency and VP = Applied peak voltage. All values are entered by the program user.

`Imped_and_Phase(X_C, X_L, Res :REAL . . .`

The meanings of the input parameters are:

`X_C = Capacitance reactance, X_L = Inductive reactance.`
`D_Out(X_Val, Y_Val:INTEGER;Show:WORDS;Value:REAL;Unit :WORDS);`

The meanings of the input parameters are:
X_Val = Starting row for output.
Y_Val = Starting column for output.
Show = String to be displayed.
Value = Real number to be displayed in the form 3:6.
Unit = String to be displayed.

`Note(Pitch, Duration :INTEGER);`

The meanings of these input parameters are:
Pitch = Frequency of the tone.
Duration = Length of the tone in milliseconds.
All of this information can be found in the `Procedures Used` part of program 4–39.

6. The procedure that uses variable parameters is

`Imped_and_Phase{...VAR Z_P, Ph :REAL};`

The meanings of these output parameters are:
Z_P = Circuit impedance.
Ph = Phase angle of applied voltage to circuit current.
This information can be found in the `Procedures Used` part of program 4–39.

7. The main sequence of the program is

`Program_Title;`
`Explain_Program;`
`Get_Values;`

This was found in the procedure

`Main_Programming_Sequence;`

8. No, it doesn't make any difference in what order the procedures or functions are defined, because of the **FORWARD** declarations.

9. These values are passed back through the variable parameters of the called procedure `Imped_and_Phase`.

10. The purpose of `User_Prompt` is to give the program user a chance to observe one screen of information for as long as needed before going on to the next screen of information.

Answers to Problems—Chapter 4

General Concepts

1. A procedure can be thought of as a mini-program used inside your total Pascal program.
3. There is no limit to the number of times a procedure can be called.
5. A procedure cannot be called through another procedure.
7. Nesting procedures is defining one procedure inside another. An example would be

```
PROGRAM Nested;
  PROCEDURE Outer;
    PROCEDURE Nested;
      BEGIN
        {This is the nested procedure.}
      END;
    BEGIN
      {This is the outer procedure.}
    END;
  BEGIN {Main program block}
  END. {of main program block.}
```

9. The term global applies to any declaration which takes place in the main program block (before any procedures are defined). An example would be

```
PROGRAM Example;
  VAR
    Global_Variable
  :INTEGER;

    {Program procedures follow...}
        .
        .
        .
```

The variable in this example, Global_Variable, may be used by any other parts of the program.
11. The term scope in Pascal refers to how far declarations such as constants and variables extend into the program.

13. An example of nested procedures using a local constant would be

```
PROGRAM Nested;
  PROCEDURE Outer;
    CONST
      LocalOne = 'This is a local constant.';
    PROCEDURE Nested;
      BEGIN
        {This is the nested procedure.}
        {The constant LocalOne is known here.}
      END;
  BEGIN
    {This is the outer procedure.}
    {The constant LocalOne is known here.}
  END;
BEGIN {Main program block}
    {The constant LocalOne is not known here.}
END. {of main program block.}
```

15. A Pascal function returns a value. This is different from a Pascal procedure in that a procedure need not return a value.

17. Yes, a Pascal function can call itself; this process is called recursion.

19. A built-in Turbo Pascal function is a function that has been defined by the developer of the Turbo Pascal system. An example is

```
SQR(Number :INTEGER or :REAL);
```

21. Passing values to a procedure means that the calling procedure sets the values of the variables to be used by the called procedure.

23. A variable parameter has its value passed from the called procedure back to the calling procedure. An example would be

```
PROCEDURE Do_Calculation(VAR PassItBack :INTEGER);
    {Body of procedure goes here.}
            .
            .
            .

PROCEDURE Call_It;
  VAR
     AValue
   :INTEGER;
    BEGIN
      Do_Calculations(AValue);
        WRITELN('Value from called procedure = ',AValue);
    END;
```

25. A library routine is a procedure or function that is routinely used in different programs.

27. Some of the most common Pascal built-in procedures are the **WRITELN** and the **READLN**.

29. The parameter list of a **FORWARD** declaration must be defined when used in the **FORWARD** declaration part. It is not defined again when the procedure or function is defined in the program.

Program Analysis

31. Program 4–40 will not compile because there is no **BEGIN** for the main program block.

33. Program 4–42 will not compile because it is calling a procedure through another procedure. To correct this, call the outer procedure from the main program block and then call the nested procedure from the called outer procedure.

35. The program will compile; no changes are necessary.

37. The program will not compile, hence nothing will be displayed on the screen.

39. You will not get the correct answer because the main programming block displays the global value of the variable **Results**. The calculations are done in the procedure **Do_Some_Calculations** and a locally defined variable by the same name is used to contain the answer for the calculations. Effectively, these are two separate variables using the same identifier, although they do occupy different places in memory.

Answers to Section Reviews—Chapter 5

Section Review 5–1

1. A relational operator is a comparison of two quantities while a Boolean operator is an expression that uses a Boolean operator such as **NOT, OR, AND** or **XOR**.

2. The relational operators compare two quantities. EQUAL is true only when the two quantities have the same value. GREATER THAN is true only when the quantity on the left is larger in value than the quantity on the right. LESS THAN is true only when the quantity on the left is smaller in value than the quantity on the right. NOT EQUAL is true only when the value of the two quantities are not equal. EQUAL TO OR LESS THAN and EQUAL TO OR GREATER THAN, are combinations of two operational operators.

3. These are the Boolean operators. **NOT** changes the value of an expression to its opposite Boolean value. **AND** is used in statements which are evaluated as TRUE only if all the expressions are TRUE. **OR** is used in statements which are evaluated as TRUE if any one expression or combination of expressions is TRUE. **XOR** (exclusive **OR**) is used in statements which are evaluated as TRUE only if one expression or the other is TRUE.

4. An example of a Pascal statement using both relational and Boolean operators can be taken from program 5–4.

```
IF (User_Input >= 8) AND (User_Input <= 10) THEN
WRITELN('Input is OK!');
```

Section Review 5–2

1. A branch block is a programming block where a different programming sequence may be used and the program will continue forward.
2. A closed branch forces a selection of two or more actions, while an open branch may or may not select an action.
3. The **IF-THEN** statement is an open branch and the **IF-THEN-ELSE** is a closed branch.
4. A sequenced branch consists of multiple selections that give more than two choices. An example would be selecting a formula from a list of formulas.
5. A semicolon is not used to separate **END** statements from **ELSE** statements. It is, however, used in compound statements within the **IF-THEN-ELSE**.

Section Review 5–3

1. The main advantage of the **CASE-OF** statement is the programming ease it affords when there are several choices available within the program.
2. The **IF-THEN-ELSE** statement can do the same thing as the **CASE-OF** statement but it is more awkward to use in programs with multiple selections.
3. The term case selector is a variable whose value must be ordinal; it selects the particular statement that will be activated.
4. The term case clause contains the group of statements that can be selected from the case selector. It can contain several case selectors for the same statement, a range of case selectors for the same statement, and an **ELSE** statement.

Section Review 5–4

1. A string is a group of characters.
2. A character string is a series of one or more characters taken from the ASCII code, enclosed by apostrophes. An example is 'Hello!'
3. A null string is a string that does not contain any characters.
4. To declare a string variable in Turbo Pascal use **STRING[N]**, where N is the maximum length of the string and may be any value from 1 to 255.
5. String concatenation is the joining together of two or more strings into one string. This is done in Turbo Pascal by using the reserved word **CONCAT**.

Section Review 5–5

1. The **COPY** function can be used to store the middle name of a person in another string variable.
2. A function is used to return the length of a string. It is the **LENGTH** function.
3. The purpose of the **POS** function is to return the value of the location of a substring inside a target string.
4. The **VAL** procedure converts a string into its numeric representation while the **STR** procedure converts a numeric value into its string representation.

Section Review 5–6

1. The word library in Pascal refers to special programs that contain useful program features and can be used by other programs.
2. Libraries can be developed in Turbo Pascal by developing programs that are called units.
3. A Turbo Pascal unit contains the Turbo reserved words **UNIT**, **INTERFACE** and **IMPLEMENTATION**. It exists on the disk in compiled form with the extension **.TPU**.
4. For a program to use a unit the Turbo Pascal reserved word **USES** followed by the name of the unit must be placed before the declaration block in the Turbo Pascal program.

Answers to Self-Test—Chapter 5

1. The procedure that performs the light check is procedure **Light__ Check**. This test is done using the **CASE** statement.
2. The procedures **Light__Check** and **Arm__Service** both use a **CASE** statement.
3. Not counting the last **BEGIN-END.**, there are ten procedures.
4. The **USES CRT** means that the program is using a built-in Turbo Pascal procedure that is contained in the unit that is automatically called by the Turbo program.
5. There are no open branches in the program.
6. There are three procedures that use the **IF-THEN-ELSE; Start__ Testing__Sequence; TP__1__Measurement,** and **Arm Drive__Disconnect.**
7. The procedure **TP__1__Measurement;** passes the value **Next** back to the calling procedure. This kind of variable is called a variable parameter.
8. A **CASE** statement was not used in **TP__1__Measurement** because the variable **Value** is of type REAL and this will not support the **CASE** statement.
9. There are no semicolons following the two **END** statements because they are part of an **IF-THEN-ELSE** series and no semicolons are allowed.
10. The procedures that use logic statements are **Start__Testing__ Sequence** and **TP__1__Measurement**. The logic statements are OR and AND respectively.

Answers to Problems—Chapter 5

General Concepts

1. Relational operators are symbols that indicate a relation between two quantities. For example, the equal sign (=) is a relational operator.

3. A) is TRUE because 4 is in the range of 1 to 12. B) is FALSE because the lowercase letter "d" is not in the range of any uppercase alphabetical characters.

5. The unique feature of Boolean algebra is that it is a two-value system. Variables are either TRUE or FALSE; there is nothing in between. Boolean multiplication is represented by the `AND` operation, while addition is represented by the `OR` operation.

7. Yes, relational and Boolean operators can be combined. An example would be

    ```
    IF (ThisValue > ThatValue) AND (ThisValue = 5) THEN . . .
    ```

9. For the command

    ```
    IF This THEN That ELSE Something_Different
    ```

    the Boolean operator is `This`; the procedures are `That` and `Something_Different`.

11. The rule for using semicolons in the **IF-THEN-ELSE** statement is that no semicolon is allowed after the last statement just before the **ELSE**. This is because **ELSE** acts as a statement terminator.

13. In the **CASE** statement
    ```
    CASE (Expression) OF (CaseClause)
    ```
    the Expression is any Pascal ordinal type. This is sometimes referred to as the case selector. The CaseClause is a list of statements, each preceeded by one or more constants. The case selector (Expression) must be of a Pascal ordinal type (cannot be REAL or STRING).

15. An example of a procedure, a function, and a compound statement in a case clause statement would be

    ```
    CASE OrdinalValue OF
      1 : This_Procedure;
          {Procedure example.}
      2 : NewValue := Operation(OldValue);
          {Function example.}
      3 : BEGIN
            This_Procedure;
            NewValue := Operation(OldValue);
          END;
          {Compound statement example.}
    ```

17. A character string is a series of one or more characters taken from the ASCII code.

19. The first Pascal statement

    ```
    WRITELN('Value := 2 + 5');
    ```

    will produce the string

    ```
    Value := 2 + 5
    ```

when executed. The second Pascal statement

```
WRITELN('Value := ', 2 + 5);
```

will produce

```
Value := 7
```

when executed.

21. A string constant is declared with

```
CONST
    St_Constant : STRING[N] = 'The_String';
```

An example would be

```
CONST
    This_will_be : STRING[79] = 'Something different.';
```

Thus, using this assignment

```
WRITELN(This_will_be);
```

will produce, when executed

```
Something different.
```

23. The built-in Turbo Pascal function that will return a selected part of the string is

```
COPY(String_Name, From, To);
```

where `COPY` = the Turbo Pascal reserved word to call the copy function

`String_Name` = the string from which the selected part is to be taken

`From` = an INTEGER that tells Pascal how many character places to count from the beginning to where the selected part of the string is to start

`To` = an INTEGER that tells how many characters of the string are to be selected

25. The **INSERT** procedure is the built-in Turbo Pascal procedure that will insert one string inside another. An example would be

```
CONST
    String1 : STRING[79] = 'This answer.';
    String2 : STRING[79] = 'is the ';
BEGIN
    INSERT(String2, String1, 6);
    WRITELN(String1);
    { Body of program...}
```

which, when executed, will produce

```
This is the answer.
```

27. The **STR** procedure is a built-in Turbo Pascal procedure that will convert a type INTEGER to a type STRING. The **VAL** procedure is a built-in Turbo Pascal procedure that will convert an INTEGER into a type STRING. An example of each is

    STR(IntegerType, StringType);

    In the above, **IntegerType** is the INTEGER to be converted to a string and **StringType** is the string variable that will contain the STRING of **IntegerType**.

    VAL (StringType, IntegerType, Success);

    In the above, **IntegerType** is the INTEGER that will be converted into the STRING **StringType**. The variable **Success** is of type INTEGER that will be zero if the conversion was a success.

29. Yes, Turbo Pascal does have a user-created library available that is called a unit.

31. A unit may contain procedures, functions, and declarations that may be used by other Turbo Pascal programs.

33. The **INTERFACE** part of a unit contains all of the information that is accessible to programs using the unit. The **IMPLEMENTATION** part contains the definition of all the procedures and functions as well as other information that will be used by the unit but is not directly accessible to other programs.

Program Analysis

35. Program 5–41 will not compile because the Boolean statements must be put into brackets.

    IF (8 > 7) AND (3 < 5)
    THEN

37. Program 5–43 will not compile because there is a semicolon (**;**) before the **ELSE** in the **IF–THEN–ELSE** statement.

39. An **END;** statement is needed to terminate the **CASE**. This **END;** statement should come after the **WRITELN** statement following the **ELSE**.

41. Yes, there is a compile-time error in the program because **STRING[0]** is an invalid STRING length.

Answers to Section Reviews—Chapter 6

Section Review 6–1

1. A data type is a Pascal variable or constant. An example is the CHARacter constant 'a'.

2. A scalar is a data type that has a distinct set of possible values, which are considered to be ordered in a specific way. An example is the letters of the alphabet.

3. The term ordinal as used in Pascal means order and refers to one-to-one correspondence with an integer sequence starting at 0.

4. An example of a non-ordinal variable would be a type REAL such as 2.35.

5. A user-defined type cannot be displayed directly on the monitor. The only types that can be displayed are those already defined by Pascal to do so.

Section Review 6–2

1. The three different kinds of programming blocks are action, branch, and loop blocks.

2. Program 6–15, presented in this section, is not practical because there must be at least one program line for each calculation of the problem.

3. A program loop has the potential of repeating a part of the program.

4. The difference between a loop block and a branch block is that a loop block has the potential of going back and repeating a part of the program over again, while a branch block always moves the program forward.

Section Review 6–3

1. A loop structure must contain a starting value, the process that is to be repeated, the value of the increment, and the conditions for termination.

2. The meaning of `Counter := Counter + 1;` is: to the value stored in the memory location called `Counter`, add one, thereby increasing the number stored in that memory location by one.

3. It is poor practice to use an equality to terminate a **REPEAT-UNTIL** loop because the value may be "skipped" by the increment and the loop will continue well beyond the intentions of the programmer.

4. The minimum number of times a **REPEAT-UNTIL** loop will be executed is one. This type of loop is always executed at least once.

5. The preferred way of terminating a **REPEAT-UNTIL** loop is with the use of the **<** or **>**. Doing this prevents the loop from continuing in case the increment skips the terminating value.

Section Review 6–4

1. The structure of a **WHILE-DO** loop is **WHILE**(expression)**DO**(statement)—as long as (expression) is TRUE, the (statement) will be executed.

2. The main difference between a **WHILE-DO** loop and **REPEAT-UNTIL** is that the body of the **REPEAT-UNTIL** will always be performed at least once, while the **WHILE-DO** might not be performed even once.

3. An error-trapping routine is a part of the program used to attempt to force a particular input from the program user.

4. The precaution to be taken when constructing a **WHILE-DO** loop is to take care not to allow the loop to repeat beyond bounds. Thus the use of the **>** or **<** is encouraged for the loop terminator.

Section Review 6-5

1. The three different kinds of loops available in Pascal are the **REPEAT-UNTIL**, **WHILE-DO**, and **FOR-TO** loops.
2. The unique properties of the Pascal **FOR-TO** loop are that it must use an ordinal variable as its counter, and that the counter will always be incremented or decremented by one ordinal value.
3. To make the **FOR-TO** loop decrement, use the reserved word **DOWNTO**:
   FOR(variable) := (expression$_1$)DOWNTO(expression$_2$)DO
4. If the beginning value of the variable is larger than the ending value and the loop is to increment, the loop will do nothing, and the Pascal program will continue forward.
5. **FOR-TO** loops are available in Pascal as a matter of convenience.

Section Review 6-6

1. A nested loop is one loop contained inside another.
2. The program structure that should be used when developing a nested loop is a block structure that makes it clear which loop is the outer loop and which one is the inner loop.
3. When using a **REPEAT-UNTIL** in a nested loop, you should make sure that the loop variables are initialized and that the inner loop variable is reset each time the inner loop is completed.
4. The advantage of using a **FOR-TO** in a nested loop is that you don't need to worry about initializing or resetting the loop variables.

Section Review 6-7

1. The applications program for this section will calculate and display the impedance of a series LRC circuit for a range and increment of frequencies selected by the program user.
2. There are five procedures used in this program: `Main_Programming_Sequence`, `Explain_Program`, `Get_Values`, `Display_Values`, and `Program_Repeat`.
3. There is only one type of loop used in this program, the **REPEAT-UNTIL**.
4. The whole program is not repeated when the program user repeats the program because it is assumed that the user now understands what the program will do, so the part explaining the program to the user is omitted the second time around.
5. No, there are no nested loops in this program.

Section Review 6-8

1. The program user could input characters other than numbers, causing a run-time error.
2. A run-time error is caused when a string is input for a number.
3. A STRING-type variable should be used for inputting the value of a number in order to prevent a run-time error caused by the input of other characters.

4. The built-in Turbo Pascal procedure for converting a string into a number is the **VAL** procedure. The converted number may be either REAL or INTEGER depending upon the type given to it by the programmer.

Answers to Self-Test—Chapter 6

1. Yes, the program will compile.
2. `Loop_7` will continue until stopped by the program user, because the variable `Counter_1` will never equal the value of 5.
3. `Loop_12` has the greatest number of loops; `Loop_3` and `Loop_9` will never loop at all.
4. The **REPEAT-UNTIL** loop in the procedure `User_Loop_Selection` acts as an error-trapping loop to insure that the user enters a whole number from 1 to 12.
5. The nested loops are `Loop_11` and `Loop_12`. `Loop_11` has three levels of nesting and so does `Loop_12`.
6. `Loop_12` uses REAL, INTEGER and user-defined types.
7. `Loop_6` gets incremented by the program line
   `Second_1 := SUCC(Second_1);`
   To decrement the loop, this line would be changed to
   `Second_1 := PRED(Second_1);`
8. `Loop_10` will loop only once, because its counting variable is set to the end count (`Counter_3 := 5;`) within the loop.
9. `Loop_4, Loop_8` and `Loop_12` decrease the value of the counter variable.
10. There are no **FOR-TO** loops that use the variables `Stepper_1`, `Stepper_2`, or `Stepper_3` as counting variables because these are of type REAL. In Pascal, the **FOR-TO** loop requires an ordinal value.

Answers to Problems—Chapter 6

General Concepts

1. The six Pascal data types presented up to this point are REAL, INTEGER, BOOLEAN, BYTE, CHAR, and STRING.
3. The types that are scalar are INTEGER, BOOLEAN, CHAR, and BYTE.
5. The **ORD** function in Pascal returns the ordinal number of an ordinal-type expression. As an example, **ORD(a)** = 97.
7. The **SUCC** function returns the ordinal value of the following piece of data, while the **PRED** function returns the ordinal value of the preceeding piece of data.
9. A program loop has the option of going back and repeating part of the program. The difference between a program loop and a branch is that a branch will always cause the program to move forward while a loop will not.

11. An example of the need for program loops is the need to solve a given equation for many different values.

13. It is considered good programming practice to insure that a loop will not go on forever. This may be prevented by setting the exit condition for the loop to values larger (for an increasing value loop) or smaller (for a decreasing value loop) than a predetermined value.

15. Any legal Pascal statements may be placed between the **REPEAT** and **UNTIL** parts of a **REPEAT-UNTIL** loop. It isn't necessary to identify the beginning and end of this type of loop with **BEGIN** and **END** statements because the **REPEAT** and **UNTIL** commands serve that purpose.

17. The Pascal command

WHILE (expression) **DO** (statement)

means while the (expression) is TRUE, continue to do the following (statement).

19. An error-trapping routine is a part of a program that attempts to prevent the program user from inputting data that may cause the program to malfunction or produce incorrect results. An example would be to ask the user to input numbers within a specified range of values.

21. A precaution that should be taken with a **WHILE-DO** loop is to insure that the loop does not continue forever by using a larger than (or less than, as the case may be) a specified exit value.

23. The counting variable in a **FOR-TO** loop must be of an ordinal type.

25. The incremental change of a **FOR-TO** loop in Pascal is 1.

27. The main reason for using a **FOR-TO** loop in Pascal is as a programming convenience.

29. The type of program structure that should be used for nested loops makes it clear where each loop begins and ends. One suggested method of doing this is to have the **BEGIN** and **END** parts (or **REPEAT** and **UNTIL** parts) of each loop line up vertically.

31. For programs with nested loops, the innermost loop is completed first, the outermost last.

Program Analysis

33. Program 6–57 will not compile because you cannot print data of this type directly to the monitor.

35. Program 6–59 does compile, because Turbo Pascal allows a BOOLEAN type to be directly output to the screen.

37. Program 6–61 will not compile because you cannot print data of this type directly to the monitor. To correct the problem you need to use the **ORD** function to print the ordinal value of the data to the monitor.

39. There is no bug in program 6–63.

41. The bug in program 6–65 is the semicolon after the **DO** in the **FOR-TO** statement.

Answers to Section Reviews—Chapter 7

Section Review 7–1

1. The word array means arrangement.
2. The floor, room, and bed number of a patient in a hospital could be thought of as subscripts.
3. Using subscripts is an easy way of making a distinction between otherwise similar objects (such as patients in a hospital).
4. Initializing an array means setting all of its elements to some predetermined value.

Section Review 7–2

1. The dimension of an array is determined by the number of index elements used for the array.
2. The values used to index elements of a Pascal array can be of any scalar or subrange type except INTEGER (they can be a subrange of INTEGER).
3. One of the advantages of declaring an array type is that arrays can be passed between procedures.
4. Arrays are congruent if they are of the same type and dimensionality.
5. The significance of having congruent arrays is that they may have values passed between them and they may be compared to see if all of their elements are equal.

Section Review 7–3

1. Allowing the program user to enter patient data in any order makes it easier for the user; any ordered listing may thus be used to copy the data into the computer.
2. Program 7–14 allowed the program user to input patient data in any order because the program allowed the user to enter the values of each index in the array.
3. Program 7–15 illustrated getting data that was arrived at by using many separate elements of data.

Section Review 7–4

1. A bubble sort is the process of taking two sequential quantities at a time from a consecutive list of quantities, comparing them, and switching their sequence depending upon their relative value. This process is continued until no further switching is required in the consecutive list. The result will be a consecutive list of sorted quantities in ascending or descending value depending upon the requirements for the switch.
2. A sorting program must go through a list at least one time. When the list is already sorted, no switch is required, and so the program would only go through once.
3. For descending order, a switch is performed if the first number is smaller than the second. For ascending order, a switch is performed if the first number is larger than the second.

4. Four passes are required: 7 3 1 8, 3 1 7 8, 1 3 7 8, and one last pass with no switch.

Section Review 7–5

1. All string characters are represented by a number assigned by the ASCII code; it is these numbers that are compared in relational statements.
2. The string sorting program uses string variables and the numerical sorting program does not.
3. The sorted string data would be: 100, 35, Capacitor, and then Transistor.
4. The data is treated as a string variable and is sorted by its ASCII code value, not its indicated numerical value. The ASCII code for a string of number characters is smaller than that for a string of alphabetical characters.

Section Review 7–6

1. A run-time error is an error that will take place during program execution.
2. No, a compiler does not catch run-time errors, because by definition a run-time error is not an error in the programming syntax; it is an error in program design.
3. A debug routine usually contains a visual display of some data and the ability to step through the procedure.
4. An auto-debug routine is a convenient way of activating or deactivating the debug feature. Usually this is done with a global BOOLEAN statement.

Section Review 7–7

1. The maximum number of shares that can be entered into this program is 10.
2. There are 11 procedures used in the final program design.
3. Array types were used so that local variables could all have the same type.
4. The program user can enter the name, price, number of shares, and whether a stock is blue chip. The user can also change the price, delete a stock, add new stock, and have all of this information displayed in alphabetical order including the value of the stock and the amount of gain or loss on the stock.

Answers to Self-Test—Chapter 7

1. The statement

```
IF NOT ((UserChoice IN ['A'..'E']) OR
        (UserChoice IN ['a'..'e']));
```

is part of an error-trapping routine to insure that the program user enters either an uppercase or lowercase letter from A through and including E.

2. `Index` is set equal to `IndexMax` so that new stock items may be added to an existing list of stocks without taking the place of any of the previously entered stocks.

3. In the procedure `Update_Stock_Price` it is necessary to have two variable parameters because the new value and difference in value of the stock is passed on to the calling procedure.

4. The **GOTOXY** command is a screen format command that determines the horizontal and vertical position of the displayed text. The `1 + Index` causes each line of text to be displayed below the other.

5. If the name of the stock must be swapped, then everything else related to that stock must also be swapped. This is necessary so that all items relating to a specific stock will have the same subscripts. If they don't, then there will be no way of sorting, entering, or displaying this related information in a consistent manner.

6. The purpose of the procedure `Continue` is to allow a pause when the text is displayed on the screen in order to give the program user time to read it. The procedure also displays the message "Press RETURN/ENTER to continue." and waits for a user response. In this fashion, the program user may take as long as he wishes to read the displayed text.

7. The purpose of the procedure `Program_Repeat` is to give the program user a "second chance" just in case the E (for Exit the program) was mistakenly pressed during the display of the main menu. This feature makes the program more user-friendly.

8. The maximum number of stocks that can be entered into this program is 10. This was determined by the **CONST MaxStocks = 10;**.

9. The stock is alphabetized every time a new stock entry is made.

10. The variable `GainLoss[Index]` is set to 0 so this variable can be initialized when it is returned back to the main programming sequence.

Answers to Problems—Chapter 7

General Concepts

1. An array can be thought of as an arrangement.

3. An example of the use of a subscript would be distinguishing one resistor from another: R_1, R_2, etc.

5. The general form for a Pascal array is
    `Identifier:ARRAY[`$index_1$`,` $index_2$`,...`$index_N$`] OF TYPE`

7. A three-dimensional array has three indices.

9. Any ordinal value except of type INTEGER may be used as an index in a Pascal array.

11. Pascal will accept an array index that is a user-defined type as long as it is ordinal.

13. Yes, arithmetic operations are allowed with an array index.

15. The significance of having congruent arrays is that the arrays may be compared.

17. Entering data interactively means entering information into the program while the program is active.

19. The rules for bubble sorting numbers are given on the first page of section 7–4.
21. The numerical operation that must be performed on an arrayed index when sorting is to increment the value of the index by one. This is done so that the order of the data can be changed on the array list.
23. A temporary value is used when switching data in a sorting program in order to perform the switch. First one quantity is stored in the temporary variable, then that quantity is made equal to the second. Now, the second quantity is made equal to the temporary variable and the switching is then complete.
25. Numbers are sorted by their numerical value, while strings are sorted by their ASCII value.
27. A run-time error is an error that is detected after the program has compiled and is being executed.
29. A debug routine is put into a program to help find run-time errors.
31. An auto-debug routine allows an easy method for activating or deactivating the debug routine.

Program Analysis

33. Program 7–32 will compile.
35. Program 7–34 will not compile because a type INTEGER is used for an index.
37. Yes, program 7–36 will compile.

Answers to Section Reviews—Chapter 8

Section Review 8–1

1. A Pascal array treats grouped data as separate elements with the same subscripts while a record can treat the separate elements as a whole unit or as separate elements.
2. The different elements of a Pascal record are called fields.
3. An individual element of a record may be accessed by using the record identifier followed by a period and then the name of the field identifier. An individual element may also be accessed by the **WITH** statement.
4. A Pascal record can be passed between procedures either as a variable parameter or as a value parameter.

Section Review 8–2

1. A record can contain an array, but arrays and records are different. An array can be an array of records for a previously defined record.
2. Yes, a record can contain another previously defined record as one of its field types.
3. A variant is a selection of record fields, each of which requires different information.
4. Mutually exclusive fields are fields within variant records.

Section Review 8–3

1. A Pascal file is an external place for getting or receiving data.
2. I/O means Input/Output.
3. Some Pascal input files are the keyboard and a location on the floppy disk. Some Pascal output files are the console monitor, printer, and a location on the floppy disk.
4. The Pascal commands that actually output data to a file are the **WRITE** and **WRITELN** procedures.
5. The Pascal commands that actually input data from a file are the **READ** and **READLN** procedures.

Section Review 8–4

1. A text file is a string of ASCII characters.
2. The name of a disk file means a physical location on the disk to the Pascal program.
3. A legal DOS file name may specify the disk drive, must have a name starting with a letter of the alphabet up to 8 characters, and may have an extension of up to three characters (that may include number characters).
4. The four possible conditions you may encounter when working with a disk file are: 1. A new file needs to be created and information needs to be put into it. 2. A file already exists and information needs to be taken from it. 3. A file already exists and information needs to be added without destroying the old information. 4. A file exists and new information needs to be added while the old information is deleted.
5. **REWRITE** creates a new disk file and opens it for receiving information. **RESET** opens an existing disk file for getting information from the file. **APPEND** opens an existing disk file and readies it for new information without destroying information already in the file. **REWRITE** will destroy an existing disk file by the same name and create a new one by that name and ready it for inputting new information.

Section Review 8–5

1. The difference between the **WRITELN** and the **WRITE** is the same for text files as it is for outputting to the screen. The **WRITELN** causes a carriage return at the end of the line, while the **WRITE** does not.
2. The difference between the **READLN** and the **READ** is the same for text files as it is for inputting from the keyboard. The **READLN** forces a return to the next line while the **READ** does not.
3. The basic structure of a text file consists of characters structured into lines, each line terminated by an end-of-line marker. The file is also terminated by an end-of-file marker.
4. A sequential file has one piece of information immediately following another. The significance is that these files cannot be randomly accessed.
5. The data types that may be put into a text file are STRING, INTEGER, REAL, and

BOOLEAN. The data types that may be read from the same file are all of those mentioned except BOOLEAN.

5. The **EOF** marker in a text file can be used in a program to determine when the file is completed.

Section Review 8–6

1. A Pascal random-access file consists of components, each of which is accessible through the action of a file pointer.
2. The Turbo Pascal procedure that is used to locate the position of a record within a random-access file is the **SEEK(FileName, Pointer);** procedure.
3. No, there is no difference in the use of the **ASSIGN**, **REWRITE**, or **RESET** procedures when used in sequential or random-access files.
4. The built-in Turbo Pascal procedure that can be used to determine the number of files in a random-access file is **FILESIZE(FileName);**

Section Review 8–7

1. A compiler directive is an instruction to the Pascal compiler.
2. The syntax for a compiler directive is **{$D−}** with no spaces.
3. If the program user attempted to cause a Pascal program to open a disk file that did not exist, a run-time error would occur and the program would terminate.
4. The Turbo Pascal compiler directive that turns off automatic I/O error checking is **{$I−}**. It is turned back on by the compiler directive **{$I+}**.
5. When automatic I/O checking is disabled, the built-in Turbo Pascal procedure **IORESULT** will have a value of 0 if there was no error and a non-zero value if there was an I/O error.

Section Review 8–8

1. The main points of the problem statement were to modify the stock market program presented as the case study in the previous chapter so that it made use of Pascal's record structure and ability to use a disk file for storing and retrieving data.
2. A recommended process to use when modifying an existing program is to make one major modification at a time and ensure that the program is working correctly before introducing the next change.
3. The first modification of the program presented in this section was developing a record structure for the data.
4. Using a record structure in this program modification simplifies the passing of parameters and the sorting of data.
5. For the program to automatically interact with a disk file to store and retrieve data, it needs a method of checking to see if a data file already exists. If such a file does not exist, the program needs to create one. If a file does exist, the program needs to open it, and retrieve the data.

Answers to Self-Test—Chapter 8

1. The name of the disk file used by the program is **STOCKS.DTA**.
2. The program checks to see if a file already exists by using the function **FileExists(FileName : Words):BOOLEAN;**.
3. The program knows the number of records already in the file by using the built-in Turbo command **FILESIZE**. This is done in the procedure **Main_Programming_Sequence;**.
4. The program will save new information to the file when the procedure **SaveRecords** is activated. This is the procedure that contains the **WRITE** command that will cause data to be written to the disk file.
5. There are five fields in each record file. This was determined by the **RECORD** declaration under the **TYPE** at the beginning of the program.
6. The maximum number of stocks that may be entered into this program is ten. This is determined by the **CONST** declaration at the beginning of the program.
7. When parameters about the stock are passed between individual procedures all of the stock information in the fields is also passed. This can be done because a type RECORD is being used as the data structure for the stock information.
8. Individual records are saved to the disk as random-access files, because the built-in Turbo Pascal **SEEK** command is used.
9. There is no provision in the program for the program user to determine the name of the file for storing the stock data.
10. The stocks are alphabetized when new stocks are entered into the list. This was determined by observing when the alphabetizing procedure **Alphabetize_Stock** is called. It is called in the procedure **Main_Programming_Sequence** immediately following the **CASE** statement, only if the program user has selected the **Input_New_Stock** procedure.

Answers to Problems—Chapter 8

General Concepts

1. A Pascal record is identified by an identifier and contains a group of separate elements that may be of different types. These elements may be treated together or separately.
3. The elements of a record may be accessed by using the record identifier followed by a period, then the name of the field identifier. An individual element may also be accessed by the **WITH** statement.
5. A Pascal array treats grouped data as separate elements with the same

subscripts while a record can treat the separate elements as a whole unit or as separate elements.

7. A variant record represents a selection of record fields, each of which requires different information.

9. A Pascal file is an external place for getting or receiving data.

11. Some Pascal input files are the keyboard and disk. Some output files are the monitor, printer, and disk.

13. A text file is a file that contains a string of characters.

15. A legal DOS file name may specify the disk drive, must have a name starting with a letter of the alphabet up to 8 characters, and may have an extension of up to three characters (which may include number characters).

17. A] **REWRITE(FileName)** creates a new disk file and opens it for receiving information. If a file by that name already exists it is first destroyed. B] **RESET (FileName)** opens an existing disk file for getting information from the file. C] **Append(FileName)** opens an existing disk file and readies it for new information without destroying information already contained in the disk file.

19. In text files, the **READLN** command forces a return to the next line while the **READ** does not.

21. A sequential file has one piece of information immediately following another. It is important to make this distinction because this type of file cannot be randomly accessed.

23. The characteristics of a Pascal random-access file are that it consists of components, each of which are accessible through the action of a file pointer.

25. There are no differences between sequential and random-access file commands.

27. The built-in Turbo Pascal procedure that will determine the number of files in a random-access file is
FILESIZE(FileName);

29. The general format for a Turbo Pascal compiler directive is **{$D-}** with no spaces.

31. The built-in Turbo Pascal statement used to check if a disk file already exists when automatic I/O error checking is disabled is procedure **IORESULT**. It will return a value of 0 if there is no error and a non-zero value if there is an error.

Program Analysis

33. The program will compile and run successfully.

35. The program will not compile because the type identifier has been used instead of the variable identifier.

37. It is the record that is the array, not the element of the record. Also, the record identifier used is the wrong one.

Answers to Section Reviews—Chapter 9

Section Review 9–1

1. To obtain color, a color graphics card must be installed and a color monitor attached.
2. In order to change the text from 80-column to 40-column, an installed color graphics card is required.
3. Text can have 16 different colors. All these colors can be displayed at the same time on the screen.
4. There are 8 different background colors available. The **CLRSCR** command is used to bring the full screen to the desired color.
5. There are basically two choices available to indicate the desired color: the color number or the color name. For example, **TEXTCOLOR(0)** or **TEXTCOLOR(BLACK)** would both produce the same affect.

Section Review 9–2

1. The two different kinds of screens that can be available to you on your monitor are text and graphics.
2. Text mode has only predefined images, while graphics allows you to define your own.
3. The one piece of hardware is an installed graphics adapter, while the two pieces of software are **GRAPH.TPU** and the appropriate graphics driver.
4. The number of different colors available with color graphics depends upon the installed color graphics hardware in your system and its mode of operation.
5. A pixel is the smallest element possible on the graphics screen.

Section Review 9–3

1. It's important to know about your system's modes and colors so you don't mistakenly try to look for lines that are drawn in the same color as the graphics screen (you won't see them).
2. A color palette means the colors available for the current active mode.
3. There are 4 color palettes available for the IBM CGA color graphics card. Three colors per palette are available.
4. Yes, there are modes where color is not available. This information can be found in table 9–3.
5. You can find out the number of modes for your system by referring to table 9–3 or by using a program that contains the built-in Turbo Pascal function **GETGRAPHMODE**.

Section Review 9–4

1. Changing a graphics mode could change the color palette and/or the number of pixels available on the graphics screen.
2. There are four line styles available in Turbo Pascal, plus the ability to have two thicknesses of each. They are a solid, dashed, dotted, or center line style each available in two thicknesses.

3. The built-in Turbo Pascal shapes presented in this section were the rectangle and the circle. The circle's line is limited to two line thicknesses.
4. A flood fill will fill an enclosed area or the surrounding area with a given color. The area to be filled is determined by whether the screen coordinates used in the flood-fill procedure are within or without the area in question.

Section Review 9–5

1. Turbo Pascal has a built-in procedure for creating bars because bar graphs are commonly used to present various kinds of technical information.
2. Flat or 3–D bars can be displayed using Turbo Pascal built-in procedures.
3. The options available for displaying a 3–D bar using Turbo Pascal built-in procedures are placing a top on the bar or omitting it. This feature is available in case you wish to stack 3–D bars.
4. You can distinguish one bar from the other with built-in Turbo Pascal procedures that use the available area-fill command and fill the bars with different patterns as well as different colors.
5. Some of the options available when using text in Turbo Pascal graphics are the text style (font), text size, and text direction (horizontal or vertical).

Section Review 9–6

1. Scaling is a process that uses numerical methods to insure that all of the required data appears on the graphics screen and utilizes the full pixel capability.
2. Coordinate transformation is the process of using numerical methods to cause the origin of the coordinate system to appear at any desired place on the graphics screen.
3. The values of the scaling factors are determined by the number of horizontal and vertical pixels, and by the minimum and maximum values of X and Y.
4. The coordinate transformation values are determined by the horizontal and vertical scaling factors, and by the minimum value of X and the maximum value of Y.

Section Review 9–7

1. The guiding principle for developing Pascal programs is to make the program easy to understand and modify.
2. One method of writing user-defined identifiers for procedures and functions is to use the __ as part of the identifier.
3. In developing identifiers for variables, you should consider using standard identifiers commonly found in specific formula relationships. The important point is to make standard formulas easily recognizable in the program.
4. Try to consider the record, **CASE**, and user-defined scalar types when creating your Pascal programs; they will make the programs easier to understand and modify.

Answers to Self-Test—Chapter 9

1. Program 9–26 graphs a sine wave. The amplitude, number of cycles, and phase angle are selected by the program user.
2. The graphics screen is initialized in the procedure `Sine_Wave`.
3. The purpose of the **ROUND** procedure is to convert a REAL number into an INTEGER. This is necessary because the **PUTPIXEL** procedure requires INTEGER values.
4. Program 9–26 would not work for every graphics adapter because it uses the built-in Turbo Pascal procedure **SETGRAPHMODE** and sets it to mode 3. Not all graphics adapters have a mode 3.
5. The statement is
   `PutPixel(XValue, 75-YValue, 2);`
6. The maximum amplitude allowed for the sine wave is 75 units.
7. No, program 9–26 does not automatically adjust itself for different size graphics screens. It doesn't use the built-in Turbo Pascal **GETMAXX** and **GETMAXY**.
8. The sine wave will always start at an *X* value of 0. The *Y* value is determined by the phase angle input by the program user. The *X* value is determined by the starting value of 0 in the **FOR–TO** loop. The *Y* value is determined by the values placed in the **SINE** function by the program user.
9. The number of colors displayed by program 9–26 will depend on the kind of graphics adapter used in the system that runs the program. At most there will be two colors; one will be the default drawing color, and the other will be the background color determined by
   SETBKCOLOR(1);

Answers to Problems—Chapter 9

General Concepts

1. To produce text in color using Turbo 4.0 Pascal, your system must contain a color monitor, an installed color graphics system, and be an IBM PC, AT, PS/2, or true compatible.
3. The maximum number of text colors you may have is 16. Yes, they can all be displayed at the same time.
5. A text screen can only display text and use text commands. A graphics screen can have individual pixels manipulated by the program user.
7. A pixel is the smallest element possible on the graphics screen.
9. The built-in Turbo Pascal procedure for determining the modes available on your system is
   GETMODERANGE(GraphDriver, LoMode, HiMode**);**
11. The purpose of the built-in Turbo Pascal functions **GETMAXX** and **GETMAXY** is to get the maximum size of the current graphics screen.

13. The palette on the graphics screen can be changed by using the built-in Turbo Pascal procedure **SETGRAPHMODE(Mode);**.

15. You can create a rectangle using Turbo Pascal with the built-in Turbo Pascal **LINE(X1,Y1,X2,Y2)** procedure to create the four sides of the rectangle, or the **RECTANGLE(X1,Y1,X2,Y2)** procedure.

17. There are only two line styles for creating circles in Turbo Pascal. They are either normal width or thick width, set by the last **SETLINESTYLE** procedure.

19. A Turbo Pascal fill style represents different methods of filling in the area of a bar in order to help distinguish one bar from another.

21. There are 12 different fill styles available for bars in Turbo Pascal.

23. Scaling is a process that uses numerical methods to insure that all of the required data appears on the graphics screen and utilizes the full pixel capability.

25. Coordinate transformation is used so that the full range of the resulting graph generated by the mathematical function may be viewed on the graphics screen.

27. The coordinate transformation values are determined by the horizontal and vertical scaling factors, and by the minimum value of X and the maximum value of Y.

29. Good programming style makes a source code easy to read, understand, and modify.

31. You should keep in mind the record, **CASE**, and user-defined scalar types when developing Pascal programs.

Program Analysis

33. What happens depends upon the installed graphics system in your computer. For some systems, this may cause a display that does not produce any legible characters.

35. Program 9–30 will not compile for two main reasons: there is no **USES GRAPH** and the graphics screen has not been initialized.

37. Program 9–32 will not compile because a variable identifier is expected in the parameter list of **INITGRAPH**.

Index